AF805344

Functional Organic Materials for Tumor Diagnostics and Therapy

Pengfei Shi • Huairong Zhang
Lei Yang • Shujuan Sun • Shusheng Zhang
Editors

Functional Organic Materials for Tumor Diagnostics and Therapy

Springer

Editors
Pengfei Shi
School of Medicine
Linyi University
Linyi, China

Huairong Zhang
School of Medicine
Linyi University
Linyi, China

Lei Yang
School of Medicine
Linyi University
Linyi, China

Shujuan Sun
School of Medicine
Linyi University
Linyi, China

Shusheng Zhang
School of Medicine
Linyi University
Linyi, China

ISBN 978-981-96-9477-8 ISBN 978-981-96-9478-5 (eBook)
https://doi.org/10.1007/978-981-96-9478-5

© The Editor(s) (if applicable) and The Author(s), under exclusive license to Springer Nature Singapore Pte Ltd. 2026

This work is subject to copyright. All rights are solely and exclusively licensed by the Publisher, whether the whole or part of the material is concerned, specifically the rights of translation, reprinting, reuse of illustrations, recitation, broadcasting, reproduction on microfilms or in any other physical way, and transmission or information storage and retrieval, electronic adaptation, computer software, or by similar or dissimilar methodology now known or hereafter developed.
The use of general descriptive names, registered names, trademarks, service marks, etc. in this publication does not imply, even in the absence of a specific statement, that such names are exempt from the relevant protective laws and regulations and therefore free for general use.
The publisher, the authors and the editors are safe to assume that the advice and information in this book are believed to be true and accurate at the date of publication. Neither the publisher nor the authors or the editors give a warranty, expressed or implied, with respect to the material contained herein or for any errors or omissions that may have been made. The publisher remains neutral with regard to jurisdictional claims in published maps and institutional affiliations.

This Springer imprint is published by the registered company Springer Nature Singapore Pte Ltd.
The registered company address is: 152 Beach Road, #21-01/04 Gateway East, Singapore 189721, Singapore

If disposing of this product, please recycle the paper.

Foreword

Cancer, an ailment characterized by uncontrolled cell proliferation, has emerged as a leading cause of mortality globally and is likely to persist as a major health threat. In tandem with the rapid advancement of conventional medical technologies for tumor treatment, functional materials as theranostic agents, which integrate diagnostic and therapeutic functions, are gaining prominence. The theranostics approach enables precise therapy through visual imaging and targeted delivery. Unlike traditional diagnostic techniques such as MRI, X-ray, and PET, functional materials as theranostic agents provide benefits, such as minimal invasiveness, reduced radiation, low toxicity, real-time responsiveness, and superior spatial resolution.

Professor Pengfei Shi, Huairong Zhang, Lei Yang, Shusheng Zhang, and Dr. Shujuan Sun have kindly provided this book, which focuses on functional organic materials in tumor diagnosis and therapy. The overall aim of this book is to bring to readers a complete scope of functional organic material design and applications in tumor diagnosis and therapy, toxicological analysis of nanomaterials, and future development trends. Hence, the content of this book covers various functional organic materials, such as self-assembled DNA nanomaterials, small-molecule fluorescent probes, metal-organic framework nanomaterials, and nano-drug carriers in tumor diagnosis and treatment applications. More specifically, this book is divided into five chapters. Chapter 1 provides an overview of self-assembled DNA nanotechnology and the properties of self-assembled DNA nanomaterials. It summarizes their roles in tumor diagnosis, tumor sensing, and tumor imaging and introduces the use of self-assembled DNA nanomaterials in tumor therapy, including chemical drug delivery, gene therapy, phototherapy, immunotherapy, and multimodal therapy. Chapter 2 explores the latest developments in the creation and implementation of theranostic fluorescent probes, particularly highlighting fluorogenic theranostic agents that are activated in specific tumor conditions and their role in enhancing chemotherapy, photodynamic therapy, photothermal therapy, and combined therapeutic approaches. Chapter 3 describes the notable benefits of nanoscale MOFs in tumor diagnosis and therapy, particularly highlighting the preparation of the MOFs, fluorescence-based tumor diagnosis and multimodal imaging, radiation therapy, chemical therapy, chemodynamic therapy, photodynamic therapy, photothermal therapy, hunger therapy, immunotherapy, and multimodal therapy based on MOFs. Chapter 4 gives an overview of single-atom nanocatalysts, mesoporous nanoparticles, liposomes, and polymeric nanoparticles in tumor diagnosis and treatment and outlines the future development trend of using suitable nanomedicine carriers to encapsulate active ingredients with poor solubility but strong curative effect, so that they can maintain activity and stability before reaching the site of action, so as to play the curative effect. Chapter 5 focuses on the toxicological analysis of nanomaterials for tumor diagnosis and treatment and introduces the toxicity profiles of titanium-based nanoparticles, graphene-based nanomaterials, and two-dimensional theranostic nanomaterials and the safety profiles of liposomal nanocarriers and MOF nanoparticles and discusses the new challenges for nanomaterials in nanomedicine.

Functional organic materials for tumor diagnosis and therapy have been widely developed. Relying on the deep understanding and fruitful research achievements of Prof. Pengfei Shi and her research group, the authors present to readers a comprehensive overview of the functional organic materials in tumor diagnosis and therapy. It is for sure that many readers will be able to learn about the preparation of functional organic materials and various tumor imaging and therapy methods from this book. Readers are also encouraged to explore and contribute to the growing literature in this evolving field. We earnestly hope that this book will promote scientific excitement and contribute to the advancement of functional organic materials in biomedicine.

School of Chemistry and Chemical Engineering Hongyuan Chen
Nanjing University, Nanjing, China

Preface

The writing and publication of a book on functional organic materials in tumor diagnosis and therapy are a matter of course of the vigorous development of functional organic materials in biomedicine. This book contains the following three aspects of significance.

First, unlike traditional diagnostic techniques such as MRI, X-ray, and PET, fluorescence imaging based on functional organic materials provides benefits such as minimal invasiveness, reduced radiation exposure, low toxicity, real-time responsiveness, and superior spatial resolution. Consequently, numerous fluorescence-based methods have been developed for cancer detection and treatment, including targeted probes for surgical guidance, responsive probes for early cancer detection, and near-infrared probes for vascular imaging. Novel tumor diagnosis and imaging methods are needed, and the development direction of tumor diagnosis techniques is also reflected in the topic of this book.

Second, functional nanomaterials for tumor-targeted therapy have advanced due to their designability and modifiability. The application progress and prospects of self-assembled DNA nanomaterials, small-molecule fluorescent probes, metal-organic framework nanomaterials, and nano-drug carriers of nanomaterials in radiation therapy, chemical therapy, chemodynamic therapy, photodynamic therapy, photothermal therapy, hunger therapy, immunotherapy, and multimodal therapy are summarized and analyzed in detail in this book, providing new strategies for tumor treatment.

Third, various cancer nano-biomaterials have demonstrated success in preclinical studies, improving therapeutic outcomes, prolonging survival, and/or decreasing side effects. However, the translation from bench to bedside remains challenging. While a number of nano-biomaterials have entered clinical trials, only a few have been approved for clinical applications. Before the application of any nano-biomaterial, it is necessary to evaluate its biological toxicity and enable its safe use in patients. Further clinical trials are carried out under the premise of ensuring that the material exhibits low toxicity in vivo and in vitro. Thus, we focused on six types of nanomaterials and introduced their toxicity. The current challenges and future perspectives for the next stage of toxicity research in nanomedicine will help promote the clinical application of these nanomaterials.

In this book, we cover the design and application of functional organic materials in tumor diagnosis and therapy, the toxicological analysis of nanomaterials, and future development trends. Each chapter is written by scientists in the field, summarizing advances in tumor diagnosis and therapy based on functional organic materials and providing a systematic introduction to help readers understand the achievements in biomedicine.

Linyi, China Pengfei Shi
Linyi, China Huairong Zhang
Linyi, China Lei Yang
Linyi, China Shujuan Sun
Linyi, China Shusheng Zhang

Acknowledgments

It is a special privilege to express our heartfelt gratitude and respect to all the people who have helped us with this book. First and foremost, we would like to give our deepest gratitude to the Fund Committee. Our work has been financially supported by the National Natural Science Foundation of China and the Natural Science Foundation of Shandong Province. Second, we are greatly indebted to all the professors in our institute who have achieved interesting and meaningful research work. Last but not least, we would like to thank the teachers and students at Linyi University. With their help and cooperation, we were able to complete this book successfully.

Contents

Contributors

Yun Luo Shandong Provincial Key Laboratory of Tumor Imaging Equipment Development and Tumor Diagnosis and Treatment Integration Technology, School of Medicine, Linyi University, Linyi, Shandong, People's Republic of China

Pengfei Shi Shandong Provincial Key Laboratory of Tumor Imaging Equipment Development and Tumor Diagnosis and Treatment Integration Technology, School of Medicine, Linyi University, Linyi, Shandong, People's Republic of China

Shu Juan Sun Shandong Provincial Key Laboratory of Tumor Imaging Equipment Development and Tumor Diagnosis and Treatment Integration Technology, School of Medicine, Linyi University, Linyi, Shandong, People's Republic of China

Chuanchen Wu Shandong Provincial Key Laboratory of Tumor Imaging Equipment Development and Tumor Diagnosis and Treatment Integration Technology, School of Medicine, Linyi University, Linyi, Shandong, People's Republic of China

Lei Yang Shandong Provincial Key Laboratory of Tumor Imaging Equipment Development and Tumor Diagnosis and Treatment Integration Technology, School of Medicine, Linyi University, Linyi, Shandong, People's Republic of China

Huai Rong Zhang Shandong Provincial Key Laboratory of Tumor Imaging Equipment Development and Tumor Diagnosis and Treatment Integration Technology, School of Medicine, Linyi University, Linyi, Shandong, People's Republic of China

Shusheng Zhang Shandong Provincial Key Laboratory of Tumor Imaging Equipment Development and Tumor Diagnosis & Treatment Integration Technology, School of Medicine, Linyi University, Linyi, Shandong, People's Republic of China

Chapter 1
Application of Self-Assembled DNA Nanomaterials in Tumor Diagnosis and Treatment

Shu Juan Sun

Abstract DNA nanostructures have emerged as pivotal tools in biomedicine. In this chapter, we review the methodologies for constructing DNA nanostructures and their applications in tumor biomarker detection, imaging, and tumor therapies.

Keywords DNA nanostructures · DNA self-assembly · Tumor biomarker detection · Tumor therapies · Multimodal therapy

1.1 Introduction

Self-assembled DNA nanomaterials possess numerous advantages, including structural stability, excellent biocompatibility, programmability, addressability, and ease of modification, making them highly versatile in biomedical applications, particularly in the diagnosis and treatment of malignant tumors. In recent years, DNA nanostructures have been combined with a variety of nucleic acid signal amplification methods to construct a series of nanoplatforms for tumor-related biomarker analysis and detection through a variety of signal output methods, significantly improving the sensing specificity and sensitivity. At the same time, a variety of nanomaterials based on self-assembled DNA nanostructures have been used in cancer therapy, including chemotherapy, gene therapy, phototherapy, immunotherapy, and multimodal therapy, further advancing the precise and efficient delivery of therapeutic agents and improving overall treatment outcomes.

S. J. Sun (✉)
Shandong Provincial Key Laboratory of Tumor Imaging Equipment Development and Tumor Diagnosis and Treatment Integration Technology, School of Medicine, Linyi University, Linyi, Shandong, People's Republic of China

© The Author(s), under exclusive license to Springer Nature Singapore Pte Ltd. 2026
P. Shi et al. (eds.), *Functional Organic Materials for Tumor Diagnostics and Therapy*, https://doi.org/10.1007/978-981-96-9478-5_1

1.2 Overview of Self-Assembled DNA Nanomaterials

1.2.1 DNA Nanotechnology

DNA is the genetic material that carries biological information. In 1953, Watson and Crick identified the double helix structure of DNA and proposed the principle of complementary base pairing, which laid the foundation for the development of nucleic acid nanotechnology. In the early 1980s, Nadrian C. Seeman pioneered the use of DNA molecules as the basic building blocks of self-assembled nanomaterials (Seeman 1982), which subverted the traditional understanding of the role of DNA and created a new field of structural DNA nanotechnology. Over the following four decades, the continuous development of DNA nanotechnology, researchers using nucleic acids to self-assemble molecular structure exploration, a variety of nanoscale structures and devices have been constructed, has had a significant impact on the development of materials science, nanoscience and biomedicine fields.

1.2.1.1 Principle of DNA Self-Assembly

The DNA molecule consists of nucleotides connected by phosphodiester bonds, and the nucleotide monomer consists of a phosphate group and a deoxyribose connected to a nitrogenous base. These four nitrogenous bases, adenine (A), thymine (T), guanine (G), and cytosine (C), can form four different nucleotide monomers. One end of the DNA strand possesses a phosphate group on the 5′ carbon of the sugar ring, which is defined as the 5′ end, while the other end possesses a hydroxyl group on the 3′ carbon of the sugar ring, which is defined as the 3′ end. DNA molecules can exist in both single-stranded DNA (ssDNA) and double-stranded DNA (dsDNA) forms. According to the principle of Watson–Crick base pairing, A pairs with T and C pairs with G by hydrogen bonds (Fig. 1.1), so that the two complementary ssDNA hybridization along antiparallel directions forms the double helix structure. These structural features of DNA enable programmable self-assembly and structural prediction.

Unique physical and chemical properties make DNA an ideal building block for self-assembling nanomaterials. A helix of dsDNA consists of ten base pairs and is 3.4 nm long, and the width of the double helix is about 2 nm. The rigidity of DNA structure can be regulated by changing the number of base pairs. dsDNA less than 150 bp in length is rigid, while most ssDNA is flexible (Roh et al. 2011; Tinland et al. 1997). In addition, DNA hybridization is reversible. The process of separating dsDNA into two ssDNA is called "denaturation," and the process of recombining dsDNA is called "renaturation." This reversibility is the basis of using temperature to control the state of DNA. Chemically, DNA is water-soluble, which facilitates modification of DNA through a variety of chemical reactions in the water environment. Both the 5′ terminal phosphate group and the 3′ terminal hydroxyl group are commonly used for chemical modification. DNA is biodegradable and can be

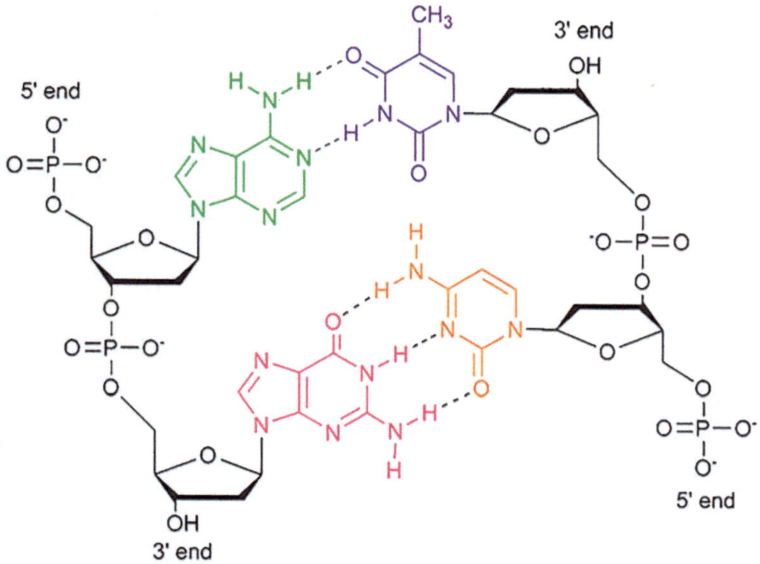

Fig. 1.1 The Watson–Crick base-paring rules. Molecular structures of adenine (A) pairs with thymine (T), and guanine (G) pairs with cytosine (C). Reprinted with the permission from Roh et al. (2011)

degraded into nucleotides under the action of nuclease. It is a natural metabolite in the human body and is nontoxic. These unique physical and chemical properties enable DNA to show higher precision and accuracy in the construction of nanomaterials than other synthetic and natural polymers (Roh et al. 2011).

1.2.1.2 Strategies for DNA Self-Assembly

DNA nanotechnology makes it easy to design and assemble DNA sequences to obtain desired structures. Common DNA self-assembly strategies are divided into three types: modular assembly, folded structure assembly, and dynamic assembly (Ge et al. 2014). The first strategy is modular assembly, which is the first mainstream self-assembly method. The idea is to decompose the target structure into smaller structural units, and use the direct strong force of nucleic acid strand in each structural unit and the weak force between structural units to build the overall structure, which can generally be used to assemble periodically repeated structures. In 1983, Kallenbach et al. (1983) first used four short DNA strands to hybridize with each other to form a four-arm structure (Fig. 1.2a). Its successful design laid the foundation for the assembly of DNA nanostructures with spatial complexity. In 1993, Fu and Seeman (1993) reported the DX module (double crossover) composed of two four-arm junctions (Fig. 1.2b). The specific binding between different DX modules can be achieved by designing viscous ends, and then the periodic

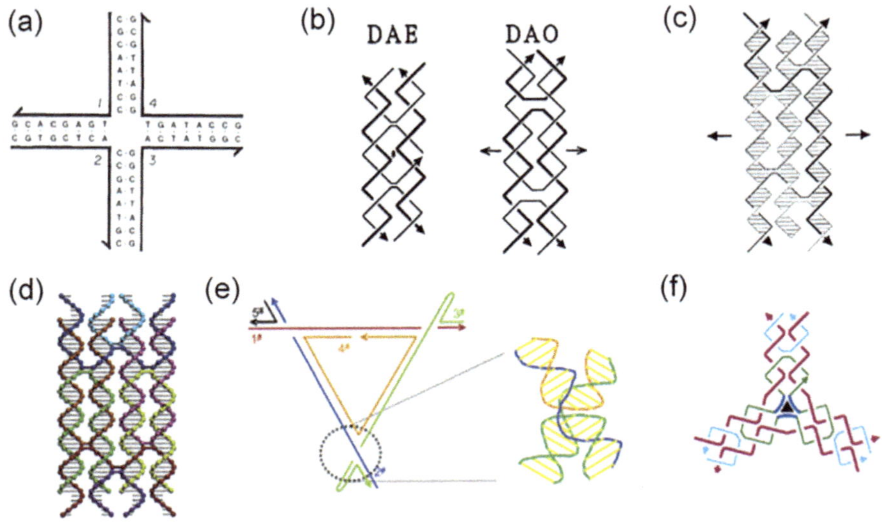

Fig. 1.2 Modular Assembly of DNA Nanostructure Models. (a) A four-arm structure composed of four short DNA strands. Reprinted with the permission from ref. [Kallenbach et al. 1983]. Copyright 1983, Springer Nature Limited. (b) Two four-arm connections composed of DX modules. Reprinted with the permission from ref. [Fu et al. 1993]. Copyright 1993, American Chemical Society. (c) Cross-junction assembly modules containing three DNA double helix regions. Reprinted with the permission from ref. [Labean et al. 2000]. Copyright 2000, American Chemical Society. (d) Cross-junction assembly modules containing four DNA double helix regions. Reprinted with the permission from ref. [Reishus et al. 2005]. Copyright 2005, American Chemical Society. (e) Rigid triangular module based on cross modules. Reprinted with the permission from ref. [Liu et al. 2004]. Copyright 2004, American Chemical Society. (f) A three-point-star DNA module. Reprinted with the permission from ref. [He et al. 2005]. Copyright 2005, American Chemical Society

one-dimensional or two-dimensional DNA lattice structure can be assembled. Subsequently, cross-junction assembly modules containing three (Labean et al. 2000) (Fig. 1.2c) and four (Reishus et al. 2005) (Fig. 1.2d) DNA double helix regions have been reported. Yan et al. (2003) designed a cross-shaped assembly module with four four-arm junctions in 2003. Later, Liu et al. (2004) extended the cross module and designed a rigid triangle module. (Fig. 1.2e). In 2005, He et al. (2005) designed a three-point-star DNA module. The module consists of three strands of DNA wrapped around each other (Fig. 1.2f). By controlling the flexibility of the three branches at the central junction, the module can be assembled into a planar array of DNA or a polyhedral structure. Further, five-point-star (Zhang et al. 2008) and six-point-star modules (He et al. 2006) have been successfully used to assemble DNA nanostructures. In 2008, Yin et al. (2008a) used two pairs of complementary sequences to form a four-domain long strand (2(a), which was self-assembled as a single-stranded tile (called single-stranded tile (SST), which could form two large structures, nanoribbons, and nanotubes.

The second strategy is folding structure assembly, mainly referring to DNA origami, which uses a long ssDNA as a template to fold, while using complementary short strands to form a characteristic structure. In 2006, Rothemund (2006) built a fully fleshed-up two-dimensional structure using a single strand of 7249-base DNA as the backbone and helper strand for binding. By designing different auxiliary strands, various two-dimensional nanostructures such as squares, rectangles, five-pointed stars, smiley faces, and two different triangles have been successfully obtained using DNA origami (Fig. 1.3a). The emergence of DNA origami is a major breakthrough in the field of DNA nanotechnology. Later, Chinese researchers used DNA origami to create the shape of a map of China (Qian et al. 2006) (Fig. 1.3b), the first asymmetrical two-dimensional nanostructure. The concept of DNA strand folding has been used to construct DNA origami structures with a wide range of structural features and complexity (Douglas et al. 2009; Veneziano et al. 2016), for example, DNA honeycombs(Wang et al. 2016), gridirons with four arm junctions (Han et al. 2013), nanocubes (Andersen et al. 2009) (Fig. 1.3c), DNA origami polyhedrons in tripods (Iinuma et al. 2014), and DNA nano-shapes with controllable curvatures (Han et al. 2011) have also been constructed.

The third strategy is dynamic assembly, which is characterized by direct control of the dynamics of self-assembly, usually using the hairpin structure as the starting point of the reaction, and the final reaction product is obtained by strand displacement continuously occurring in the single-strand region. The advantage of this strategy is manifested in the dynamic control of the nanostructure. Dynamic DNA nanoassembly mainly refers to the establishment of nucleotide system with dynamic function, and the nanomechanical device of DNA construction refers to the change of the configuration of DNA complex by changing the conditions (Ge et al. 2014). In 1999, Mao et al. (1999) induced the mutual change of B-DNA and Z-DNA configurations by changing buffers. Subsequently, molecular tweezers powered by DNA were designed (Yurke et al. 2000). In 2002, Yan et al. (2002) designed another DNA machine to define the conversion between PX and JX2 configurations. At the same time, these DNA-assembled nanodevices are also endowed with the function of transportation, through the dynamic switching of the structure to achieve the transportation of functional goods (Goodman et al. 2008; Park et al. 2006; Douglas

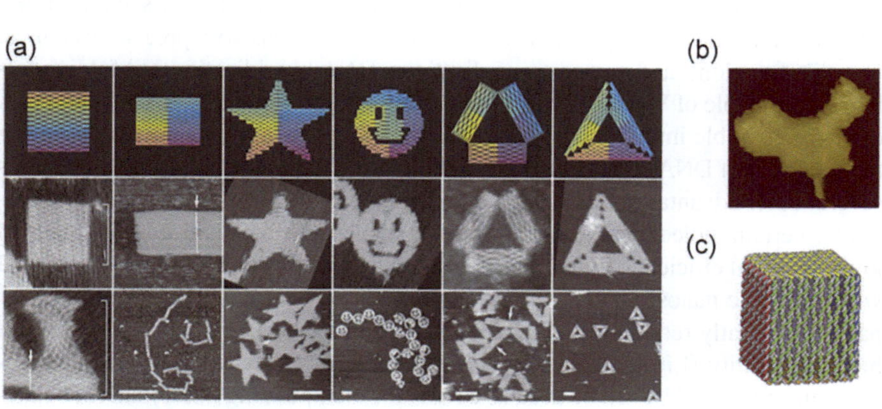

Fig. 1.3 2D and 3D DNA Nanostructures Constructed Based on Folded Structure Assembly. (a) 2D DNA origami nanostructures, including squares, rectangles, five-pointed stars, smiling faces and two different triangles. Reprinted with the permission from ref. [Rothemund 2006]. Copyright 2006, Springer Nature Limited. (b) 2D nanostructure in the form of a China map. Reprinted with the permission from ref. [Qian et al. 2006]. (c) 3D DNA nanostructure: a nanocube. Reprinted with the permission from ref. [Andersen et al. 2009]. Copyright 2009, Macmillan Publishers Limited. All rights reserved

et al. 2012). In addition to the two types of nanodevices, configuration conversion and analog switching, another nanodevice is the DNA walker, which refers to the DNA nanomachines walking in a specific direction in an orderly manner, and it has many walking strategies. The original designers made the DNA walker by manually adding instruction strands (Shin and Pierce 2004; Sherman and Seeman 2004). Later, strategies to assist walker in walking by deoxyribozyme or restriction endonuclease were reported (Tian et al. 2005; Bath et al. 2005). Further, the walking direction of the DNA walker is extended from the linear direction to the two-dimensional (Lund et al. 2010) and three-dimensional planes (Jung et al. 2016). A common strategy for dynamic assembly is the strand displacement reaction, which refers to the process of replacing one of the pre-hybridized DNA strands with a complete or partial complementary match to the target sequence. The reaction is a competitive process. This reaction process begins with a single strand of the target structure (toehold), where the target DNA can replace a pre-hybridized strand of DNA by branch migration. Strand displacement reaction is the most commonly used reaction mode for devices such as circuits, catalytic amplifiers, automatic molecular motors, and reconfigurable nanostructures (Zhang and Seelig 2011). Strand displacement reactions can be used to assemble dynamic nanostructures. First, the hairpin structure of DNA molecule is opened after reacting with the substituted strand, but it is still not separated from the reactant. The opened strand in the hairpin of DNA molecule can be used as a starting strand to continue to react with a new reactant (Yin et al. 2008b).

1.2.2 Properties of Self-Assembled DNA Nanomaterials

DNA nanotechnology holds great promise for future biomedical engineering and the development of novel diagnostic analyses and therapies. Self-assembled DNA nanostructures have emerged as a type of nano-biomaterials with precise structures, versatile functions, and numerous applications. DNA hybridization and folding follow the principle of Watson–Crick base pairing, so nanostructures made from DNA can self-assemble into specific preprogrammed topologies by precisely controlling the sequence of DNA strands for predictability. As a natural nanomaterial, DNA has several other advantages. First, DNA nanostructures demonstrate the ability to recognize certain molecular and membrane signals when functionalized with targeted molecules and efficient cellular internalization (Lee et al. 2016). Second, compared with inorganic nanostructures, DNA nanostructures show more favorable biocompatibility, greatly reduced cytotoxicity in vivo and in vitro, and have more reliable biodegradability (Liu et al. 2016). Third, through DNA hybridization, nanostructures made from DNA can be used as a basis for manipulating other materials functionalized by DNA, and the sub-nanoscale addressability of DNA nanostructures allows them to precisely and tailor-made modify a variety of chemical and biological entities, such as metal nanoparticles or biological components (Zhang et al. 2018). This is conducive to giving DNA self-assembled nanostructures more functionality and properties, making them suitable as precise diagnostic tools and multifunctional vectors for targeted drug delivery. With such a wide range of sizes, custom functionalities, and biological properties (Hong et al. 2017), DNA nanostructures play an important role in many fields such as catalysis (Wang et al. 2014a), biosensing (Lu et al. 2013; Chandrasekaran et al. 2016; Rajendran et al. 2012), bioimaging (Jiang et al. 2016), and drug delivery and therapy (Nishikawa et al. 2019).

1.3 DNA Self-Assembled Nanomaterials in Tumor Diagnosis

With the rapid advancements in DNA nanotechnology, a diverse array of design principles and assembly techniques have been developed to enhance the stability and complexity of DNA nanostructures. When coupled with various sensing modalities and signal amplification strategies, DNA nanostructures are increasingly pivotal in fields such as biosensing and cellular imaging. Cancer-associated biomarkers, including DNA, RNA, proteins, lipids, and metabolites, serve as quantifiable indicators of specific oncogenic states (Huang et al. 2018). The detection of these biomarkers is crucial for distinguishing patients across different clinical stages and for the development of more precise and targeted therapeutic approaches. Owing to their distinct advantages, DNA nanostructures can be engineered as versatile components in the construction of biosensors for tumor biomarker detection.

1.3.1 Overview of Nucleic Acid Probes Based on DNA Nanostructures

The underlying principle of DNA probe identification is the ability of one DNA strand to recognize and hybridize with its complementary strand, forming the characteristic Watson–Crick double helix. Leveraging this, a variety of DNA-based probing strategies have been developed, including fluorescence in situ hybridization (FISH) (Kishi et al. 2019), DNA molecular beacons (MBs) (Wile et al. 2014), and toehold-mediated double-stranded DNA (dsDNA) probes (Yang et al. 2015). Additionally, functional nucleic acids with unique sequences have been identified, encompassing DNAzymes capable of catalyzing specific reactions and aptamers capable of recognizing and binding to specific targets. Ideally, DNA aptamers can be selected to recognize virtually any type of target molecule, including nucleic acids, proteins, ions, small molecules, and even pH levels (Wang et al. 2021a). Given the ease with which DNA can be modified, functionalized, and amplified, DNA probes have found extensive applications across various fields of chemistry and biology. However, simple linear DNA cannot penetrate cells directly and is easily degraded by nucleases in the cellular environment, while DNA, as a building block, can be assembled into nanostructures of various shapes and sizes, many of which exhibit enhanced anti-degradation properties and can be internalized by cells efficiently without the assistance of any transfection reagents. The integration of DNA probes with DNA nanostructures presents a significant potential for the detection of cancer-related biomarkers.

1.3.2 Applications of DNA Nanomaterials in Tumor Sensing

Biosensor is a kind of detecting tool that is highly sensitive for biological substances and capable of converting their concentrations into measurable signals. Biosensors are regarded as powerful tools for assessing biological processes and are particularly valuable in medical diagnostics, especially for serious diseases such as cancer (Mittal et al. 2017). DNA nanostructures, as efficient probe carriers, offer advantages such as low cost, high stability, and modifiability, making them suitable for the detection of various tumor-related biomarkers.

1.3.2.1 Nucleic Acid Sensing

Nucleic acid detection plays a crucial role in disease diagnosis and monitoring. For instance, the methylation of tumor-specific DNA can be detected in the circulating cell-free DNA and has been employed in early cancer screening (Salta et al. 2018). Abnormal miRNA expression is closely linked to tumor pathogenesis, making miRNA a valuable biomarker for early tumor diagnosis (Zhang et al. 2024a; Bertoli et al. 2015). Accurate measurement and imaging of mRNA associated with tumors contribute substantially to precancer identifying (Schwarzenbach et al. 2011).

For miRNA detection, in complex biological environments, shared sequence similarities and low numbers of miRNAs present challenges for biosensing. Gao et al. designed a DNA tetrahedral nanoprobes built on fluorescence resonance energy transfer (FRET) to detect miRNA-146 sensitively, where no enzyme amplification strategy by assistant DNA was involved (Gao et al. 2020a). Liu and his colleagues have proposed a DNA tetrahedron nanostructure (DTN) capable of detecting three kinds of miRNA. In brief, TOTO-1, a type of fluorescent donor, was embedded in nucleic acid chains of DTN, whose three vertices were equipped with three organic dyes. As the target RNAs approach, oligonucleotide chains with dyes could hybridize with them resulting in a FRET between TOTO-1 and the fluorescent stains (Zhao et al. 2020b). Similarly, Yu et al. have demonstrated a self-assembly DNA tetrahedron nanomachine to evaluate intracellular miRNA content. This nanomachine employed the DNA tetrahedron nanostructure as its framework and an allosteric DNAzyme controlled by an miRNA inhibitor as its core. When the target was present, the tetrahedron was activated to undergo a configuration change and generated a fluorescence signal (Yu et al. 2022). Wang et al. (2019a) reported a 3D DNA tetrahedron-mediated hyperbranched hybridization chain reaction (HCR) for the detection and imaging of cellular miR-21. Each vertex of the tetrahedral DNA nanostructure assembles a DNA hairpin to construct a quadrivalent tetrahedral DNA nanoprobe. This strategy can extend the HCR toward multiple reaction directions, increase the collision probability of probes, and achieve a high local probe concentration, thereby greatly accelerating the reaction kinetics of the HCR.

In addition, electrochemical sensing and photoelectrochemical sensing are also used to detect miRNA. Wen et al. (2020) developed a two-phase miRNA biosensing system based on a modular framework nucleic acid (FNA) platform. In the presence of the target miRNA, the probe DNA first hybridizes with the target and the reporter DNA in a homogeneous solution, forming a DNA-RNA complex that includes a 10 nt anchor sequence. Through continuous base stacking, this complex can be effectively immobilized on the interface by means of the 10 nt anchor sequence on the FNA, and thereby generates a signal on the electrode (Fig. 1.4). Photoelectric chemical (PEC) biosensors, which combine the characteristics of light irradiation and electrochemical detection, have the potential for clinical diagnosis based on portability, cost effectiveness, and high sensitivity (Zhao and Ding 2019; Li et al. 2016; Gao et al. 2020b). Li et al. assembled DNA tetrahedrons as tracks on the surface of a TiO_2/MIL-125-NH_2-functionalized paper electrode to construct a two-in-one photoelectric sensor with an antifouling surface and backbone-rigidified interfacial tracks for detection of miRNA-182-5p (Shan et al. 2023). As shown in Fig. 1.5, the presence of miRNA-182-5p can gradually transform the DNA three-way junction into a circular DNA walker. The DNA walker can hybridize with a PDA-labeled DNA probe on the DNA tetrahedral through the multi-recognition fragments it contains. With the help of Nt.BstNB I endonuclease, PDA-labeled ssDNA and DNA walker are released, triggering the next cyclic walking and cleaving activity. The ultrasensitive and selective detection of target miRNA in complex substrates is achieved by generating significant photocurrent response.

For mRNA detection, fluorescence analysis based on nucleic acid is identified as one of the most powerful techniques in recent years. On account of the excellent biocompatibility and stability of DNA tetrahedron frameworks, several DNA tetrahedron-based nanoprobes have been assembled to detect RNA in cells (Zhou

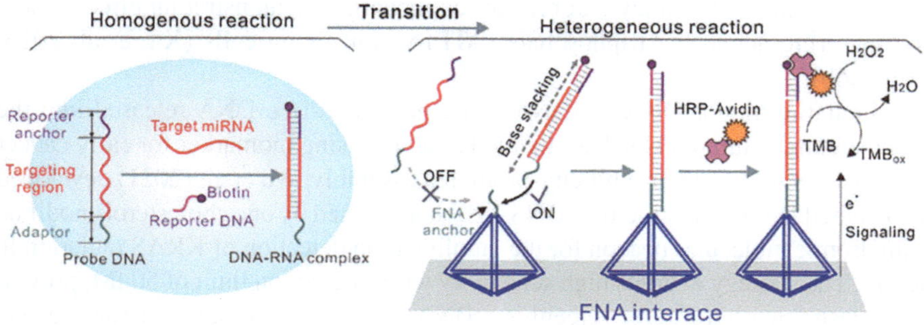

Fig. 1.4 DNA Framework-mediated electrochemical biosensing platform for microRNA analysis. Reprinted with the permission from Wen et al. (2020). Copyright 2020, American Chemical Society

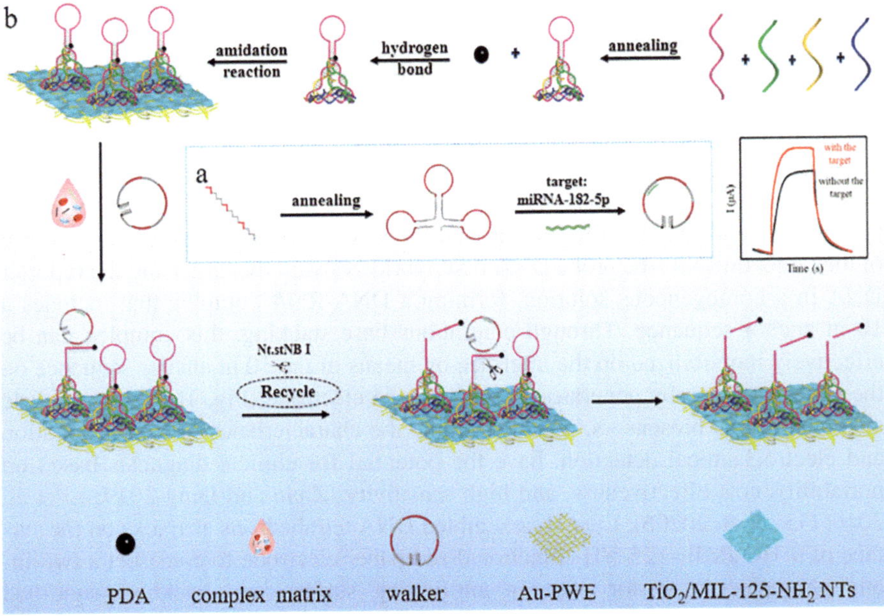

Fig. 1.5 Tetrahedral DNA nanostructure-based photoelectrochemical sensing platform for microRNA analysis. Reprinted with the permission from Shan et al. (2023). Copyright 2023, American Chemical Society

et al. 2016). He et al. had demonstrated a DNA tetrahedral nanoprobe utilizing FRET also known as DNA tetrahedral nano-tweezers, which was successfully applied in the detection of tumor-associated mRNA. This probe could sense intracellular mRNA efficiently and reliably because false positive signals could be avoided by FRET "off" to "open" probes and DNA tetrahedral nanoprobes had better cellular permeability (He et al. 2017) (Fig. 1.6a). Additionally, Xie et al. introduced a competitively mediated FRET switch-based self-assembled DNA tetrahedral molecular beacon to address the challenge of inadequate signal separation between identification element with signal module and single functional. This technique exhibited superior accuracy and broader applicability, demonstrating effectiveness in measuring adenosine triphosphate (ATP) within living cells (Xie et al. 2016) (Fig. 1.6b).

Circulating tumor DNAs (ctDNAs), a type of cell-free DNA released into the bloodstream by tumor cells, has emerged as a promising biomarker for early cancer diagnosis. Based on hydrophilicity-promoted assembly, Wu et al. (2021) developed a core/shell DNA nanostructure (DNS-MB) constructed by one cholesterol-modified palindromic molecular beacon for the amplification detection of KRAS mutation in serum. The strategy exhibits high sensitivity with a detection limit of 50 fM, providing a desirable platform for screening ctDNAs. Similarly, the same research group developed a reconfigurable lipidic moiety-attached DNA nanoparticle constructed from a cholesterol-conjugated multifunctional hairpin-type DNA probe by

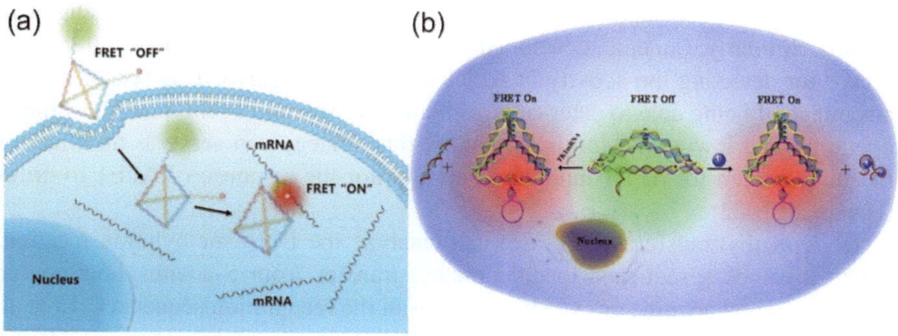

Fig. 1.6 Schematic illustrations of FRET-based DNA tetrahedron probes for intracellular mirRNA detection. (**a**) FRET-based DNA tetrahedron for mRNA detection. Reprinted with the permission from He et al. (2017). Copyright 2017, American Chemical Society. (**b**) Competition-mediated FRET-switching DNA tetrahedron for intracellular molecular detection. Reprinted with the permission from Xie et al. (2016). Copyright 2016, American Chemical Society

hydrophobicity-mediated self-assembly. The cancer-related p53 gene can be detected down to 10 pM by the DNA nanoparticle (Xue et al. 2022).

1.3.2.2 Protein Sensing

Proteins are also one of the meaningful biomarkers in terms of screening, monitoring and treatment of various diseases (Cohen et al. 2020). DNA-binding proteins recognize their specific targets through a combination of two readout mechanisms: base readout and shape readout (Slattery et al. 2014). Base readout refers to the direct interaction between proteins and DNA bases. Most protein–DNA interactions can be thought of as forming stable complexes or recognizing DNA structures to read bases indirectly. While shape readout is the ability of a protein to recognize a specific overall or local shape of DNA at its binding site (Azad et al. 2018). DNA shape readout relies on both intrinsic and protein-induced DNA deformations, including local structural features such as minor groove narrowing and base pair geometry (Lin and Guo 2019; Jolma et al. 2013).

Apurinic/apyrimidinic endonuclease 1 (APE1) plays a critical role in DNA repair and redox regulation, making tumor cells more susceptible to radiation and chemotherapy by modulating their intracellular activity. Consequently, precise sensing and imaging of APE1 hold a significant potential for advancing cancer diagnosis and treatment (Wang et al. 2009). Su et al. developed a novel strategy for probing and regulating intracellular APE1 activity by using DNA tetrahedrons. The tetrahedron with an AP site-containing antenna (OUT-tetrahedron) exhibits high sensitivity and specificity to APE1 and was used for in vitro detection of APE1 and cellular fluorescence imaging. The tetrahedron with an apurinic/apyrimidinic (AP) site on its scaffold (IN-tetrahedron) exhibits high binding affinity to APE1 but limits enzymatic catalysis, which makes it an APE1 inhibitor. It was used for suppressing the APE1 activity in living cells and sensitizing cancer cells to anticancer drugs (Zhang et al. 2019a). Similarly, Liu et al. developed a molecular beacon based-DNA tetrahedron sensor for detecting the activity of APE1 in vitro and in vivo. This sensor has the advantage of being able to enter cells without a transfection agent, featuring simple operation and low cost. More importantly, this strategy can serve as an effective means to screen for APE1 inhibitors and monitor the response of APE1 to drugs (Han et al. 2022).

DNA methyltransferase (MTase) can catalyze the DNA methylation reaction, which involves a sequence-specific methyl transfer from S-adenosylmethionine (SAM) to target adenine or cytosine residues in the recognition sequences (Li et al. 2015; Hou et al. 2019). DNA methylation plays an important role in the regulation of gene expression, cell growth, and proliferation in both prokaryotes and eukaryotes. Aberrant DNA methylation has been proven to influence gene expression and the interaction between proteins and DNA, which may be associated with various diseases, including leukemia. Therefore, it has been regarded as a new family of disease biomarkers. DNA adenine methyltransferase (Dam) is a kind of DNA MTase. Gao et al. (2020c) reported a strategy for onsite analysis of Dam activity by employing DNA tetrahedra-based hydrogel and amyloglucosidase; the DNA tetrahedrons contained recognition sites and were also used as the building blocks for constructing hydrogel, which can realize target recognition and signal transduction and be used as a simple and user-friendly POCT tool.

Zhuo et al. proposed a regulatable DNA nanodevice, termed the AND logic gate-regulated DNAzyme nanoflower, for specifically monitoring the activity of DNA repair enzymes, with the aim of enabling early evaluation of cancer mediated by genomic instability (Zeng et al. 2024) (Fig. 1.7). The DNAzyme nanoflower was rationally designed through the self-assembly of DNA duplexes modified with both AP sites and methyl lesion sites. It could be reconfigured under the repair of AP sites by APE1 and O6-methylguanine sites by O6-methylguanine methyltransferase (MGMT), producing a fluorescent signal that realizes sensitive monitoring of the activities of APE1 and MGMT.

The mucin 1 protein (MUC-1) is an important cancer biomarker associated with colon, breast, lung, and many other cancers (Huang et al. 2020; Han et al. 2021). Because its expression level is relatively low in the early stage, sensitive and rapid detection is of great significance for the clinical diagnosis of cancer. Hu et al. (2022) used tetrahedral DNA nanostructure with multiple biomolecular recognition domains to develop an ultrasensitive electrochemical biosensor for MUC-1 protein. These multiple biomolecular recognition domains enhance the sensitivity of detection. When the target MAC-1 is captured by the protein aptamer located on the DNA tetrahedron, the ferrocene-labeled DNA strands will be released from the electrode surface, thereby producing a significantly reduced electrochemical signal (Fig. 1.8). This strategy can sensitively detect MUC-1 with the limit of 0.31 fg mL^{-1}.

DNA nanostructures are not only being used to detect a single target, but are also being used to perform more complex detection tasks through rational design. For example, Zhou et al. (2020b) reported a DNA tetrahedron module as a versatile unit for diverse parallel and multiplexed sensing. Three different fluorophore-quencher

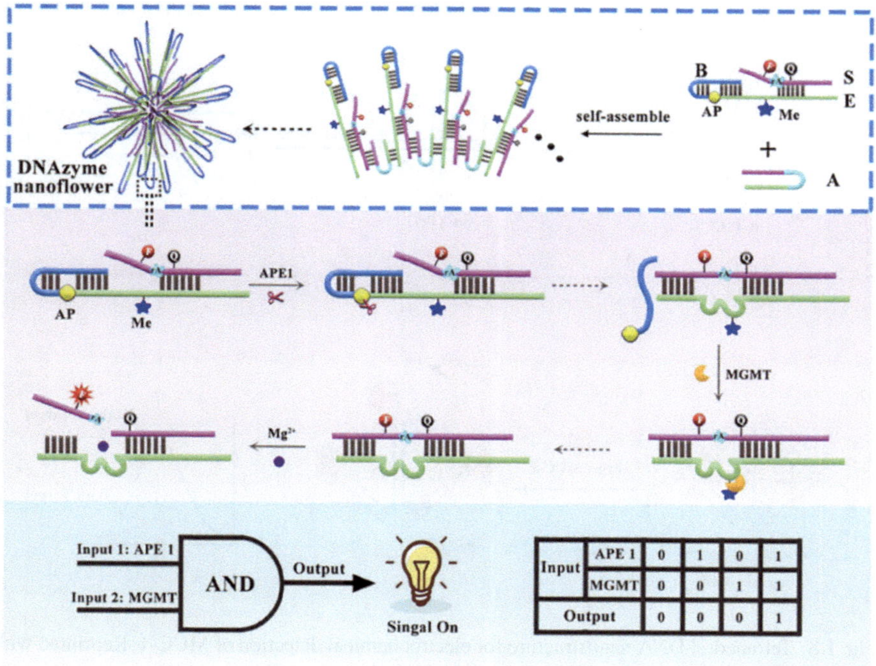

Fig. 1.7 DNAzyme nanoflower structure and its logic gate operation for APE1 and MGMT Detection. Reprinted with the permission from Zeng et al. (2024). Copyright 2024, American Chemical Society

pairs and predesigned hairpin structures were introduced onto the edges of the tetrahedra. These structures allow three different miRNAs (miRNA-21, miRNA-221, and miRNA-155) to uncage the respective hairpins, leading to the switched-on fluorescence of the corresponding fluorophores. By introducing recognition sites for different endonucleases onto the edges of the DNA tetrahedron, a module is constructed that can respond to three different endonucleases (Nt.BbvCI, EcoRI, and HindIII). Finally, a DNA tetrahedral module was constructed to detect ATP, VEGF, and Thrombin, which, unlike the first two modules, responds to the target with decreased fluorescence. This strategy can be used for imaging of MCF-7 and HepG2 cancer cells and their distinction from normal cells.

1.3.2.3 ATP Sensing

Adenosine triphosphate (ATP) is the primary energy carrier within cells and plays a critical role in various biological processes (Alberts 2002). It provides energy for cellular metabolism, ion transport, and signal transduction, while also regulating cell proliferation and apoptosis (Nelson and Cox 2000; Knowles 1980). In the tumor microenvironment, ATP metabolism and energy demands undergo significant

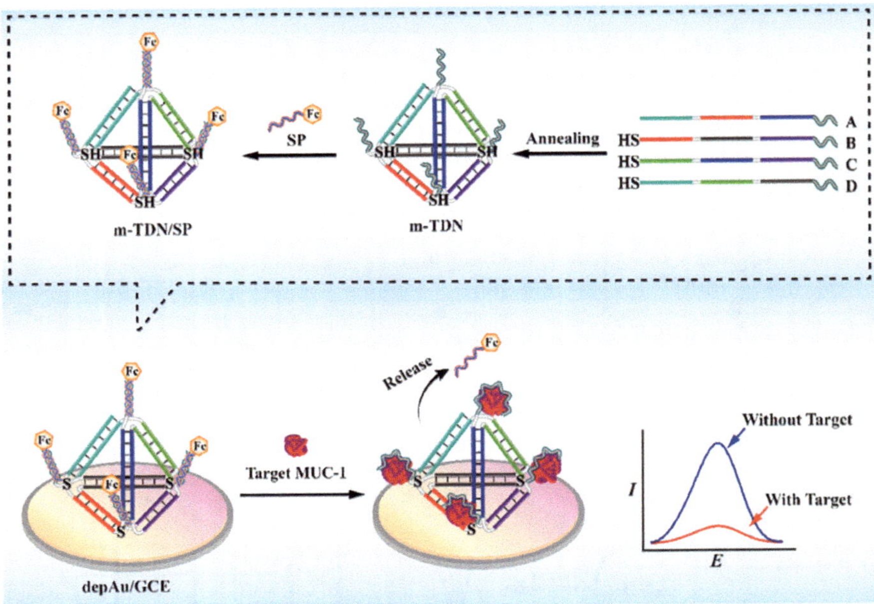

Fig. 1.8 Tetrahedral DNA nanostructure for electrochemical detection of MUC-1. Reprinted with the permission from Hu et al. (2022). Copyright 2022, American Chemical Society

changes (Vlaskovska et al. 2001; Ma et al. 2008; Gourine et al. 2005) with tumor cells preferring glycolysis for ATP production, even in the presence of oxygen—a phenomenon known as the Warburg effect (Vander Heiden et al. 2009). ATP is not only closely linked to tumor biology but also plays a pivotal role in the pathology of various diseases, including neurodegenerative disorders, cardiovascular diseases, metabolic disorders, and inflammation-related conditions (Spari and Beldi 2020; Mattson et al. 2008; Idzko et al. 2014; Gottesman et al. 2002). Thus, ATP is central to cellular energy regulation and serves as a key signaling molecule within the tumor microenvironment. In recent years, ATP sensing technologies have found increasing application in tumor research, providing powerful tools to elucidate tumor metabolism and develop novel diagnostic and therapeutic strategies (Spari and Beldi 2020; Huang et al. 2022a; Deng et al. 2020).

Traditional chemical sensing methods (Wang et al. 2014b; Moro et al. 2010; Imamura et al. 2009), including electrochemical sensors (Yao et al. 2009; Llaudet et al. 2005; Huang et al. 2010) and colorimetric methods (Tsuboi et al. 2004; Bell et al. 2007), are constrained by their inability to enable real-time ATP detection in living systems. In contrast, metal nanostructures, with their excellent optical properties, large surface area, and superior biocompatibility, provide high sensitivity and selectivity, making them ideal for ATP sensing applications. Yu et al. (2018) developed an innovative sensor using Metal-Enhanced Near Infrared Fluorescence (ME-NIRF) technology, incorporating the fluorescence signal enhancement properties of gold nanoparticles to achieve remarkable sensitivity in ATP detection. The sensor operates by specifically binding near-infrared dyes to ATP, resulting in detectable changes in fluorescence intensity. Li et al. (2022a) reported a device termed "Enzymatically Controlled Nanoflares," which integrates fluorescently labeled nanoparticles with enzyme-responsive probes. Each nanoparticle is functionalized with fluorophores capable of emitting a signal upon interaction with specific enzymes. In the presence of target molecules, these molecules activate the enzymes, triggering a reaction that either releases or modifies the fluorescence of the probes. This mechanism enhances the sensitivity and specificity of molecular detection by utilizing enzyme-mediated signal changes. The system facilitates precise molecular recognition and biosensing, enabling detailed downstream analysis based on the altered fluorescence signal (Fig. 1.9a). Ren et al. (2021) proposed a

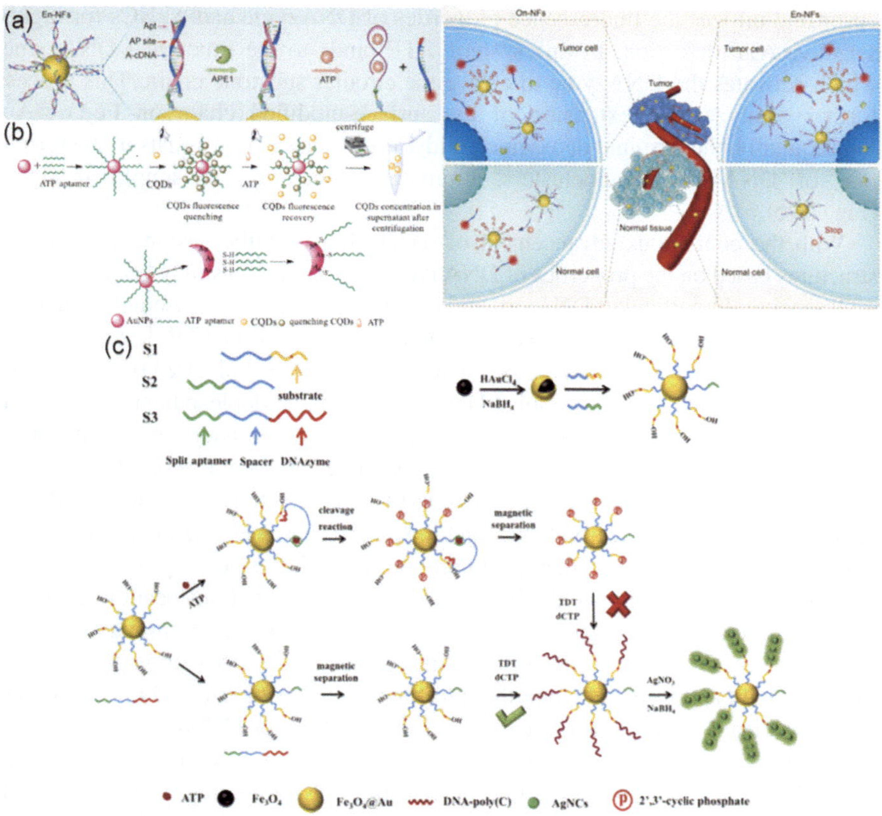

Fig. 1.9 Schematic of ATP sensing platform based on metal nanomaterials (**a**) The enzymatically ATP sensing mechanism of En-NFs in tumor cells. Reprinted with the permission from Zhao et al. (2022). Copyright 2022, American Chemical Society. (**b**) AuNPs and CQDs-based sensing platform for ATP detection. Reprinted with the permission from Ren et al. (2021). Copyright 2021, Author(s). This article is distributed under a Creative Commons Attribution (CC BY) license. (**c**) ATP fluorescence adapter constructed based on Fe_3O_4@Au-S1/S2 composites. Reprinted with the permission from Cai et al. (2023). Copyright 2023, The Royal Society of Chemistry

label-free fluorescence aptasensor for detecting ATP by combining gold nanoparticles (AuNPs) with carbon quantum dots (CQDs). In this system, ATP specifically binds to the aptamer conjugated to the AuNPs, resulting in alterations in the fluorescence of the CQDs. The sensor exhibits significantly enhanced sensitivity and selectivity due to the strong interaction between ATP and the aptamer, enabling accurate detection even at low ATP concentrations. This approach minimizes background noise and is well-suited for real-time biosensing applications (Fig. 1.9b). He and his colleagues developed a ratiometric fluorescence biosensor for detecting ATP utilizing silver nanocluster beacons (SNCs) and ligand-triggered rolling circle amplification (RCA). In this system, ATP specifically binds to an aptamer, triggering the rolling circle amplification reaction. Subsequently, the silver nanoclusters emit a fluorescence signal dependent on the presence of ATP, enabling ratiometric fluorescence detection (Qian et al. 2023). Cai et al. (2023) designed a fluorescence aptasensor that utilizes the fluorescence properties of DNAzyme and AgNCs for highly sensitive ATP detection. In this system, ATP binds to the functional DNAzyme, which activates the DNAzyme to cleave the circular substrate chain. This process further catalyzes the extension of the surface-modified chain on Fe_3O_4@Au. Subsequently, the terminal deoxynucleotidyl transferase (TdT) assists in the formation of DNA-silver nanoclusters, ultimately generating an amplified signal (Fig. 1.9c).

With the continuous advancement of nanotechnology, the design of functional structures based on the properties of DNA molecules has increasingly been applied to ATP sensing. In particular, DNA origami technology, with its precise nanostructures and enhanced specificity, demonstrates significant potential in ATP detection (Yuan et al. 2024; Puchkova et al. 2015; Ochmann et al. 2017; Li et al. 2023a). Gao and his colleagues reported a self-assembled DNA origami-based duplexed aptasensor combined with centrifugal filters for efficient and rechargeable adenosine triphosphate (ATP) detection. In this system, DNA origami technology is used to construct duplexed aptasensors with specific structures that can selectively recognize ATP molecules. With the assistance of centrifugal filters, the sensor effectively separates and concentrates ATP samples, thereby improving detection efficiency and sensitivity (Wang et al. 2022b). Ding et al. (2022) designed a nanopore translocation technique based on shaped DNA origami carriers, with surface modification through aptamers to enhance performance. In this system, the surface of the DNA origami carriers is modified with aptamers, allowing them to specifically recognize and bind ATP. Sen and his colleagues developed a highly sensitive method for ATP detection based on single-molecule fluorescence enhancement. In this system, DNA origami is used to precisely position and assemble Au@Ag nanostar (NSs) structures, which act as optical antennas to significantly enhance the fluorescence signal. When ATP binds to a specific aptamer, the fluorescence signal is amplified, enabling high-sensitivity ATP detection at the single-molecule level (Kaur et al. 2023) (Fig. 1.10).

The emergence of framework nucleic acids (FNA) has introduced a valuable approach for designing highly reliable aptasensors (Zhao et al. 2023; Zhang et al. 2021b; He et al. 2023). Among them, tetrahedral FNA (tFNA), characterized by its stable three-dimensional DNA nanostructure, has emerged as an ideal nanocarrier

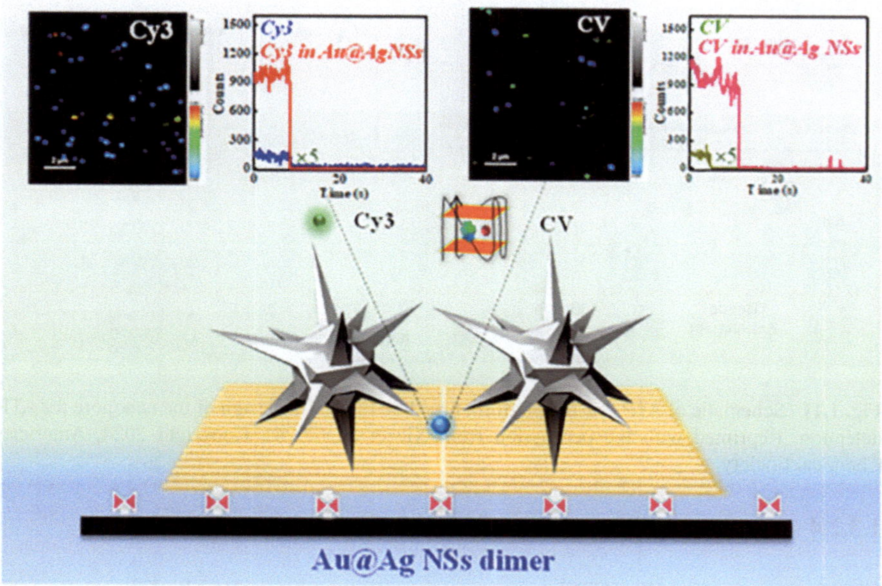

Fig. 1.10 Schematic of ATP sensing platform based on DNA Origami. Au@Ag nanopillar optical antenna assembled using DNA origami for sensitive ATP detection. Reprinted with the permission from Kaur et al. (2023). Copyright 2023, American Chemical Society

for ATP sensing (Mao et al. 2020; Hu et al. 2019; Guo et al. 2022). Xie et al. (2024) developed a highly sensitive method for ATP detection. In this approach, nucleic acid nanostructures are designed as carriers, while nanopore sensors facilitate the specific recognition and detection of small molecules. When the target small molecule interacts with specific binding sites within the nucleic acid structure, the nanopore sensor produces a change in electrical signal, enabling quantitative detection (Fig. 1.11). Xie et al. (Huang et al. 2024) developed a magneto-fluorescent nanobiosensor based on tFNA and an aptamer strategy for trace ATP detection. In this system, aptamers are immobilized on the tetrahedral carrier; when ATP binds to the aptamers, the fluorescence signal significantly increases, enabling highly sensitive detection of trace ATP. Li et al. (2024a) developed a selective electrochemiluminescence sensor based on a DNA tetrahedron molecular sieve. The DNA tetrahedral structure functions as a molecular sieve, selectively allowing the target molecule ATP to enter the sensor based on its size. This molecular sieve effect enables the sensor to precisely distinguish molecules of different sizes and detect them via electrochemiluminescence signals, facilitating highly sensitive ATP detection.

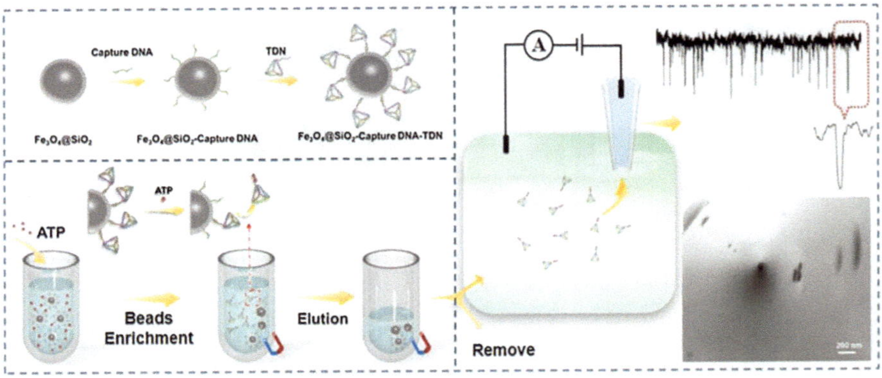

Fig. 1.11 Schematic of ATP sensing platform based on FNA. The design of the nanopore for ATP detection. Reprinted with the permission from Xie et al. (2024). Copyright 2024, American Chemical Society

1.3.2.4 Circulating Tumor Cell (CTC) Sensing

Circulating tumor cells (CTCs) are tumor cells that have detached from primary tumor tissue and are present in peripheral blood, carrying crucial markers of tumor metastasis (Micalizzi et al. 2017; Krebs et al. 2014). As an alternative source of cancer cells, CTCs have garnered significant attention for their role in the early detection, prevention, and mechanistic studies of tumor metastasis (Riethdorf et al. 2010; Pantel and Brakenhoff 2004; Fidler 2003) and the CTC analysis is increasingly recognized as a distinctive real-time liquid biopsy method for cancer patients (Nadal et al. 2013; Alix Panabières and Pantel 2013).

Due to their numerous advantages, including small size, ease of modification with various ligands, and a high surface-to-volume ratio, nanoparticles have been widely employed in the detection of CTCs. Moreover, the interaction between nanoparticles and CTCs significantly enhances enrichment efficiency and detection sensitivity (Ding et al. 2020). To achieve ultrahigh sensitivity and specificity in the capture of CTCs and to facilitate downstream cellular and molecular analysis, Chen et al. (2019a) developed a device named "NanoOctopus" by connecting magnetic microparticle with long multimerized aptamer DNA strands. Each DNA sequence contained 4500 repeating "suckers" of DNA aptamer sequences that could specifically combine with biomarker proteins on the surface of target cells. The sensitivity and specificity were enhanced through multivalent binding of the DNA aptamer to cell receptors. Moreover, captured cells could be released by DNase I, which would allow for downstream cellular and molecular analysis (Fig. 1.12a). Sun et al. (2020) developed an assay platform based on fluorescence spectrometry for the highly sensitive CTC detection by combining of immunomagnetic separation with aptamer-mediated double rolling circle amplification (Fig. 1.12b). In addition, Liu et al. (2024) developed a high-sensitivity strategy targeting breast cancer cells, utilizing a homogeneous DNAzyme and controllable cholesterol-stacking DNA machine that

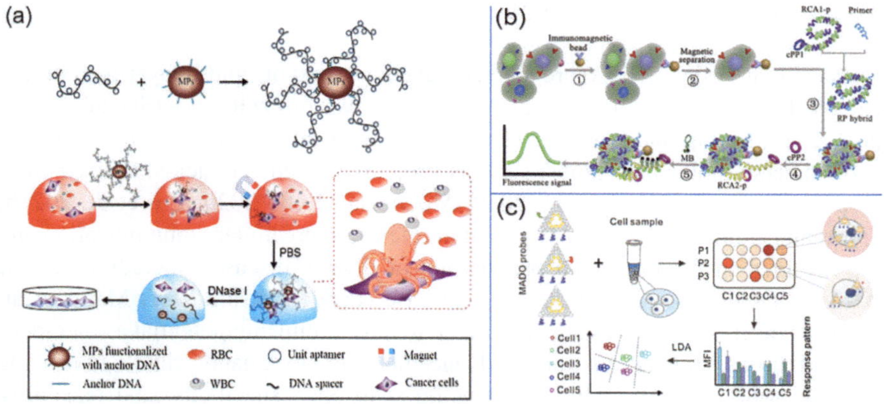

Fig. 1.12 DNA nanostructures for circulating tumor cell sensing (**a**) Schematic illustration of the preparation and application of magnetic particles (MPs) functionalized with anchor DNA for targeted cancer cell capture. Reprinted with the permission from Chen et al. (2019b). Copyright 2019, American Chemical Society. (**b**) Immunomagnetic-assisted and aptamer-mediated double rolling circle amplification for CTCs detection. Reprinted with the permission from Sun et al. (2020). Copyright 2020, American Chemical Society. (**c**) Diagram showing the use of MADD probes in combination with a linear discriminant analysis (LDA) for the identification and differentiation of various cell types within a sample based on their response patterns. Reprinted with the permission from Chen et al. (2022c). Copyright 2022, American Chemical Society

integrates a dual-target recognition aptamer system for CTCs quantitation. Chen et al. (2024) reported the use of DNA-tetrahedral frameworks anchored with aptamers as probes to specifically target receptor–ligand interactions on the cell membrane for capturing circulating tumor cells. Li and co-authors devised a novel chemical nose sensor that is based on multiple aptamer-integrated DNA origami (MADO) for the high-precision identification of cancer cells (Fig. 1.12c) (Chen et al. 2022b).

In addition to CTCs, some studies have directly identified cancer cells. For example, Li et al. (2023a) reported an extracellular ATP-activated HCR for sensitive detection of cancer cells. The Apt-trigger probe consists of an aptamer sequence for specific recognition of target cells and a trigger sequence for initiating the HCR assembly. In the presence of ATP, the ATP aptamer will bind to ATP, causing it to dissociate from hairpin H1. Subsequently, the dissociated hairpin will bind with the ATP-trigger, thereby initiating HCR with hairpin H2 on the target cell surface. The strategy enables quantitative analysis with a detection limit of 25 cells in 200 μL of binding buffer and also allows for the identification of target cells among control cells and mixed cell samples.

1.3.2.5 Extracellular Vesicles Sensing

Extracellular vesicles (EVs) are a heterogeneous group of lipid bilayer membrane-enclosed nanovesicles (30–1000 nm) released by various cell types (Sabaté del Río et al. 2022; Kalluri and LeBleu 2020; Bordanaba-Florit et al. 2021). These vesicles carry diverse molecular cargos, such as proteins and nucleic acids, derived from their parental cells (Yang et al. 2019; Möller and Lobb 2020; Liu et al. 2021b; Herrmann et al. 2021). EVs play essential roles in intercellular communication and are involved in the onset and progression of diverse diseases, including cancer, Alzheimer's disease, and Parkinson's disease (Penders et al. 2021; Marar et al. 2021). Cancer-derived EVs are considered to have promising potential as therapeutics and disease biomarkers. Rapid advancements in DNA nanotechnology provide valuable tools for EV analysis, enhancing sensitivity, specificity, and feasibility (Zhang et al. 2020; Deng et al. 2024). A wide range of DNA nanomaterials, including linear DNA probes, DNA nanostructures, and hybrid DNA nanomaterials, has been utilized for extracellular vesicle analysis.

Exosomes (EXOs) are nanosized (30–100 nm) membrane-enclosed extracellular vesicles derived from many cells, including leukocytes, epithelial cells, and cancerous cells (Valadi et al. 2007; György et al. 2011). EXOs carry a variety of biomolecules, including proteins and nucleic acids, and play a role in intercellular communication (Pisitkun et al. 2004; Kooijmans et al. 2012; Février and Raposo 2004). They have been established as important biomarkers for disease diagnosis and treatment. Currently, the separation and extraction of exosomes with minimal damage and high purity remain a challenge due to the complexities of biological media, which is crucial for downstream applications. To solve this problem, Tang et al. (2023a) developed a DNA-based hydrogel for specific and nondestructive separation of EXOs from complex biological media. In this strategy, RCA is used to synthesize super-long DNA strands containing polyvalent aptamers, and DNA hydrogels are formed by pairing complementary bases. These polyvalent aptamers can bind to receptors on EXOs, and this specific and efficient binding ensures highly specific EXOs separation during the formation of DNA hydrogels. The introduction of other DNA probes can detect EXOs pathogenic miRNAs and accurately identify breast cancer patients (Fig. 1.13). In addition, this EXOs-encapsulated DNA hydrogel can be used to treat myocardial infarction in mice. This research is expected to advance the development of EVs in the field of nucleic acid nanomedicine.

Zhang et al. (2019a) used triangular pyramid DNA nanostructures loaded with gold nanoparticles through the electrostatic interaction to obtain a novel Raman probe. By assembling Raman molecules and DNA recognizable sequences on the nanostructures, strong electromagnetic hot spots may be generated at the junctions between AuNPs, resulting in a strong surface-enhanced Raman scattering (SERS) signal. This strategy can be used to detect CTC and exosomes with high specificity and sensitivity (Fig. 1.14).

MiRNA-based liquid biopsies are subject to many limitations in clinical application, such as heterogeneous sources in plasma and cumbersome detection procedures. To address these challenges, Lei et al. (2023) developed a one-pot assay

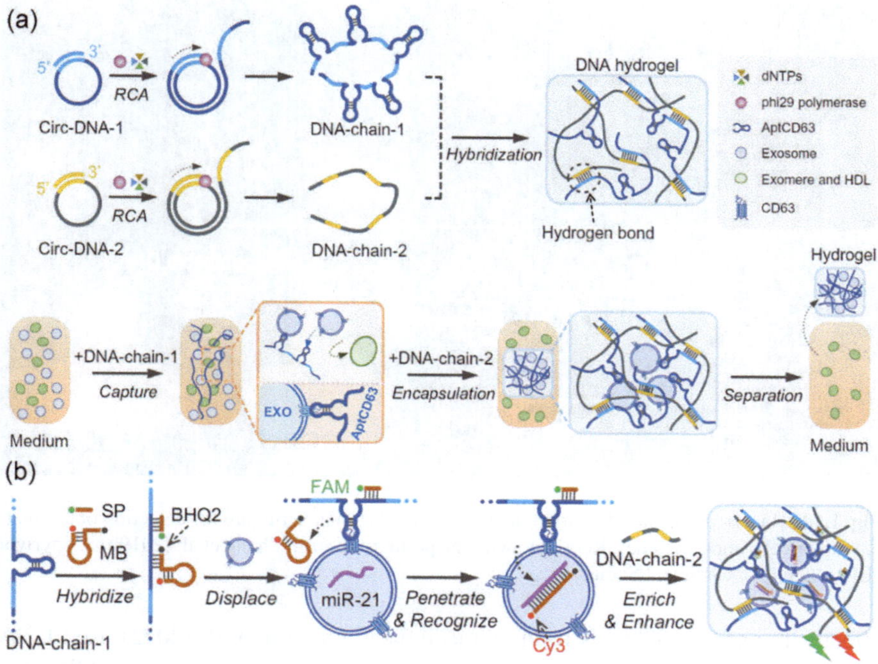

Fig. 1.13 DNA-based hydrogel for EXO separation and sensing. (**a**) The construction process of DNA-based hydrogel and the principle of EXO separation. (**b**) Molecular beacons (MB) and single probes (SP) were used to detect exosome microRNA-21 based on DNA-based hydrogel. Reprinted with the permission from Tang et al. (2023b). Copyright 2023, the Author(s)

called dual-surface protein-guided orthogonal recognition, which can achieve in situ profiling of microRNAs to detect tumor-derived exosomal miRNAs and enhance the diagnostic accuracy of prostate cancer. As shown in Fig. 1.15a, in this strategy, two allosteric aptamers targeting exosomal markers CD63 and tumor marker EpCAM were used to create an orthogonal labeling barcode, and the selective sorting of tumor-specific exosome subtypes was achieved based on the zipper-like hybridization between orthogonal barcode and cholesterol-labeled complementary DNA tags. The labeled barcode on tumor-derived exosomes mediated targeted membrane fusion with liposome probes importing gold nanoflares as miRNA detection reagents, enabling in situ profiling of tumor-derived exosomal miRNAs. Similarly, Xie et al. (2022) reported a DNA zipper-mediated membrane fusion for exosomal miRNA detection and cancer diagnosis. As shown in Fig. 1.15b, a lipid vesicle probe containing miR21-targeting molecular beacons (MBs) is constructed and loaded with zipper DNA constructs (ZDCs) on its surface. The complementary zipper DNA constructs (cZDCs) are modified on the exosome of interest. Upon mixing, zipping between ZDC and cZDC mediates the membrane fusion of EXOs and vesicle probes, triggering the recognition of exosomal miR21 by MBs. The strategy shows great potential in liquid biopsy for early cancer diagnosis. In order to

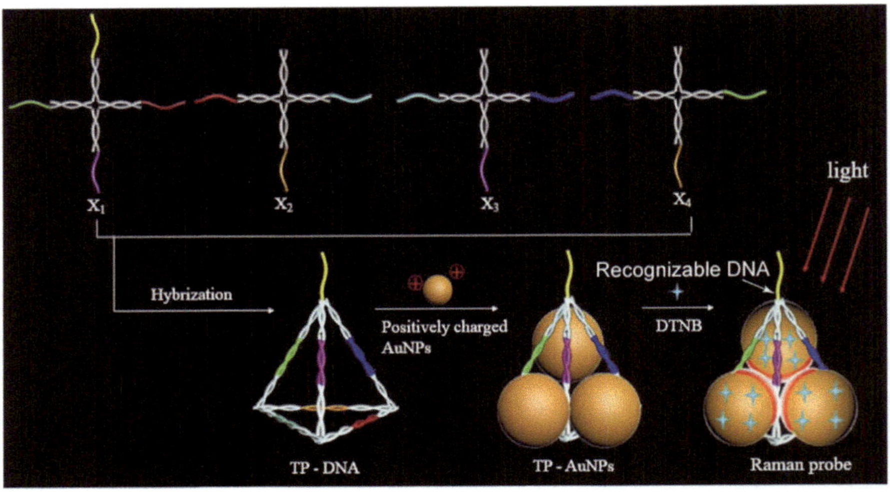

Fig. 1.14 Raman probe based on triangular pyramid DNA nanostructure for detection of circulating tumor cells and exosomes. Reprinted with the permission from Zhang et al. (2019b). Copyright 2019, American Chemical Society

effectively enrich exosomes from extracellular fluids, Lu et al. (2022) used DNA tetrahedral nanostructures as "nanoweights" during centrifugation, enabling the enrichment and quantification of cancerous exosomes in human serum. Two different DNA tetrahedral nanostructures (DTNs), each carrying a specific aptamer, were constructed for exosome biomarker recognition. Upon incubating with clinical samples, one DTN triggered the cross-linking of multiple target exosomes, thus enabling low-speed and fast centrifugation for enrichment. The other DTN further refined the target exosome subtype and initiated an HCR for sensitive signal amplification.

Wang et al. reported a method featuring successive two-stage HCR and DNAzyme for sensitive, in situ detection of miRNAs in EVs (Fig. 1.16a) (Wu et al. 2019). The DNA strands for cascade HCR were delivered into EVs through electroporation. The target miRNAs in EVs triggered the first HCR, which exposed the initiator sequences for the second HCR. The two-stage HCR brought numerous pairs of DNAzymes subunits into close proximity, activating the DNAzymes to cleave the reporter probes and generate a fluorescent signal. Zhao et al. (2023) reported a DNA cage-based thermophoretic assay for selective in situ analysis of mature miRNAs in EVs, effectively eliminating interference from pre-miRNAs. The DNA cage encapsulated pre-hybridized sequences, including a fluorophore-labeled recognition sequence (F) and a quencher-labeled complementary sequence (Q), within a confined cavity, permitting the entry of short mature miRNAs while excluding longer pre-miRNAs. After incubating EVs with the DNA cage probes designed for mature miRNA recognition, the EVs loaded with DNA cages were mixed with a PEG solution and subjected to IR laser irradiation for thermophoretic accumulation and subsequent analysis of mature miRNAs within the EVs. Liao et al. had developed a three-dimensional (3D) probe based on a DNA cube for the detection of exosomal

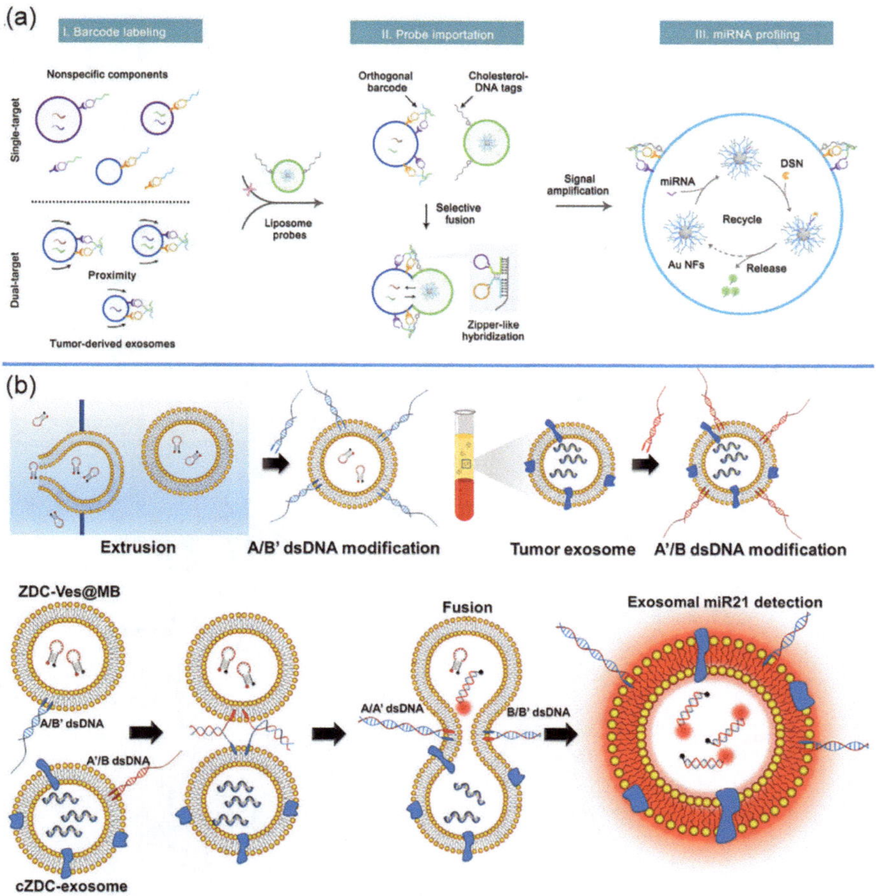

Fig. 1.15 Exosomal miRNA analysis based on DNA nanostructures (**a**) miRNA profiling of tumor-derived exosomes via dual-surface-protein orthogonal barcoding. Reprinted with the permission from Lei et al. (2023). Copyright 2023, The Authors. (**b**) DNA zipper-mediated membrane fusion for exosomal miRNA detection. Reprinted with the permission from Xie et al. (2022). Copyright 2022, American Chemical Society

miR-21 by measuring a single fluorescent signal (Xu et al. 2022). Li et al. introduced two fluorescence dyes into DNA cube and measured the FRET effect between two fluorescent signals to achieve in situ detection of exosomal miRNA (Chen et al. 2022d) (Fig. 1.16b). Wang et al. developed a Y-shaped DNA (Y-DNA) probe for in situ detection of exosomal miRNAs (Wang et al. 2020a).

In addition, various hybrid DNA nanomaterials were engaged in EV analysis, such as spherical nucleic acid (Zhao et al. 2020b), locked nucleic acid-functionalized gold nanopillars (Lee et al. 2019), DNA-graphene hybrid nanomaze (Bao et al. 2021) nanozyme-functionalized flower-like DNA structures (Zhang et al. 2023), and aptamer-functionalized metal-organic framework (Chen et al. 2022a).

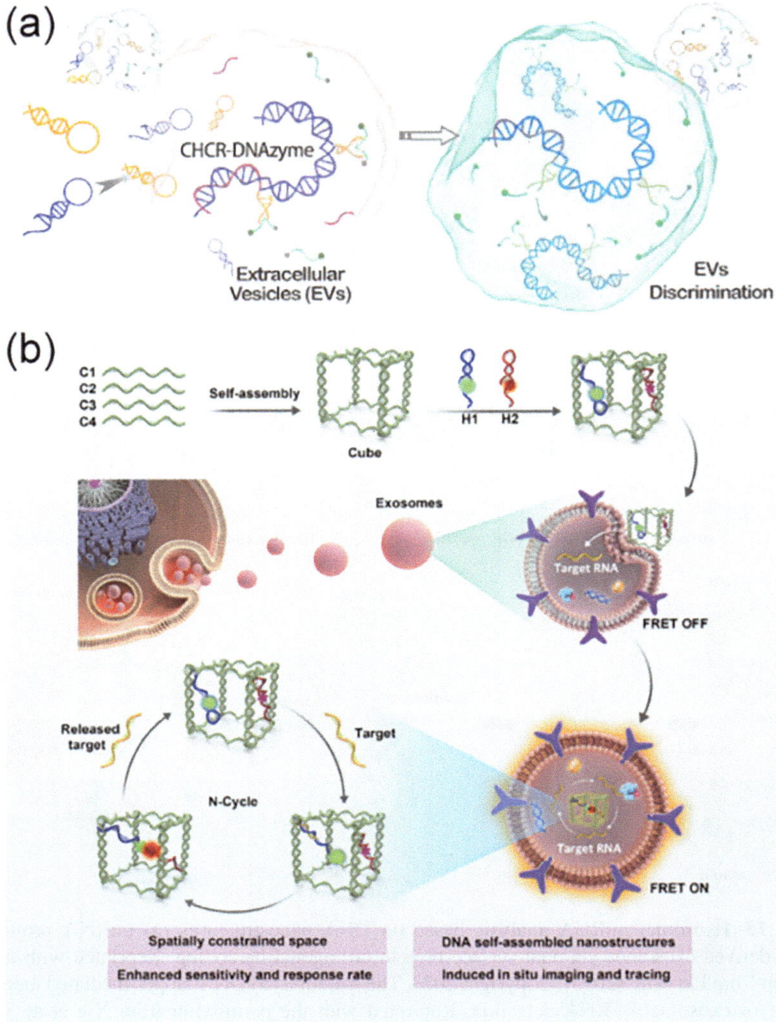

Fig. 1.16 DNA nanostructures for extracellular vesicles sensing (**a**) Schematic illustration of the CHR DNAzyme-mediated discrimination of extracellular vesicles. The DNAzyme selectively binds to and differentiates EVs based on their distinct surface markers. Reprinted with the permission from Wu et al. (2019). Copyright 2019, American Chemical Society. (**b**) Representation of the self-assembly and target-responsive behavior of the Janus DNA Cube for in situ imaging and tracking of exosomes. Upon target binding, the DNA cube undergoes structural reconfiguration, thereby amplifying fluorescence signal sensitivity and response rate, facilitating precise detection of exosomes. Reprinted with the permission from Chen et al. (2022b). Copyright 2022, American Chemical Society

1.3.3 DNA Nanomaterials in Tumor Imaging

DNA nanostructures are exceptional tools for bioimaging due to their excellent biocompatibility and stability. These nanomaterials have significantly advanced early disease diagnosis and tissue imaging. The fundamental principle of bioimaging involves the specific recognition of targets by probes, which is accompanied by changes in fluorescence signals (Wang et al. 2021b). Molecular imaging, particularly in situ fluorescence imaging of tumor-related biomarkers, enables the discrimination of cancer cells even at the single-cell level, providing pathophysiological information before observable morphological abnormalities. This capability aids in early cancer diagnosis and prognosis. Common imaging biomarkers include miRNA (Zhang et al. 2019c, 2022a; Yang et al. 2022; Wang et al. 2018, 2019b; Chen et al. 2019b), glycoproteins (Zhu et al. 2021a), telomerase (Yang et al. 2017; Chen et al. 2017), and other molecules.

1.3.3.1 RNA Imaging

For mRNA imaging, significant exploration has been conducted using DNA nanostructures. Combining the advantages of functional nucleic acids and logic circuits, Wang et al. (2020b) built two cascade logic circuits, OR–AND and AND–AND, for cellular DNA calculation and subcellular mRNA imaging. In this approach, an i-motif structure and ATP aptamers, embedded within a truncated square pyramid cage, serve as logical control units. In response to intracellular H^+ and ATP, these units trigger the release of the sensing element, enabling targeted mRNA imaging. In recent times, nonenzymatic signal amplification strategies have arisen as pivotal platforms for the visualization of intracellular biomolecules (He et al. 2020). Several enzyme-free signal amplification techniques, including HCR, catalytic hairpin assembly (CHA), and entropy-driven signal amplifiers, have been demonstrated for DNA nanostructure-mediated intracellular imaging (Wang et al. 2021c).

Inspired by the tentacles of an octopus, Yang et al. (2020a) prepared a multivalent DNA triangular prism (DTP), which is assembled by linking three biotinylated DTPs by a streptavidin for sensitive, rapid, and multiplexed imaging of miRNAs in living cells (Fig. 1.17a). In the presence of miR-155 and miR-21 targets, two CHA reactions are initiated successively, enabling precise imaging analysis of aberrant expression and dynamic changes of target miRNAs in different cells. He et al. (2018) reported an entropy-driven 3D DNA tetrahedral amplifier capable of identifying and imaging specific mRNA in living cells (Fig. 1.17b). Due to the unique entropy-driving force, this autonomous DNA circuit can be specifically activated within living cells, providing significant signal amplification for ultrasensitive detection and imaging of mRNA. Wu et al. (2020) reported a tripartite DNA probe based on a Y-scaffold for fluorescent imaging of RNA in living mice via an HCR circuit (Fig. 1.17c). This tripartite probe can be selectively and efficiently internalized into cells overexpressing folic acid receptors through endocytosis, achieving a

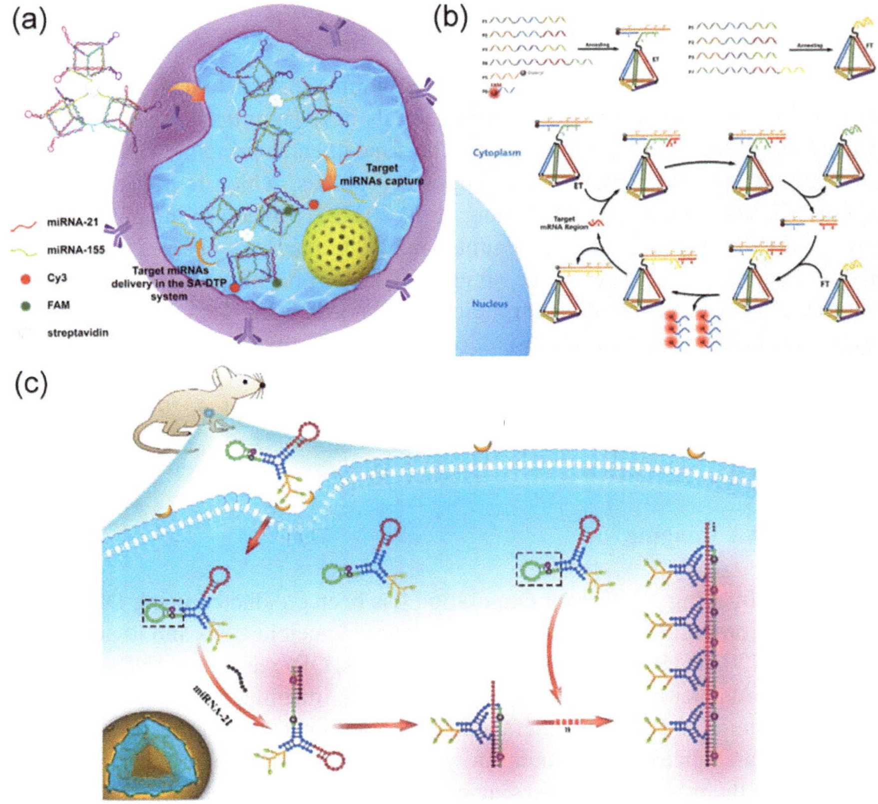

Fig. 1.17 Schematic representations of various RNA imaging techniques (**a**) Bioinspired framework nucleic acid capture microRNAs biomarkers in living cells. Reprinted with the permission from Yang et al. (2020b). Copyright 2020, American Chemical Society. (**b**) mRNA-initiated, three-dimensional DNA amplifier functional inside living cells. Reprinted with the permission from He et al. (2018). Copyright 2018, American Chemical Society. (**c**) In vivo RNA imaging in mice enabled by a hybridization chain reaction circuit with a tripartite DNA probe. Reprinted with the permission from Wu et al. (2020)

high-contrast image for miR-21 in live mice. Xu et al. (2024) designed a strategy based on a nanofirecracker probe, utilizing an intramolecular HCR amplifier for rapid, efficient, sensitive, and specific detection and imaging of surviving mRNA both in vitro and in vivo.

Huang et al. reported a triplex functionalized DNA tetrahedral nanoprobe for imaging intracellular pH and tumor-related mRNA (Zhu et al. 2019). As shown in Fig. 1.18, the mRNA-targeted H1 and H2 were stably assembled on the extended short hairpin probes of the DNA tetrahedron through Hoogsteen bonding to form DNA triplex. This DNA triplex served as a crucial element for conformational conversion. In the presence of high intracellular pH and survivin mRNA, a hybridization chain reaction was triggered between H1 and H2, which were released from the

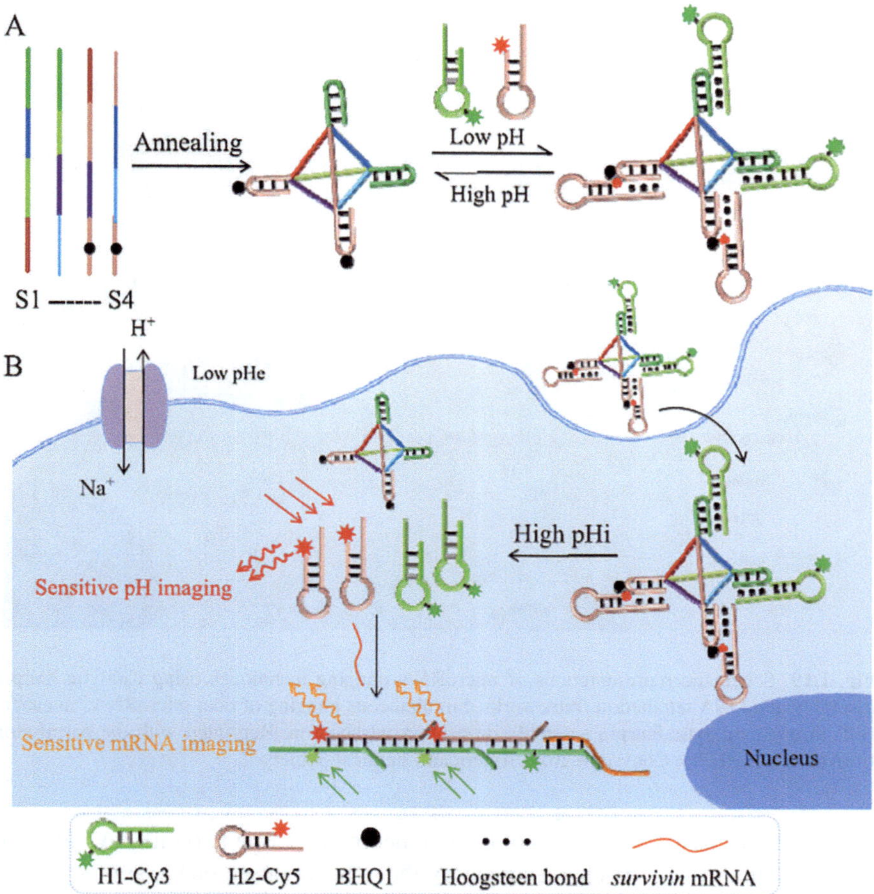

Fig. 1.18 Triplex functionalized DNA tetrahedral nanoprobe for imaging of intracellular pH and survivin mRNA. Reprinted with the permission from Zhu et al. (2019). Copyright 2019, American Chemical Society

DNA triplex. The resulting long double-stranded DNA activated a FRET signal, enabling the imaging of intracellular pH and mRNA.

For miRNA imaging, Wu and co-authors developed size-controlled DNA tiles that enhance signal amplification for sensitive and specific imaging of miRNAs in living cells (Peng et al. 2023). Mo et al. propose a new sensing platform based on tetrahedral framework nucleic acids (tFNA). In the presence of the target miRNA, FRET occurs. Moreover, the helper strands facilitate the release of the target miRNA through strand displacement, initiating a new round of reactions that generate an amplified FRET signal (Mo et al. 2021). Wang et al. (2022c) designed catalytic hairpin assembly-polymerized tFNAs for in situ imaging of miRNAs in living cells. Zhang et al. reported an APE1-assisted highly specific signal amplification biosensor for imaging of microRNA in situ (Zhang et al. 2024c). Li et al. (2022a) designed

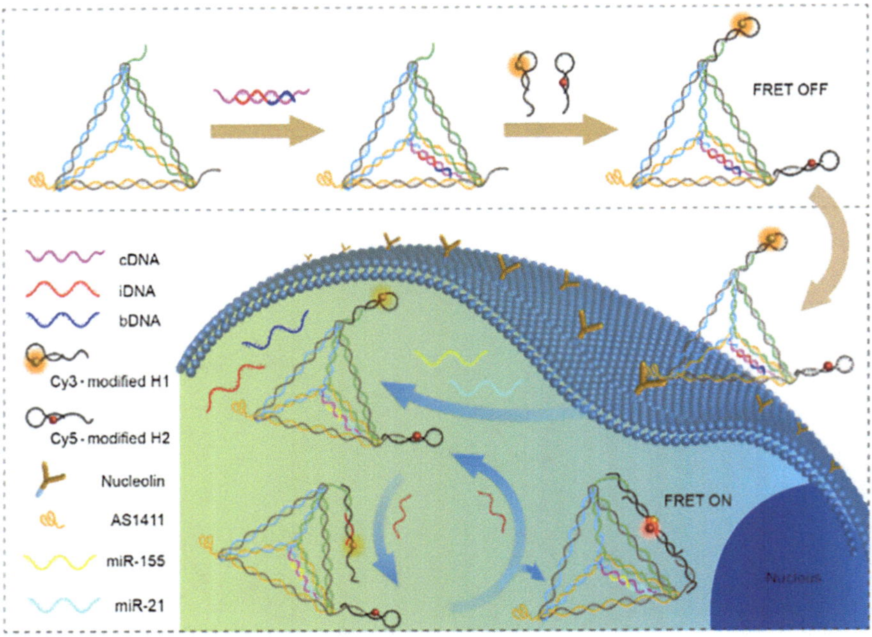

Fig. 1.19 Schematic representations of microRNA imaging techniques using catalytic hairpin assembly and DNA tetrahedral frameworks. Simultaneous imaging of dual microRNAs in cancer cells through catalytic hairpin assembly on a DNA tetrahedron. Reprinted with the permission from Li et al. (2022c). Copyright 2022, American Chemical Society

catalytic hairpins at two vertices of the tetrahedron to image two miRNAs simultaneously. This strategy is characterized by the connection on one side of specially designed tetrahedrons that capture DNA, block DNA, and start DNA. Once the probe is inside the cell, the capture DNA can hybridize with miRNA-21 and miRNA-155, releasing the start DNA that directly triggers the CHA reaction on the tetrahedron. As a result, the two fluorophores are brought closer to each other, leading to a significant FRET signal that identifies miRNAs associated with tumor cells (Fig. 1.19). Zhang et al. combined the tetrahedral DNA framework with the HCR amplification reaction to detect miRNA-21 in cancer cells, which fully reflects the advantages of the tetrahedral DNA framework in imaging (Zhang et al. 2022b). In addition, MNAzyme- and DNAzyme-mediated signal amplification is also commonly used in combination with DNA nanostructures for intracellular miRNA detection (Dan Zhu et al. 2021b; Li et al. 2019; Li et al. 2020a).

DNA nanomachines demonstrate significant potential for imaging various biomolecules in living cells (Shen et al. 2024). However, their broader application faces several limitations, including suboptimal driving factors, complex designs, and intricate operating procedures (Ma et al. 2018; Li et al. 2020a, 2023c). Li et al. (2023a) reported an endogenous enzyme-powered DNA nanomotor for microRNA imaging in living cells. As shown in Fig. 1.20, the DNA nanomotor is constructed

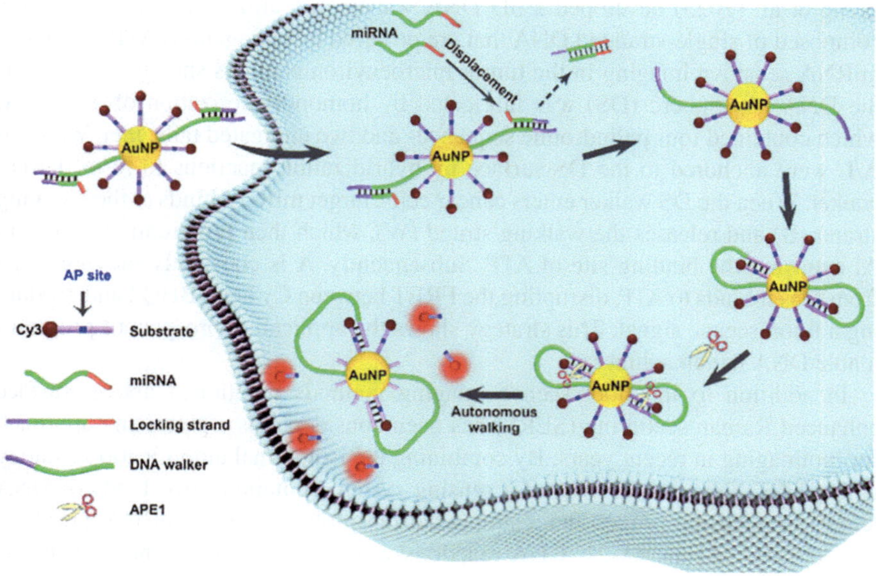

Fig. 1.20 Endogenous substance-driven DNA nanomachines based on gold nanoparticles for intracellular miRNA imaging. Reprinted with the permission from Li et al. (2023b). Copyright 2023, American Chemical Society

by modifying the surface of gold nanoparticles with different DNA strands, including the Cy3 labeled-substrate strand containing an AP site, and a DNA walker strand that can partially hybridize with the locking strand. Notably, this strategy utilizes endogenous APE1 enzymes as fuel to drive cyclic movement. Through a toehold strand-displacement reaction, the target miRNA in the cell hybridizes with the locking strand, releasing the DNA walker strand, which then partially hybridizes with the substrate to form a double strand. Intracellular APE1 recognizes and cleaves the AP site in this double-stranded region, releasing a short DNA strand labeled with Cy3, which generates a strong fluorescent signal. APE1 then drives the DNA walker to perform repeated motions, achieving signal amplification for imaging.

"One-to-one" signal-triggered model probes do not meet the requirement of sensitive imaging of low-abundance targets in living cells. Li et al. (2020b) reported an amplified FRET nanoflare, which employs endogenous mRNA as fuel to realize amplified signal detection of intracellular miRNA. The miRNA-triggered, mRNA-powered amplified FRET nanoflare was constructed by functionalizing AuNP with two types of DNA duplexes. The miRNA introduced in the form of T can replace S to generate FRET signals, and the GAPDH mRNA introduced in the form of F can release R* through strand displacement reaction, analogously, the released R* can release T, and the released T can continuously replace the S. Therefore, significantly amplified FRET signals are generated through repetition of the aforesaid process.

In addition to metal nanoparticles, soft spherical nucleic acids composed solely of DNA are also utilized in the construction of DNA nanomachines. For example,

Liang et al. (2023) developed a 3D DNA walker that utilizes DNA nanospheres composed of single-stranded DNA that are powered by endogenous ATP to achieve miRNA sensitive imaging in the tumor microenvironment. As shown in Fig. 1.21, the DNA nanosphere (DS) was fabricated by homopolymerization of a ssDNA, which contained four palindromic sequences and two elongated tails, then W/B and A/L were anchored to the DS surface by hybridization reactions to form the DS walker. When the DS walker enters cancer cells, target miR-21 binds to the blocking strand (B) and releases the walking strand (W), which then preferentially binds to A, exposing the binding site of ATP. Subsequently, A is constantly moving away from L as it binds to ATP, disrupting the FRET between Cy3 and BHQ2 and producing a fluorescence signal. This strategy shows the application prospect of programmable DNA nanomachines.

In addition to the fluorescence imaging methods mentioned above, surface enhanced Raman scattering (SERS) has been considered as a promising approach for bioimaging in recent years. By combining the isothermal amplification strategy with SERS, Yue et al. developed a sensing system containing two kinds of DNA tetrahedron-hairpin probes and a simple hairpin probe to detect miRNA-21. Upon the presence of miRNA-21, CHA reactions were triggered and generated hyperbranched three-dimensional DNA structures. These structures were captured on a

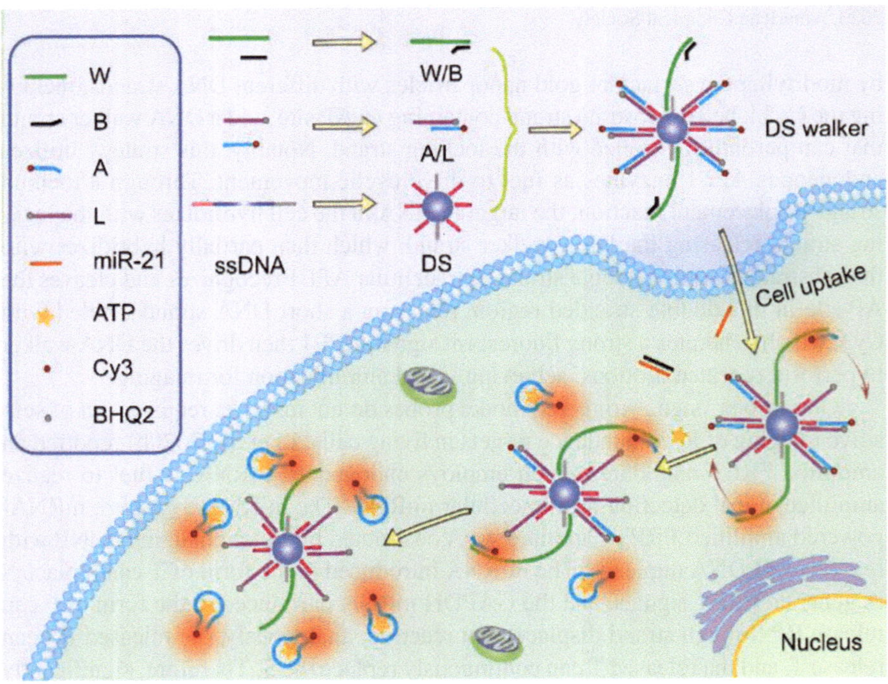

Fig. 1.21 Endogenous ATP-driven DNA nanomachine based on spherical nucleic acid for intracellular miRNA imaging. Reprinted with the permission from Liang et al. (2023). Copyright 2019, American Chemical Society

copper membrane, forming the SERS active hot spots. The enhanced SERS signal improved the efficiency of signal amplification (Yue et al. 2022). In addition to tetrahedral nucleic acid frameworks, a variety of DNA self-assembled nanostructures such as triangular DNA scaffold (Xue et al. 2019a), DNA network structure (Li et al. 2022a), DNA nanocluster (Pan et al. 2024), DNA nanoflower (Liu et al. 2020), and DNA hydrogels (Sun et al. 2022a) have been used for intracellular miRNA imaging.

1.3.3.2 Protein Imaging

Abnormal protein expression often indicates the occurrence and development of disease (Zhang et al. 2019a; Guo et al. 2019). Therefore, the localization and detection of intracellular proteins are of great significance. Beyond nucleic acid detection, DNA nanostructure-based probes have also been employed for protein imaging.

Telomerase (TE) and APE1 are two key enzymes that significantly influence tumor development and treatment outcomes (Curtin 2012). TE functions as a ribonucleoprotein that catalyzes the synthesis and extension of telomere DNA (Shay and Wright 2019). The major challenge in detecting telomerase within living cells is the complex intracellular environment, which complicates effective cell penetration, stable signal output, and precise intracellular imaging. Meng et al. (2019a)

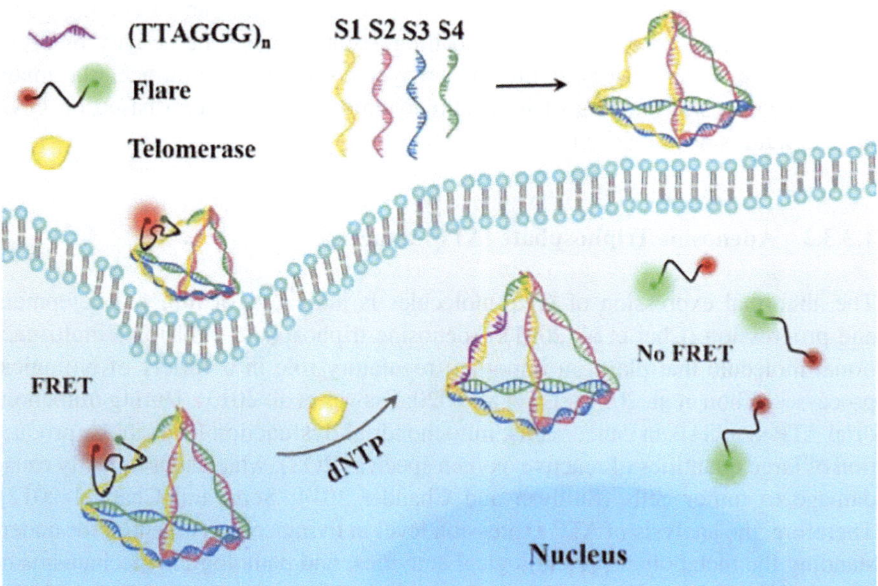

Fig. 1.22 DNA tetrahedron-based sensors for intracellular telomerase imaging. FRET investigation for DNA tetrahedron-based ratiometric analysis of intracellular telomerase activity. Reprinted with the permission from Yue et al. (2020)

introduced an innovative amplification-free method for ultrasensitive monitoring of intracellular telomerase activity based on FRET, achieving single-cell level imaging of telomerase activity (Fig. 1.22). Yue et al. (2020) designed a tetrahedral DNA nanoprobe integrated with a structure-switchable molecular beacon to detect and image telomerase in living cells. The telomerase-catalyzed extension of the nanoprobe primer strands triggers a strand displacement reaction, releasing free molecular beacons and thus giving increased FRET signals. This strategy enables precise detection of telomerase activity, distinguishing cancer cells from normal cells effectively. Dong et al. (2022) designed a multi-arm tFNAs probe to observe the entire process of apoptosis by simultaneously imaging intracellular cytochrome and telomerase through fluorescence. Zhang et al. (2022c) designed a tFNA docking assembly (DTDA) that undergoes structural changes in response to microenvironmental pH variations, enabling the measurement of telomerase activity.

Li and his colleagues have reported a DNA tetrahedron, one vertex of which has a hybrid chain consisting of a DNA strand decorated with an AP site and a DNAzyme strand, which can be used for imaging APE1. The specific recognition of APE1 and AP site renders DNAzyme activated. DNAzyme then cleave the molecular beacon (MB), restoring its fluorescence via FRET. After the MB is cleaved, DNAzyme will recognize and cleave next complete MB continuously, achieving the signal amplification effect (Zhou et al. 2020b, c). In another study, Zhang et al. (2024b) reported a self-recycled strategy for imaging of APE1 through the hybridization of molecular beacons and tFNAs. DNA damage is intricately linked to tumorigenesis, with 8-oxoguanine (8-OG) being a common marker of oxidative DNA damage (Singh et al. 2011; Ciccia and Elledge 2010). Wu et al. (2022) designed a cloning-upon-repair DNA tetrahedron nanoswitch to fluorescently image the activity of 8-OG activity in cells, demonstrating high accuracy and specificity in intracellular imaging. Hence, in situ imaging of intracellular tumor-related protein based on DNA nanostructures was achieved.

1.3.3.3 Adenosine Triphosphate (ATP) Imaging

The abnormal expression of micromolecules is indicative of tumor development and progression (Chai et al. 2021). Adenosine triphosphate (ATP) is a multifunctional molecule that plays an important regulatory role in a variety of biological processes (Zhou et al. 2021a; Lu et al. 2020; Biswas et al. 2013). During mitochondrial ATP synthesis in cancer cells, mitochondrial dysfunction leads to the production of large quantities of reactive oxygen species (ROS), which subsequently cause damage to tumor cells (Sullivan and Chandel 2014; Sena and Chandel 2012). Therefore, the analysis of ATP expression level in living cells is essential for understanding the metabolism, physiological activities, and pathological mechanisms of tumor cells (Wang et al. 2021a). Currently, ATP imaging primarily relies on fluorescence-based detection methods. Jiang et al. (2020) designed a simple fluorescent double-targeted DNA nanodevice for base excision repair (BER), enabling in

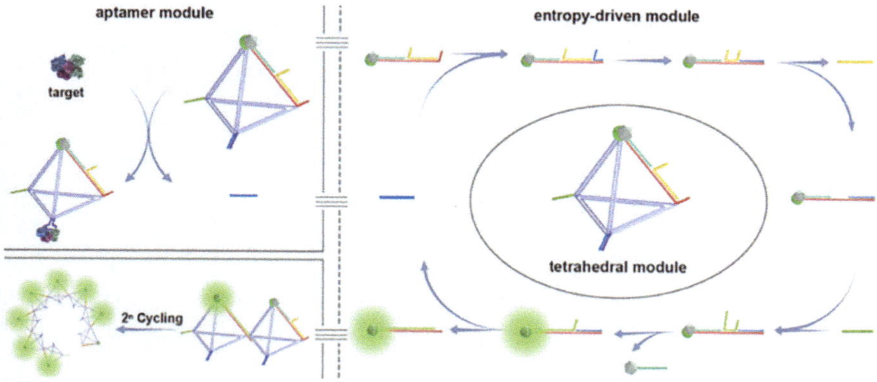

Fig. 1.23 Schematic representations of DNA tetrahedron-based systems for adenosine triphosphate imaging. Modular engineering of a DNA tetrahedron-based nanomachine for ultrasensitive detection of intracellular bioactive small molecules. Reprinted with the permission from Li et al. (2023b). Copyright 2023, American Chemical Society

situ imaging of ATP while mitigating the mutagenic effects of uracil on ATP aptamers.

Li et al. (2023c) developed a DNA tFNA-based split aptamer probe for ratiometric fluorescence imaging of ATP in living cells. Upon binding to ATP, the probe undergoes a conformational change that generates a strong FRET signal, enabling in situ imaging of ATP in tumor cells (Fig. 1.23). Yang et al. (2023b) introduced a modular tFNA-based nanomachine that can measure and image intracellular small molecules. The nanomachine exhibits sufficient sensitivity to ATP with a low detection limit of 0.40 pM and possesses anti-interference ability against other analogous small molecules. In situ imaging of ATP was realized. To avoid nonspecific interference, Wen et al. (2024) reported an enzymatically controlled DNA tetrahedron nanoprobes for biosensing and specific imaging of ATP in tumor cells. Based on a tetrahedron of DNA, the nanoprobe is formed by binding to a DNA sequence that has modified the AP site and contains an ATP aptamer. In the presence of APE1, the AP site will be recognized and cleaved, producing short DNA strands and less stable double-stranded structures. ATP aptamers dissociate due to binding ATP, resulting in detectable fluorescence. This strategy shows potential in the precise detection of multiple tumor biomarkers.

Chai et al. (2021) developed a DNA nanodevice for spatially selective, AND-gated imaging of ATP and glutathione in mitochondria. As shown in Fig. 1.24, the DNA nanodevice named A-G/NT consists of two modules: a redox-activatable aptamer probe (A-G) for sensing and a nanoparticle (NP) modified with TPP for mitochondria-targeted delivery. The A-G probe was constructed through the hybridization of an ATP aptamer-based A-strand and a G-strand that contains a chemically modified disulfide bond. When the A-strand hybridizes with the G-strand, the Cy3 labeled on the A-strand is brought close to the black hole quencher label on the G-strand, resulting in a low-fluorescent background. Upon cleavage of the disulfide

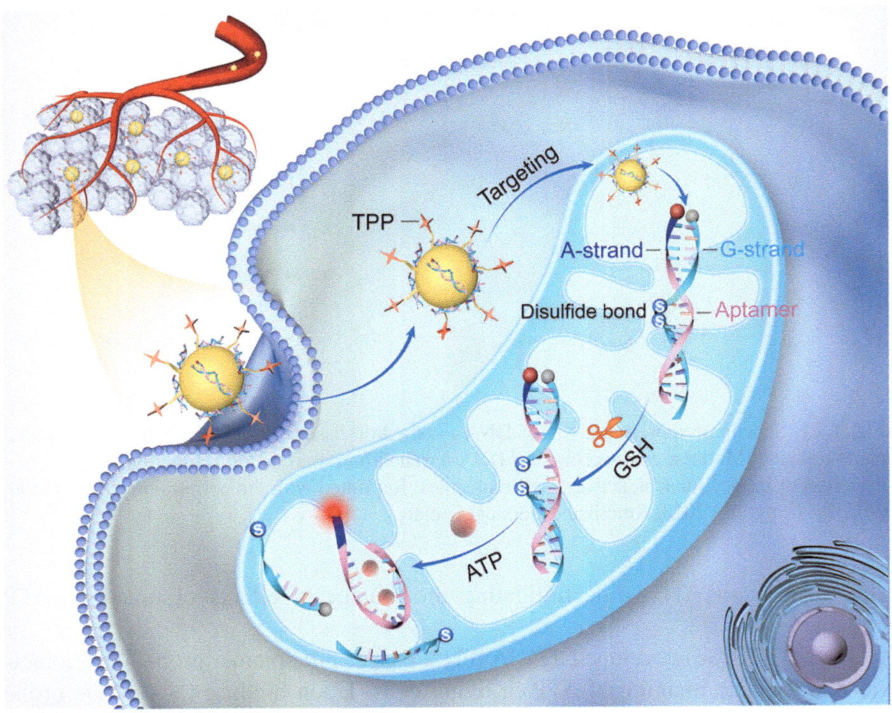

Fig. 1.24 A redox-activatable DNA nanodevice for AND-gated imaging of ATP and glutathione in mitochondria. Reprinted with the permission from Chai et al. (2021). Copyright 2021, American Chemical Society

bond by glutathione (GSH), the binding affinity between the resulting G-strand and the A-strand is reduced. In the presence of ATP, the A-strand binds to ATP and separates from the cleaved G-strand, resulting in an enhanced fluorescence signal. This strategy provides a DNA nanodevice capable of responding to two targets simultaneously, offering potential for multi-target imaging of organelles.

1.3.3.4 Cancer Cell Imaging

Accurate imaging of cancer cells is both desirable and challenging. A molecular platform that meets the requirements for in situ fluorescence imaging of biomarkers can identify cancer cells at the single-cell level, thereby revealing pathophysiological information prior to any observable morphological abnormalities. This capability is crucial for supporting early cancer diagnosis and prognosis (Calin and Croce 2006) Imaging of tumor-associated miRNAs can provide abundant information for cancer diagnosis. However, trace amounts of miRNA in normal cells will inevitably produce false positive signals, for example, miR-21 is found in normal cells (Calin and Croce 2006), which will affect the discrimination of cancer cells during miRNA

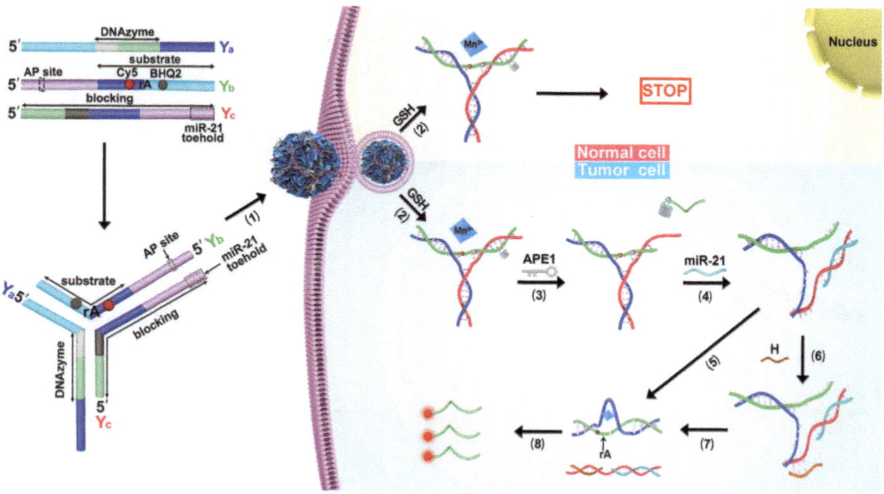

Fig. 1.25 Dual-locked DNAzyme platform for discrimination of cancer cells. Reprinted with the permission from Huang et al. (2022a). Copyright 2022, American Chemical Society

imaging. To overcome the problem of nonspecific interference with single biomarker imaging, Huang et al. (2022b) reported an enzyme (APE1)/miRNA-responsive DNAzyme-based imaging platform for in vitro and in vivo discrimination of cancer cells. As shown in Fig. 1.25, the imaging platform consists of a double-locked DNAzyme sensor and the honeycomb MnO_2 nanosponge (hMNS). The DNAzyme sensor can be released inside the cell, and in the presence of APE1 and miR-21, the DNAzyme cleavage capability was recovered and a fluorescence signal was generated. This strategy enables accurate identification of cancers in vitro and in vivo, showing great potential for biomedical applications.

In order to accurately identify different subtypes of cancer cells, Wu et al. (2023a) reported a dual-miRNA-propelled three-dimensional DNA walker for discrimination of breast cancer cell subtypes in clinical tissue samples. As shown in Fig. 1.26, the DNA nanomachine was constructed by gold nanoparticle functionalized with ssDNA and a DNAzyme probe, which can be specifically initiated by intracellular miR-21 and miR-31, the sensitivity was improved by DNAzyme-mediated signal amplification. This nanomachine can be used for effective detection of breast cancer subtypes in vivo and in clinical samples, showing that this strategy is suitable for precise management of cancer.

DNA logic gates are widely accepted as an intelligent biocomputing operator that can confer special cell-recognition properties by simultaneously analyzing multiple cancer biomarkers (Zhou et al. 2020a; Zhang et al. 2021a; Song et al. 2019; Okamoto et al. 2004; Liu et al. 2021a; Jiao et al. 2020; Chen et al. 2022d). Using DNA logic gates to modulate DNA circuits can help improve the specificity and sensitivity of cancer recognition. Many DNA nanosystems have been reported that use two cancer biomarkers (microRNA, mRNA, or membrane proteins, etc.)

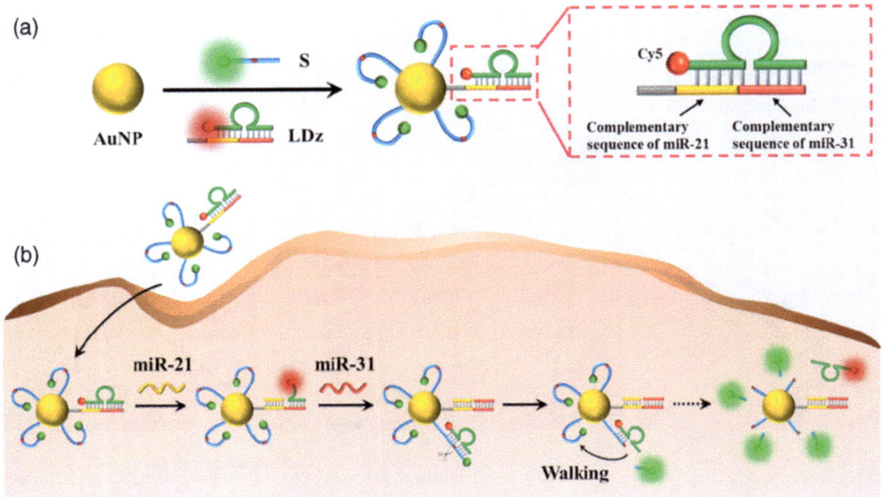

Fig. 1.26 Dual-MiRNA-propelled dual-color DNA walker for discrimination of breast cancer cell subtypes. (**a**) The construction principle of DNA walker and its intracellular response to miR-21 and miR-31; (**b**) Dual-color using two miRNAs for breast cancer cell subtype identification. Reprinted with the permission from Wu et al. (2023b). Copyright 2022, Chinese Chemical Society

as inputs to perform imaging operations based on AND logic gates, and have made great progress in improving the accuracy of cancer imaging (Wu et al. 2023b; Yu et al. 2021; Peng et al. 2018; Li et al. 2021; Cai et al. 2022). Chao et al. (2023) developed a compute-and-release logic-gated DNA cascade circuit for cancer cell imaging. As shown in Fig. 1.27, the system comprises a compute-and-release (CAR) logic gate, a double-amplified DNA cascade circuit (termed CHA-HCR), and a MnO_2 nanocarrier. The MnO_2 nanoflower was used as a nanocarrier to deliver CAR logic gates and CHA-HCR circuits, and it can release DNA in response to intracellular GSH. In detail, endogenous biomarker (miR-892b) was set as the dynamic threshold and another one (miR-21) was the target. This CAR-CHA-HCR circuit enables the comparison of the relative concentrations of the two miRNAs during sensing, and then executes an intelligent output. It executes YES in a positive condition (CmiR-21 > CmiR-892b) and INHIBIT in a negative condition (CmiR-21 < CmiR-892b). By utilizing this strategy, which is capable of comparing the relative concentrations of two biomarkers during sensing, it allows for accurate identification of positive cancer cells, even in mixed cell populations.

In addition, HCR, as an efficient means of enzyme-free signal amplification, has also been used for imaging cancer cells. For example, Yang et al. (2023b) reported a dual-responsive DNA tetrahedron nanomachine (drDT-NM) for cancer cell imaging. As shown in Fig. 1.28, the nanomachine was constructed by fixing two recognition modules, the MUC1 aptamer (MA) and the hairpin H1* encoding the nucleoprotein-specific G-rich AS1411 aptamer, on two independent vertices

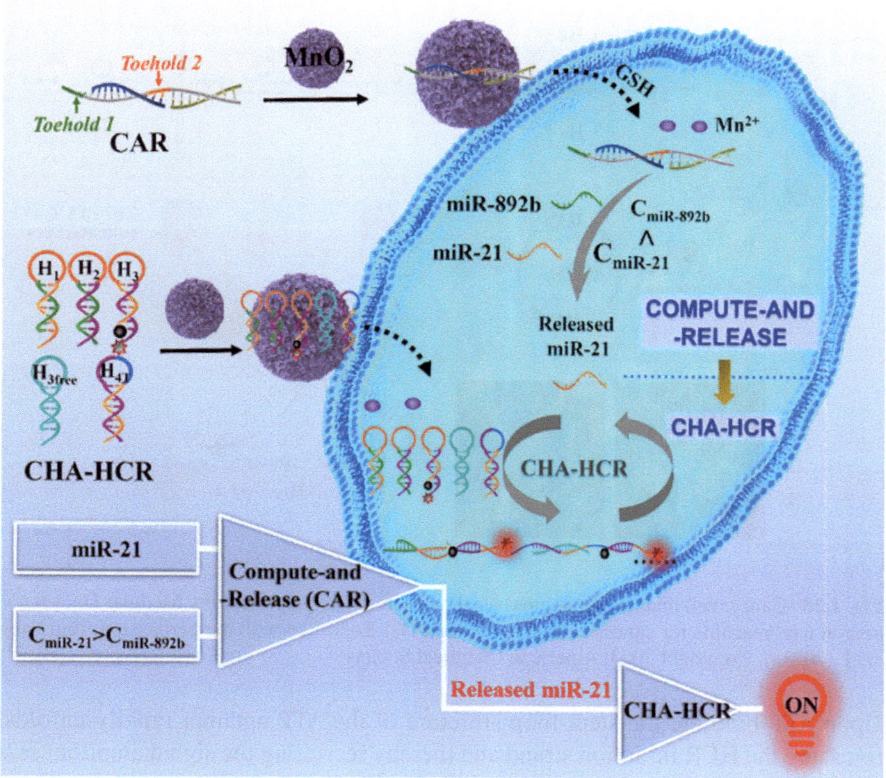

Fig. 1.27 Compute-and-release logic-gated DNA cascade circuit (CAR-CHA-HCR) for cancer imaging. Reprinted with the permission from Chao et al. (2023). Copyright 2023, American Chemical Society

of the DNA tetrahedron. In the presence of MUC1 and nucleolin, two independent hybridization chain reactions (HCRM and HCRN) are initiated with two sets of four functional hairpin reactants. In the shared HCRN duplex products, the parent AS1411 aptamers are cooperatively merged and folded into G-quadruplex concatemers to embed Zn-protoporphyrin IX (ZnPPIX/G4) for fluorescence signaling readout, achieving intracellular assay and highly discernible cell imaging. Notably, the tandem ZnPPIX/G4 units can also act as therapeutic cargos for photodynamic therapy (PDT) of cancer cells. This study provided a strategy for bivariate detection and PDT of cancer. Similarly, Li et al. (2024b) developed a membrane protein targeted and ATP-fueled "AND" logic gate nanodevice for amplified and accurate identification of cancer cells. The nanodevice (H1) is functionalized with dual aptamer: a cell membrane anchoring module (P-apt) and the ATP-gated signal amplifier switch module (Aapt). The 5′ end of the ATP aptamer is structured as a stem-loop to prevent the release of the HCR initiation strand until ATP is present. In the presence of the target membrane protein, the membrane protein aptamer navigates and binds to the surface of the cell membrane.

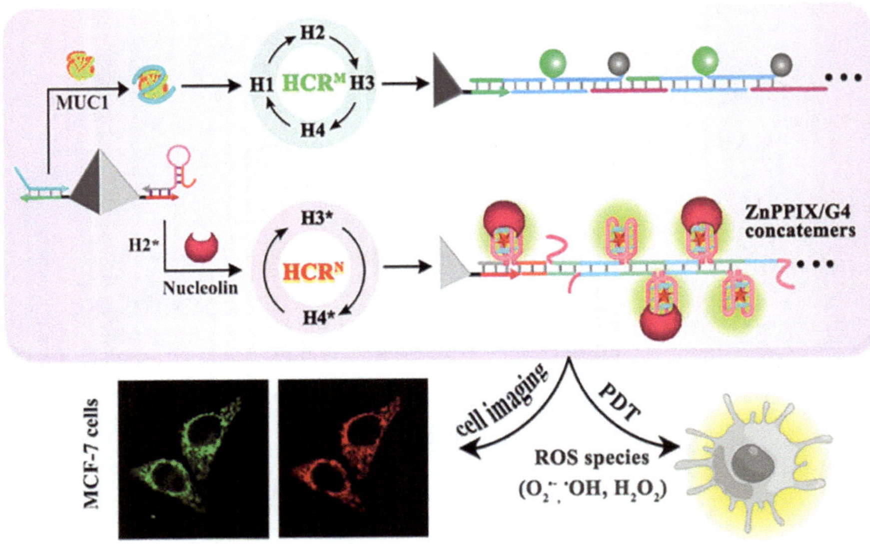

Fig. 1.28 Cancer cell imaging strategies based on HCR signal amplification. Modular DNA tetrahedron nanomachine for cancer cell imaging and PDT. Reprinted with the permission from Yang et al. (2023a). Copyright 2023, American Chemical Society

Upon ATP binding, the stem-loop structure of the ATP aptamer rapidly unfolds, releasing the HCR initiation strand and thereby recruiting the signal amplifier H2/H3 to produce sufficient fluorescent signals. The fluorescent signal of the device can only be detected when both the membrane protein and ATP are present. This strategy provides a promising means for precise imaging of tumors and ATP-regulated tumor treatment.

1.3.3.5 In Vivo Imaging

Rapid advances in DNA nanotechnology and optical imaging provide opportunities for accurate visual monitoring of cancer occurrence and progression. An ideal in vivo target sensing platform should possess several key characteristics: good stability, excellent biocompatibility, extended circulation time, specific targeting capabilities, and enhanced tumor accumulation. Utilizing probes based on self-assembled DNA nanostructures, various imaging strategies have been reported for the sensitive and real-time imaging of tumor biomarkers or tumor tissues in vivo (Meng et al. 2019b).

Wei et al. (2020) reported an autocatalytic DNAzyme (ACD) biocircuit for in vivo amplified microRNA imaging based on honeycomb MnO_2 nanosponge (hMNS)-sustained HCR and DNAzyme biocatalysis. The versatile hMNS scaffolds were used as carriers for delivering DNA probes and providing Mn^{2+} as cofactors for DNAzyme. Additionally, they served as intelligent magnetic resonance imaging (MRI) agents. Through the subsequent synergistic cross-activation between HCR and DNAzyme amplicons, the system transforms the limited miRNA recognition event into an amplified fluorescence signal readout, enabling in vivo microRNA imaging and demonstrating great promise for accurate clinical tumor diagnosis.

Multiple RNA-based biomarkers have been identified and play an important role in biology (Tyagi 2009). Fluorescent probes targeting RNA can serve as an effective tool for in-depth understanding of tumor biology and cancer therapy. Wu et al. (2020) reported a new cell-targeting fluorescent probe that enables RNA imaging in living mice via an in vivo HCR. This is the first time that fluorescent probes have been used for RNA imaging in living mammalian animals. As shown in Fig. 1.29, the Y-shaped probe was assembled through sticky-end hybridization of three DNA motifs: two hairpin structures and a cell-targeting DNA-peptide conjugated folate probe (FAP), which was decorated with multiple folate (FA) ligands for targeting folate receptors that are overexpressed in tumor cells. In the presence of miR-21, hairpin probes H1 and H2 can undergo HCR assembly and activate a fluorescence signal. The Y-shaped probe can be selectively taken up into tumor cells of living

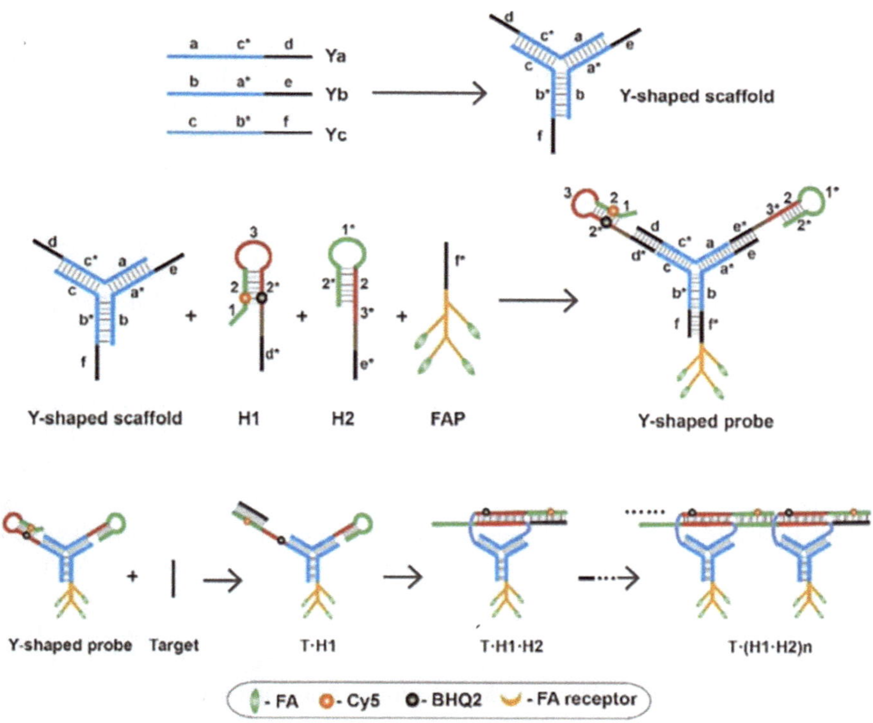

Fig. 1.29 HCR-based strategies for in vivo tumor imaging. The stepwise assembly process of tripartite Y-shaped DNA probe and in vivo HCR circuit for RNA imaging via intravenous injection. Reprinted with the permission from Wu et al. (2020)

mouse via intravenous injection, which permits an activated HCR circuit-based near-infrared (NIR) fluorescence response for in vivo RNA imaging in the living mouse. Similarly, based on the principle of HCR signal amplification, Zhang et al. (2022b) reported a tetrahedral DNA framework-enhanced HCR probe for in situ tumor imaging. Two hairpins (H1 and H2) are assembled at two vertex positions of the DNA tetrahedron. In the presence of target miRNA-21, an HCR is initiated, forming connections between multiple DNA tetrahedra and resulting in a significant fluorescence signal. The DNA tetrahedron improves the reaction efficiency of the hairpin probes in vitro, while enabling accurate and rapid identification membrane and intracellular cancer-related targets in living cells. Additionally, the system, based on the spatial confinement of the DNA tetrahedron, exhibits a high detection rate. Superior fluorescence was shown in vivo after 15 s of peritumoral injection and 10 min of caudal vein injection. This strategy demonstrates great potential for clinical diagnostic and therapeutic applications.

DNA circuits can autonomously program the temporal and spatial regulation of biological systems with predictability and are therefore considered a powerful molecular tool for bioanalysis and biomedicine (Liao et al. 2020; Hu et al. 2020; He et al. 2018). Endogenous activated DNA circuits ensure specificity and sensitivity of in vivo imaging. Shang et al. (2022) constructed an endogenously activated DNA circuit by integrating CHA and tumor-specific enzymes for imaging intracellular miRNAs. The CHA reactants are designed to have long biphasic regions to enable initial effective plugging. In addition, specific recognition sites for endogenous DNA repair enzymes are also incorporated into the long chain. Once this blocking double strand is catalytically cleaved and removed by a DNA repair enzyme, the CHA reactant can be released to recognize the target miRNA. As the target catalyzes the cross-opening of the two CHA reactants, a large number of duplex DNA products are produced, generating amplified readout signals. This strategy facilitates high-precision, high-reliability in vivo imaging of tumor cells, thus providing a powerful toolbox for cancer diagnosis and treatment.

DNAzyme is easy to synthesize and has high catalytic activity and programmability (Xu et al. 2019; Wu et al. 2019). Many probes built with DNAzyme have been used to detect metal ions in vitro and in vivo (Xiong et al. 2020; Shi et al. 2020; Yao et al. 2009; Gao et al. 2019). Wu et al. (2021) developed an imaging system based on DNAzyme. As shown in Fig. 1.30, an intracellular miRNA-unlocked entropy-driven catalysis module and an endogenous metal ion-responsive DNAzyme module were integrated for specific in vivo imaging. The activity of the DNAzyme is inhibited by blocking sequences. Overexpressed miR-21 in cancer cells triggers the entropy-driven catalysis reaction, releasing various promoters that compete with the locking strand, resulting in the activation of the DNAzyme. The free DNAzyme, in the presence of intracellular Na^+, cleaves the substrate and produces an amplified fluorescent signal. Gao et al. (2021) developed a DNAzyme walker combined with gold nanoparticle (AuNP) and DNAzyme for imaging intracellular miRNAs. In order to solve the problem of short circulation time due to the fragility of DNAzyme catalytic core, a circular bulging DNA shield was introduced into the middle of the

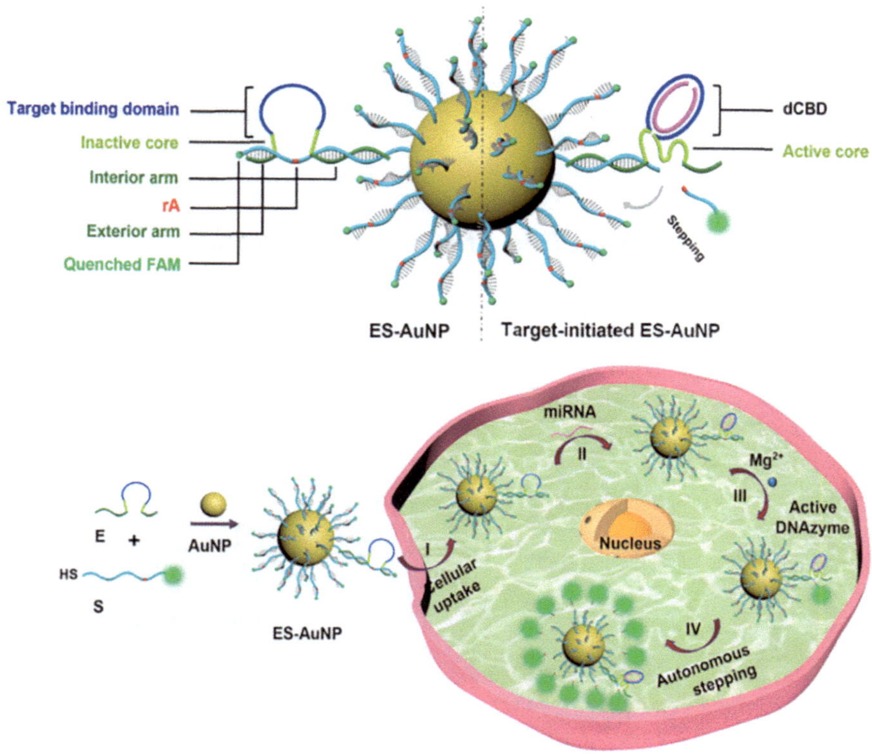

Fig. 1.30 DNAzyme-based strategies of amplified imaging of miRNAs in vivo. Self-protected DNAzyme walker with a circular bulging DNA shield for amplified imaging of miRNAs in vivo. Reprinted with the permission from Wu et al. (2021). Copyright 2021, American Chemical Society

catalytic core, thereby constructing a self-protected DNAzyme walker capable of fully stepping on the substrate-modified AuNPs. The catalytic activity of DNAzyme walker is activated by target miRNA and its stepwise movement is powered by the specific hydrolysis of substrates. The DNAzyme walker can be used for tumor imaging in living mice and represents a promising tool for cancer diagnosis and prognosis.

In addition to in vivo tumor imaging strategies based on intracellular miRNA responses, strategies that respond to other factors have also been developed. For example, Yuan et al. (2022) developed a dual-driven DNA nanorobot based on a DNA triangular prism, which is responsive to both extracellular pH and membrane receptors. By utilizing two distinct types of biomarkers as inputs, this strategy not only distinguished target tumor cells from a similar mixture of cells but also enabled accurate tumor imaging in living mice via intravenous injection (Fig. 1.31). Dong et al. (2024) reported a protease-responsive bitext editing peptide-loaded multifunctional DNA nanoflower (ppDNF) for breast cancer imaging

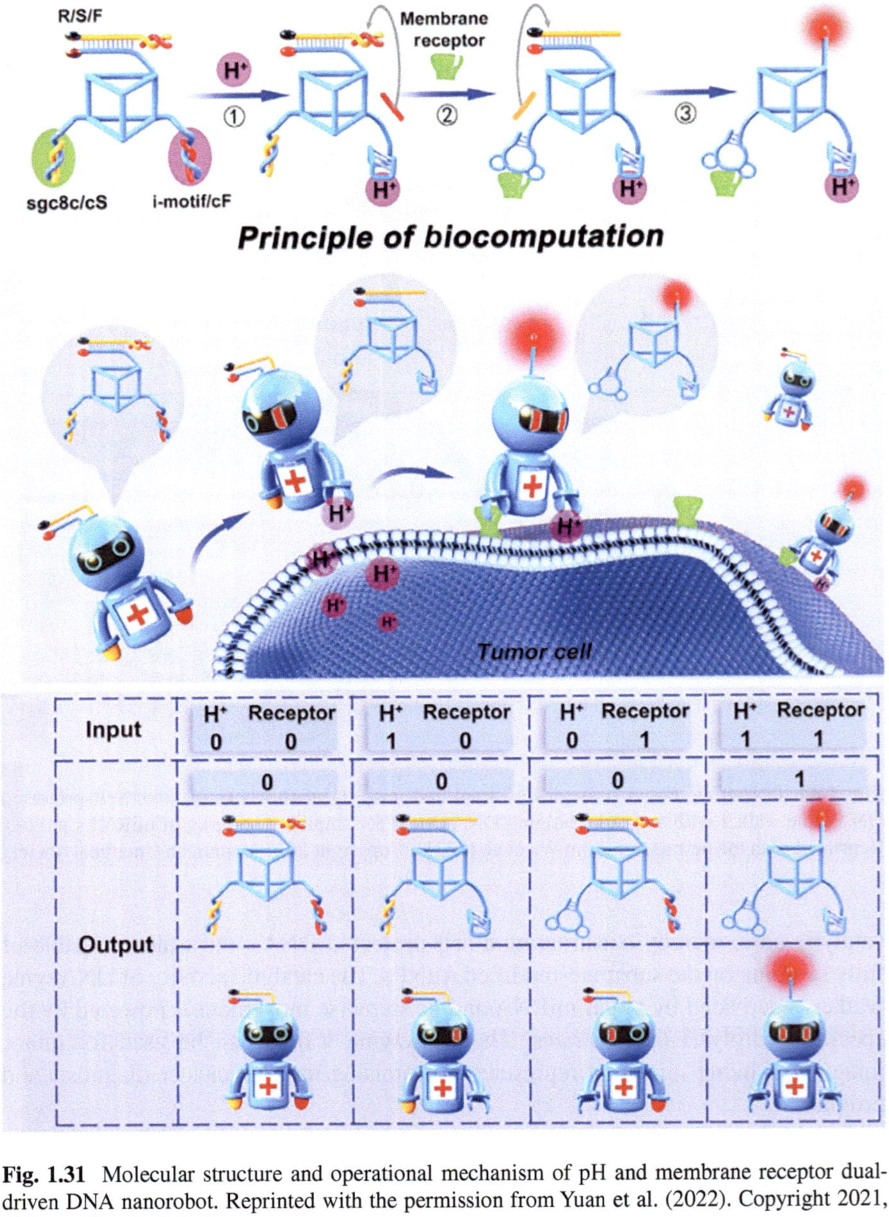

Fig. 1.31 Molecular structure and operational mechanism of pH and membrane receptor dual-driven DNA nanorobot. Reprinted with the permission from Yuan et al. (2022). Copyright 2021, Chinese Chemical Society

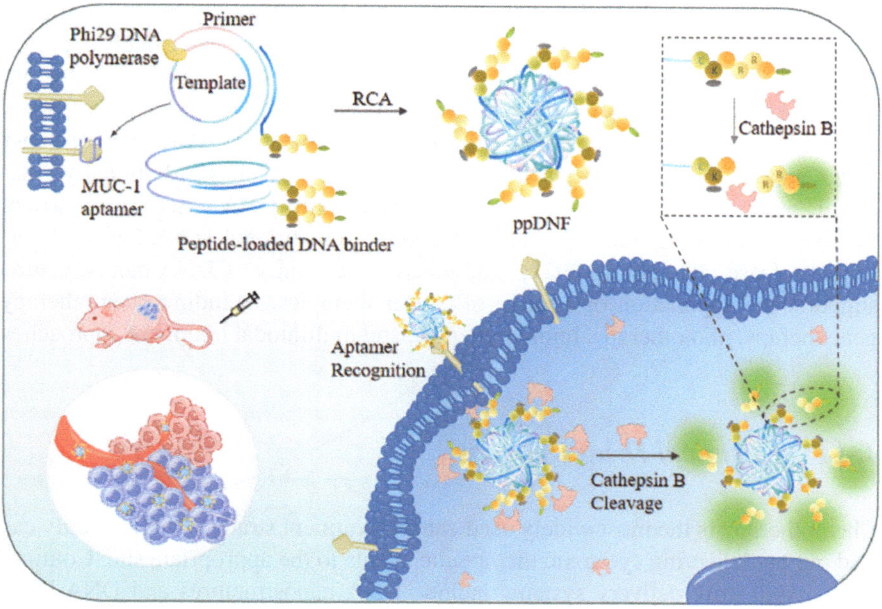

Fig. 1.32 Protease-responsive bitext editing peptide-loaded multifunctional DNA nanoflower for breast cancer imaging in vivo. Reprinted with the permission from Dong et al. (2024). Copyright 2024, American Chemical Society

in vivo. As shown in Fig. 1.32, the ppDNF was constructed using RCA and comprises two functional domains: the MUC1 aptamer and a peptide-binding site. The peptide-binding site consists of a DNA sequence complementary to the binding site and a peptide substrate probe, which is labeled with FAM at one end and DABCYL at the other. This ppDNF can respond to Cathepsin B, which is overexpressed in breast cancer cells, and can be used for robust tumor imaging in living cells and mice.

1.4 Applications of Self-Assembled DNA Nanomaterials in Cancer Therapy

In recent decades, DNA nanotechnology has advanced significantly, with the applications of self-assembled DNA nanomaterials in biomedicine expanding rapidly, particularly in drug delivery and cancer therapy. DNA-based nanostructures have the unique ability to penetrate cell membranes without the need for transfection. More importantly, they offer greater potential as drug delivery candidates due to their high biocompatibility, structural diversity, low cytotoxicity, and superior accessibility compared to traditional nanoscale carriers like gold nanoparticles and liposomes (Xia et al. 2018; Pei et al. 2014; Liang et al. 2014). The potential of DNA nanostructure-based formulations for cancer therapy has been extensively explored, with various therapeutic strategies successfully demonstrated both in vitro and in vivo. DNA nanostructures primarily serve as carriers for delivering therapeutic agents, including anticancer drugs, antisense oligonucleotides (ASOs), photosensitizers, nanoparticles, and proteins, directly to pathological sites. Additionally, functional modules can be seamlessly integrated into DNA-based nanocarriers, facilitating real-time tracking, precise targeting, conditional activation, and controlled release (Jiang et al. 2021). The inherent versatility of DNA nanostructures supports the application of a range of cancer therapies, including chemotherapy, gene therapy, phototherapy, immunotherapy, and multimodal treatment approaches.

1.4.1 Chemical Drug Delivery

Chemotherapy is the most widely used cancer treatment strategy and is mainly carried out by delivering cytotoxic therapeutic agents to the appropriate site. Common DNA-based drug delivery systems include DNA nanostructures and DNA-based nanoparticles, and their functions as carriers mainly include improving the biological distribution, specificity, and efficacy of traditional drugs. Numerous self-assembled DNA nanocarriers have been developed to act as multifunctional platforms, integrating drug loading, targeting, and delivery modules, while exhibiting minimal cytotoxicity. Doxorubicin, a widely used chemotherapeutic agent, is commonly employed as a model drug due to its ability to intercalate into double-stranded DNA and be released upon DNA strand breakage. Daunorubicin, another intercalating agent, follows a similar delivery mechanism. Other model drugs, such as fluoridine (FU), paclitaxel (PTX), and camptothecin (CPT), are typically loaded onto DNA carriers through conjugation or encapsulation (Jiang et al. 2021).

Various targeting molecules have been conjugated to DNA nanostructures to create targeted drug delivery systems that enhance tissue specificity and minimize adverse side effects resulting from nonspecific interactions. Aptamers, known for their high affinity and specificity to targets, are widely employed as targeting agents to modify DNA nanostructures. Other commonly used targeted molecules include folic acid (Zhang et al. 2017a), cell-penetrating peptides (Xia et al. 2016; Tian et al. 2018), and Affibody molecules (Zhang et al. 2017b).

Liu et al. (2019b) utilized an sgc8c aptamer to specifically deliver doxorubicin (Dox) to protein tyrosine kinase 7 (PTK7)-overexpressing cells, demonstrating that aptamer modification can enhance cytotoxic effects in PTK7-positive cells. The same research group also used a DNA tetrahedron as a novel DNA-based nanocarrier modified with aptamer AS1411 to target the delivery of the antimetabolite drug molecule 5-fluorouracil (5-FU) to breast tumor cells. Enhanced nanocarrier uptake efficiency into breast tumor cells compared to normal cells indicated that the drug-loaded nanostructure was more effective than free 5-FU at killing breast tumor cells (Zhan et al. 2019) (Fig. 1.33a).

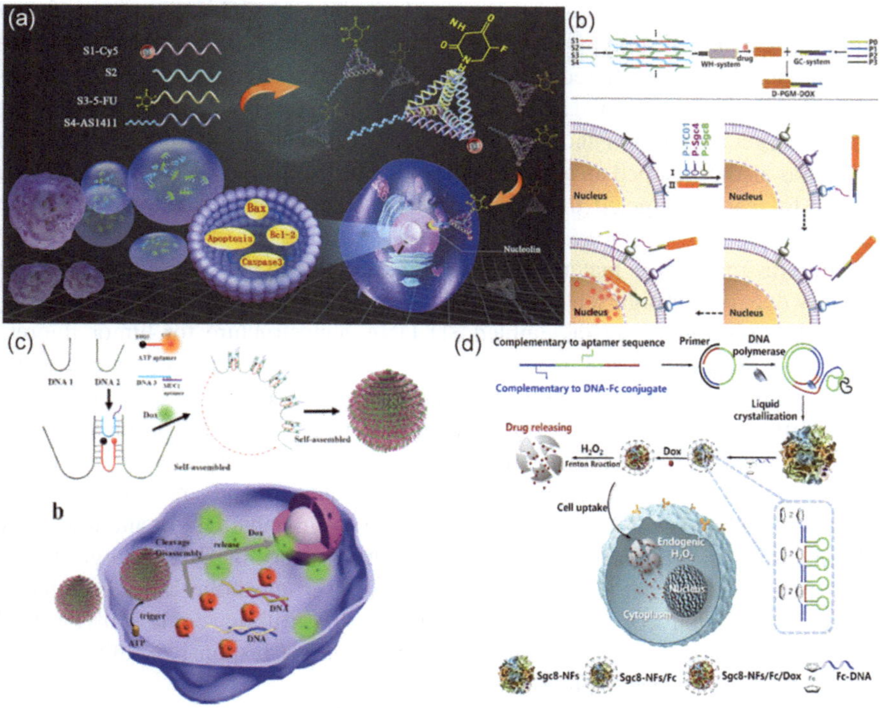

Fig. 1.33 Schematic representations of DNA nanostructures for targeted therapy and imaging in cancer cells. (**a**) DNA-based nanomedicine for targeting and enhancing therapeutic efficacy in breast cancer cells. Reprinted with the permission from Zhan et al. (2019). Copyright 2019, American Chemical Society. (**b**) missile-like DNA nanostructure with warhead and guidance control for aptamer-based targeted drug delivery into cancer cells. Reprinted with the permission from Wang et al. (2022c). Copyright 2020, American Chemical Society. (**c**) ATP-triggered drug release from self-assembled 3D DNA nanostructures for fluorescence imaging and tumor therapy. Reprinted with the permission from Jiang et al. (2022a). Copyright 2022, American Chemical Society. (**d**) Engineering of self-degradable cancer-targeting DNA nanoflowers with an artificial sandwich base. Reprinted with the permission from Zhang et al. (2019d). Copyright 2019, American Chemical Society

To address the susceptibility of DNA nanostructures to nuclease degradation, Xue et al. (2020) reported a periodically ordered nick-hidden DNA nanowire (NW) with high serum stability and active targeting functionality for imaging and therapy of cancers. The core of the nanowires consists of a plurality of connected DNA double helices, and the shell consists of a regularly arranged upright hairpin aptamer sgc8. All ends of these components are hidden from attack by nucleases, while the target binding site is exposed, allowing delivery to the cancer target.

In order to overcome the challenges of nonspecific adsorption and off-target effects encountered by DNA nanomaterials as drug carriers, the same research group also developed a DNA nanostructure to function as a DNA nanoscale precision-guided missile (D-PGM) for highly efficient loading and precise delivery of chemotherapeutic agents to cancer cells (Ouyang et al. 2020). As shown in Fig. 1.33b, the D-PGM consists of a warhead (WH) and a guidance/control (GC) system. The WH functions as the drug carrier, while the GC is an aptamer-based logic gate capable of subtype-specific cell recognition through sequential disassembly mediated by cell-anchored aptamers. This strategy achieves effective identification of target cancer cells and active targeted Dox delivery, laying the foundation for specific target recognition and precision cancer treatment. In addition, this research group has also developed an aptamer-embedded hierarchical DNA nanocluster as a carrier for cancer-targeted drug delivery to overcome off-target toxicity (Wang et al. 2022b).

Stimulus-responsive materials are ideal carriers for precisely controlled drug delivery due to their high selectivity. Jiang et al. (2022b) designed a highly concentrated ATP-responsive self-assembled 3D DNA nanostructure for tumor fluorescence imaging analysis and therapy. The 3D DNA nanostructure was prepared by self-assembly of DNA containing aptamers and specific sequences, and Dox can be loaded efficiently (Fig. 1.33c). This strategy effectively solves the problems of stable nanocarrier transport, targeted delivery, and drug localization and release, which effectively increases the antitumor effect of drugs and avoid adverse effects on normal tissues. Zhang et al. (2019a) developed H_2O_2-degradable DNA nanoflowers by incorporating an artificial sandwich base for the delivery of Dox to target tumor cells (Fig. 1.33d). Wang et al. (2019c) reported a self-catabolic DNAzyme nanosponges drug delivery system consisting of the rolling circle polymerized DNAzyme–substrate scaffolds and the encapsulated pH-responsive ZnO nanoparticles (NPs), which enable precise drug release.

In conclusion, DNA nanotechnology provides abundant structural and functional modules that significantly enhance the safety, specificity, and efficacy of chemotherapy.

1.4.2 Gene Therapy

Gene therapy aims to prevent or treat diseases by introducing healthy genes or gene regulators into abnormal cells. Self-assembled DNA nanomaterials can be used as nonviral vectors to deliver gene regulators, such as antisense DNA (ASD), DNAzyme, and small interfering RNA (siRNA), which are primarily used for post-transcriptional gene silencing. These regulators degrade target messenger RNA (mRNA) through different mechanisms, thereby preventing the transcription of undesirable proteins. ASD induces targeted mRNA hydrolysis via RNase H, DNAzyme recognizes and cleaves mRNA at specific sites, and siRNA promotes mRNA degradation by forming the RNA-induced silencing complex (RISC) (Jiang et al. 2021). The main obstacle to gene therapy at present is the insufficient uptake of these nucleic acid gene regulators by cells. DNA nanotechnology provides a variety of vectors, showing advantages in loading nucleic acid gene regulators and improving cell uptake efficiency.

Zhou et al. (2021b) utilized an aptamer-modified DNA tetrahedron to effectively deliver small interfering RNA to glioma cells to induce apoptosis. The downregulation of survivin protein and mRNA expression levels indicated that the strategy effectively silenced the target gene (Fig. 1.34a). To address the challenge of low cellular uptake efficiency of DNAzymes, Meng et al. (2019a) developed tetrahedral DNA nanostructures (TDN-Dz13) for delivery of DNAzymes to regulate target genes. Dz13 is a DNAzyme that cleaves c-Jun mRNA, inhibiting the growth of squamous cell carcinoma. The researchers incorporated the Dz13 sequence into the 5′-end of single-stranded DNA to form a modified TDN-Dz13. Dz13 delivered by the TDNs showed high cell uptake efficiency and gene silencing activity in cells

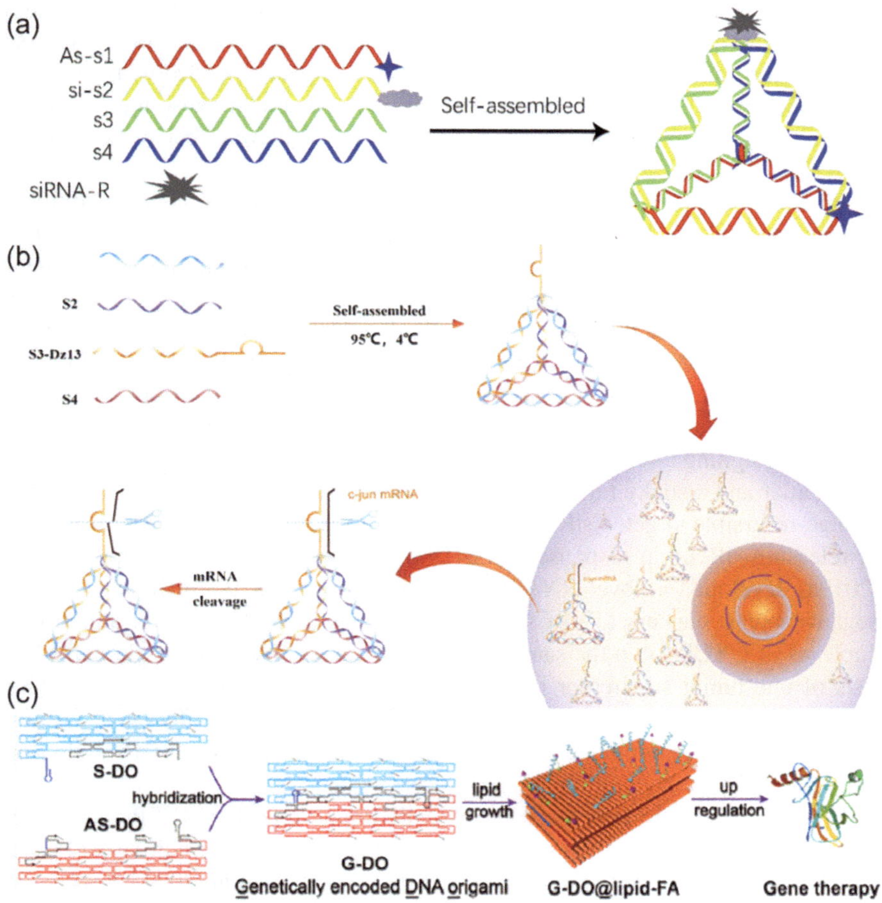

Fig. 1.34 Schematic representations of DNA nanostructures for gene silencing in cells (**a**) Self-assembled DNA nanostructure as a carrier for targeted siRNA delivery in glioma cells. Reprinted with the permission from Zhou et al. (2021a). Copyright 2021, The Author(s). (**b**) Tetrahedral DNA nanostructure-delivered DNAzyme for gene silencing to suppress cell growth. Reprinted with the permission from Meng et al. (2019b). Copyright 2019, American Chemical Society. (**c**) Genetically encoded DNA origami for in vivo gene therapy. Reprinted with the permission from Wu et al. (2023b). Copyright 2023, American Chemical Society

to cut target c-Jun mRNA, thereby reducing cell proliferation (Fig. 1.34b). Common strategies often expose siRNA to the outside of the DNA nanostructure, increasing the risk of enzymatic degradation during delivery. Therefore, Xue et al. (2019b) employed DNA tetrahedrons as building blocks to construct a novel DNA-based nanogel for intracellular siRNA delivery. This self-assembled nanogel can efficiently enter cells and knock down the expression of target gene. To fully exploit the long DNA scaffolds used in DNA origami, Ding et al. used two complementary DNA strands of functional genes as DNA scaffolds to construct DNA origami encoded by genes for gene therapy (Wu et al. 2023a). The complementary sense and antisense chains can be directly folded into two DNA origami monomers through their corresponding short chains. After hybridization, the assembled gene encoding DNA origami has precisely organized lipids on the surface, which can be used as a template for lipid growth. This lipid-coated DNA origami encoded with antitumor gene (p53) can effectively penetrate the cell membrane and cause the significant upregulation of p53 protein in tumor cells, so as to achieve effective tumor treatment (Fig. 1.34c).

In order to endow the delivery system with high stability and sensitivity to endogenous stimuli, Xue et al. (2021a) developed a three-dimensional (3D) DNA core/shell nanostructure for siRNA delivery using Y-shaped backbone-rigid triangular DNA bricks and gold nanoparticles. SiRNAs are internally encapsulated in 3D DNA shell, protecting them from enzymatic degradation. The siRNAs are released in response to endogenous miRNAs, enabling effective gene silencing within tumor cells (Fig. 1.35). In 2024, the same research group proposed a four-sticky end-contained 3D structural unit (PS-unit), constructed from two identical DNA components by designing three palindromic fragments. Building on this, they incorporated an aptamer and a sticky tail reservoir into one of the components to create a size-change-reversible targeted siRNA delivery nanocarrier. This nanocarrier is capable of releasing siRNA in response to GSH (Wu et al. 2024). In addition, Liu et al. (2019a) developed a novel RNA-triple-helix hydrogel for the treatment of triple negative breast cancers (TNBCs) by incorporating RNA-triple-helix and siRNA duplexes of CXCR4 into the same RNA nanoparticles. The RNA-triple-helix consists of one tumor suppressor miRNA (miRNA-205) and one oncomiR inhibitor (miRNA-221), both of which showed a remarkable synergistic effect in abrogating tumor. This strategy provided a potential treatment with high specificity and selectivity for TNBCs.

Clustered regularly interspaced short palindromic repeats (CRISPR)/CRISPR-associated protein 9 (Cas9) system has evolved as a broad gene-editing tool based on RNA-guided nuclease activity (Jinek et al. 2012; Sun et al. 2024). Leveraging DNA nanotechnology to construct delivery vectors with optimized packaging, dosage control, targeting, and gene-editing system release characteristics is crucial for reducing off-target effects and enhancing gene-editing efficiency. Significant progress has been made in this area, with numerous explorations underway. Li et al. (2022a) developed a nanocarrier that can deliver CRISPR/Cas9 and DNAzyme at the same time, yielding good results in breast cancer treatment by regulating gene expression. In this work, long ssDNA chains containing multiple sgRNA recognition sequences, DNAzyme sequences, and HhaI enzyme cleavage sites are synthesized through the rolling circle amplification, and compressed by Mn^{2+} to form DNA nanoparticles (DNC); finally, the acid-degradable polymer polyglycerol dimethacrylate (PGDA)-coated HHAI enzyme was assembled on its surface to form the H-DNC. Upon entry into cancer cells, the system can be activated by endogenous H^+ to efficiently release CRISPR/Cas9 and DNAzyme, thus reducing off-target problems. In addition, Liu et al. (2019b) reported a branched DNA-based nanoplatform to efficiently incorporate sgRNA/Cas9 complex and antisense via nucleic acid-based coassembly. Further, Tang et al. (2023b) used DNA origami as the carrier of CRISPR/Cas9 systems for gene therapy. The DNA origami loads the sgRNA/Cas9 complex through the PAM-rich region and DNA/RNA hybridization, and achieves targeted delivery based on DNA aptamers and influenza hemagglutinin (HA) peptides. It is opened through GSH reduction when entering the target cell, and then the sgRNA/Cas9 complex is released by RNase H cleavage, enabling the editing of tumor-related genes. These successful explorations of delivering CRISPR-Cas9 by DNA nanocarriers indicate new possibilities for combining molecular biology techniques with DNA nanotechnology.

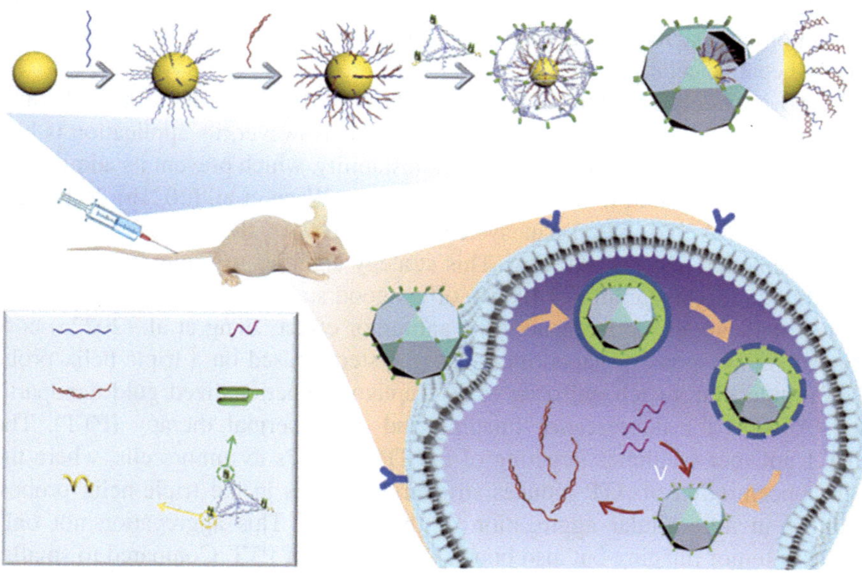

Fig. 1.35 Schematic representations of DNA-based nanostructures for targeted delivery and gene therapy. Programmable tiling of rigidified DNA bricks on gold nanoparticles to create a multifunctional shell for cancer-targeted delivery of siRNAs. Reprinted with the permission from Xue et al. (2021b). Copyright 2021, The Author(s)

1.4.3 Phototherapy

Photodynamic therapy (PDT) and photothermal therapy (PTT) are two common forms of light therapy for tumors. The integration of DNA nanotechnology in these therapies has yielded significant advancements. DNA nanocarriers enhance the solubility, specificity, and cellular uptake efficiency of photosensitizers and photothermal agents, thereby promoting the production of reactive oxygen species (ROS) or heat upon irradiation, leading to the destruction of malignant cells. Notably, compared to chemotherapy, phototherapy offers the distinct advantages of minimizing drug resistance, long-term toxicity, and side effects (Jiang et al. 2021).

PDT is highly regarded for its potential in tumor therapy, though its current limitations include the challenges posed by the hypoxic tumor microenvironment and safety concerns related to the biological specificity of photosensitizers or carriers. To address these issues, Pan et al. (2020) designed DNA nanosponges capable of efficiently loading and delivering photosensitizers, precisely targeting tumors, and significantly alleviating hypoxia-induced resistance, thereby enhancing the efficacy of PDT. To solve the problem of poor selectivity of traditional photosensitizers, Meng et al. (2021a, b) constructed MUC1 aptamer-modified tetrahedral nanostructures for targeted delivery of the photosensitizer 5, 10, 15, 20-tetrakis (1-methylpyridinium-4-yl) porphyrin (TMPyP4). These nanostructures allowed for the selective delivery of TMPyP4 to MUC1-positive cancer cells, promoted the production of reactive oxygen species, and exhibited a weak killing effect on MUC1-negative cells.

IR780 is widely recognized as a potential photosensitizer for cancer therapy, exhibiting a good response to both PTT and PDT. However, its application is hindered by poor water solubility and low photostability, which prevent its administration via parenteral routes. To address these issues, Wang et al. (2021b) designed a tetrahedral DNA (Td)-based nanosystem to load IR780 (IR780@Td) via electrostatic interaction and π-π stacking. This strategy significantly improves the water solubility and light stability of IR780, offers good stability and a suitable size for in vivo applications, and enhances the antitumor effect. Jiang et al. (2022a) constructed ATP-responsive nanodrug delivery systems based on a triple-helix probe (THP) molecular switch, utilizing MUC1 aptamer-functionalized gold nanoparticles (AuNPs) for fluorescence imaging and photothermal therapy (PTT). The MUC1 aptamer facilitates targeting of the THP-AuNPs to tumor cells, where the high concentration of ATP induces structural changes in the triple-helix probes, resulting in intracellular aggregation of the AuNPs. This aggregation not only enhances tumor imaging but also boosts the efficacy of PTT. Compared to smaller AuNPs, the aggregated AuNPs generate higher temperatures under NIR irradiation, effectively inducing apoptosis in tumor cells (Fig. 1.36).

In summary, DNA nanotechnology primarily enhances the specificity and efficacy of phototherapy. However, compared to chemotherapy, the application of phototherapy in vivo remains relatively underexplored. Future research should focus on combining phototherapy with other therapeutic strategies, such as chemotherapeutic agents or gene therapy, to harness the synergistic effects of multimodal cancer treatment.

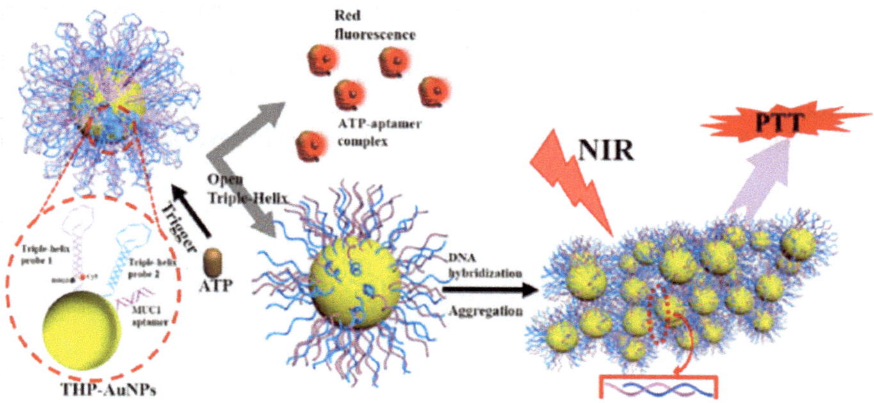

Fig. 1.36 Schematic representations of DNA-based nanostructures for photodynamic and photothermal therapy. ATP-triggered intracellular aggregation of a gold-nanoparticle-equipped triple-helix molecular switch for fluorescence imaging and photothermal tumor therapy. Reprinted with the permission from Jiang et al. (2022b). Copyright 2022, American Chemical Society

1.4.4 Immunotherapy

Immunotherapy has gained significant attention due to its low potential side effects and high specificity. With their exceptional structural programmability, permeability, and biocompatibility, DNA nanostructures are among the most promising candidates for delivering immune components to facilitate effective immunotherapy. The applications of DNA-based nanoplatforms in cancer immunotherapy can be categorized into four major areas based on their targets: (1) inducing immunogenic cell death in tumor cells, (2) modulating antigen-presenting cells (APCs) with adjuvants and nanovaccines, (3) engineering effector immune cells, and (4) manipulating the interactions between T cells and tumor cells by regulating immune checkpoint inhibitors and bispecific engagers (Sun et al. 2022b).

Unmethylated cytosine-phosphate-guanine (CpG) oligonucleotides are commonly used as immunotherapeutic agents and vaccine adjuvants, which contain certain sequence motifs that can be recognized by toll-like receptor 9 (TLR9) and stimulate cytokine release and an immune response (Krieg 2006). TLR9 is expressed in antigen-presenting cells (APCs) and is generally located in the endosome. However, CpG oligonucleotides face challenges such as instability and low efficacy. Multiple vectors based on self-assembled DNA nanostructures have been developed for CpG delivery-based immunotherapy, such as Y-shaped tiles (Nishikawa et al. 2008), dendrimers (Rattanakiat et al. 2009), hydrogels (Nishikawa et al. 2011), DNA tetrahedra (Li et al. 2011), and DNA origami tubes (Schuller et al. 2011). Yu et al. (2020) reported polymer DNA nanowires and nanospheres encoding multivalent CpG motifs by HCR to enhance immunostimulatory activity and trigger a potent immune response. The self-assembled polymer DNA nanoassemblies, which contain multivalent CpG motifs, effectively induce strong immune responses and subsequently result in cancer cell death (Fig. 1.37a).

In addition to CpG motifs, double-stranded RNA (dsRNA), which activates endosomal TLR3, can also serve as a potent immune adjuvant in immunotherapy (Alexopoulou et al. 2001). Qi et al. (2020) used the newly developed single-strand

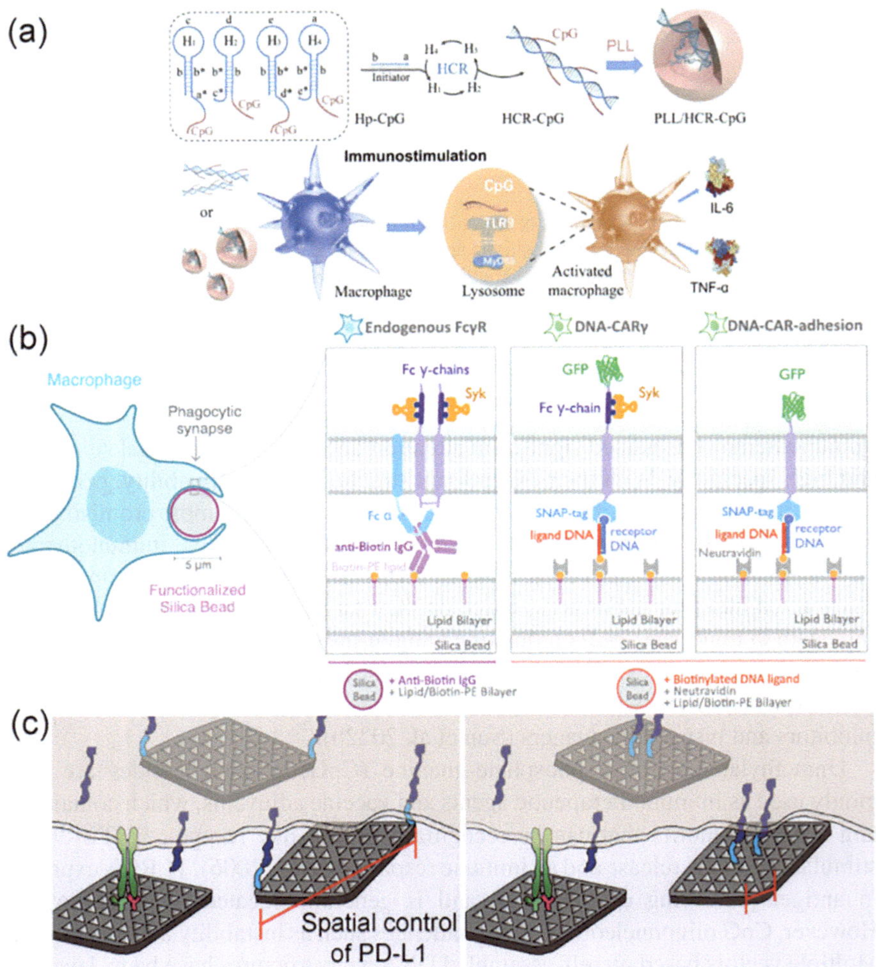

Fig. 1.37 Schematic representations of DNA-based platforms for cancer immunotherapy (**a**) Enhanced immunostimulatory activity of cytosine-phosphate-guanine immunomodulator by the assembly of polymer DNA wires and spheres. Reprinted with the permission from Yu et al. (2020). Copyright 2020, American Chemical Society. (**b**) Tight nanoscale clustering of Fcγ receptors using DNA origami to promote phagocytosis. Reprinted with the permission from Kern et al. (2021). Copyright 2022, The Authors. (**c**) Spatial regulation of T-cell signaling by programmed death-ligand 1 (PD-L1) on wireframe DNA origami flat sheets. Reprinted with the permission from Fang et al. (2021). Copyright 2021, American Chemical Society

RNA origami technology to construct RNA origami nanostructure serving as a novel TLR3 agonist for potent and safe anticancer immunotherapy. It was demonstrated that the RNA origami therapy induced durable antitumor immunity, reduced the immunosuppressive microenvironment in the peritoneal compartment, and provided a potent line of synthetic immune adjuvants with high structural stability (Tian et al. 2023). Cyclic dinucleotides, such as c-di-GMP (CDG) and 2′,3′-cyclic GMP-AMP (cGAMP), are agonists for stimulator of interferon genes (STING), representing another promising candidate as an immune adjuvant for cancer immunotherapy. Yang et al. (2020a) employed cytosine-rich i-motif DNA to efficiently load CDG at physiological pH. Chen et al. (2021) utilized polymerized guanidine-containing disulfides to assemble with cGAMP and then coordinate with Mn^{2+} ions to build a robust DNA nanoparticle.

DNA origami holds immense potential in immunotherapy, significantly enhancing immune stimulation and improving targeting (Quyang et al. 2017). It plays a crucial role in both adoptive immunotherapy and cancer vaccines. Sun et al. (2022b) developed DNA origami structures for use in nanoscale artificial antigen-presenting cells (aAPCs). By anchoring costimulatory ligands that target CD28 antibodies and T cell receptor (TCR) ligand peptide-major histocompatibility complex (pMHC) on its three vertices and three edges, their DNA-origami-based aAPCs demonstrated higher tumor growth inhibition in adoptive immunotherapy. The strategy exhibits remarkable potential in adoptive cell therapy. Liu et al. (2021c) constructed tubular DNA nanostructures for loading molecular adjuvants and antigenic peptides inside the lumen. These structures can be activated in a subcellular environment, triggering T cell activation and cancer cell toxicity. The strategy can reduce tumor size and prevent recurrence, and is a potential candidate for the development of cancer vaccines. Kern et al. (2021) proposed a DNA-origami-based phagocytic system loading FcγR receptor ligands, which enhanced macrophage phagocytosis by regulating these ligands' density on the DNA origami structure (Fig. 1.37b). Fang et al. (2021) modified DNA origami flat sheets with CD3 and CD28 activating antibodies (FS-α-CD3-CD28), which can induce strong T cell activation. In addition, the spatial organization of Programmed Death-1 (PD-1) determines its ability to regulate T cell signaling, and this finding may guide the development of future nanomedicine-based immune modulation therapies (Fig. 1.37c).

Sun et al. reported a hydrogel based on branched-strand DNA tile as a nanovaccine for synergistic immunotherapy (Sun et al. 2020). The self-assembled DNA nanogel, with its optimal size, demonstrated effective lymphatic drainage and high vaccine efficacy. When embedded with CpG motifs, this DNA nanogel significantly inhibited tumor growth and enhanced immunotherapy effectiveness following vaccination with free OVA antigen. Additionally, Yang et al. introduced a method for constructing bispecific engagers using circular aptamers. Their study demonstrated that circular bispecific aptamers can effectively guide naive T cells in recognizing and eliminating cancer cells (Zhang et al. 2020).

In summary, DNA nanotechnology has greatly promoted the progress of tumor immunotherapy.

1.4.5 Multimodal Therapy

Therapeutic agents kill tumor cells through chemical toxicity, biological interference, or physical damage. However, inherent limitations and side effects often hinder the success of single treatment strategies. Increasing evidence suggests that monotherapy is insufficient for addressing the complexities of most cancer therapies. For instance, tumor cells may develop resistance to chemotherapy, phototherapy suffers from limited tissue penetration and reduced efficacy in metastatic cancers, and gene therapy typically offers only short-term benefits while potentially triggering immune responses (Jiang et al. 2021). Self-assembled DNA nanostructure-based materials present a versatile and promising solution, serving as multifunctional drug delivery systems. By employing DNA nanotechnology to integrate multiple treatment strategies into a single platform, the synergistic effects of different therapies can be maximized, offering a more effective approach to cancer treatment. Recent research has explored these possibilities, highlighting the potential of multimodal therapy in overcoming the limitations of single-agent treatments.

As one of the most widely used DNA nanostructures, DNA tetrahedrons have become increasingly important in the field of multimodal therapy. Numerous studies have focused on the simultaneous delivery of chemotherapeutic agents like Dox and nucleic acid drugs using DNA tetrahedrons for enhanced tumor treatment. Zhong et al. proposed aptamer-modified DNA tetrahedron for targeted chemogene combination therapy based on co-delivery of Dox and DNAzyme (Ren et al. 2021). Aptamers and DNAzyme extend from two vertices of the DNA tetrahedron, and one edge of the DNA tetrahedron is a pH-responsive i-motif sequence that could release Dox. Compared with single drug therapy, the DNA nanocarriers can greatly improve the therapeutic effect on cancer cells. Tang et al. (2021) reported an aptamer-modified DNA tetrahedron (TET)-based DNA nanogel can be loaded with ASO and Dox for chemotherapy/gene combination therapy against multidrug resistant tumors (Fig. 1.38). Sun et al. (2022a) used a DNA tetrahedron to co-load Dox and ASO, combined it with gold nanoclusters (AuNCs), and self-assembled them into a programmed disassembly nanohybrid delivery system for combination gene-chemotherapy. In addition, Li et al. (2020b) developed a triple combination nanosystem for gene therapy, chemotherapy, and phototherapy, delivered by multifunctional RNA nanohydrogels, which contain three lung cancer inhibitor microRNAs for silencing three targeted mRNAs, a chemotherapy drug Dox, and a photosensitizer TMPyP4. This strategy has achieved an ideal synergistic therapeutic effect and shown a great potential in the field of multifunctional synergistic cancer treatment.

Cancer chemical immunotherapy is the combination of chemotherapy and immunotherapy, which can play a synergistic role and improve the antitumor efficiency (Salas-Benito et al. 2021). Wang et al. (2022a) used DNA tetrahedron as a carrier to co-deliver Dox and CpG for synergistic cancer therapy. This strategy enhances the immunostimulatory effect while preserving the tumor-killing activity of Dox, and this dual-modality therapy shows effective synergistic antitumor effects (Fig. 1.39). In addition, Shen et al. (2024) used DNA tetrahedron to co-deliver Dox

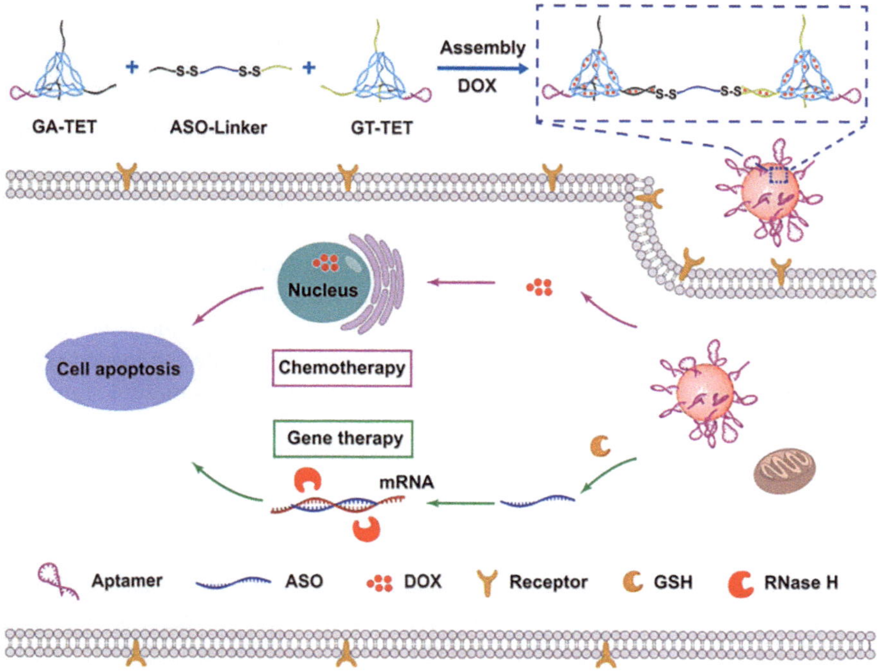

Fig. 1.38 Schematic representations of DNA systems for multimodal therapy. Aptamer-modified DNA tetrahedron-based nanogel for combined chemo/gene therapy of multidrug-resistant tumors. Reprinted with the permission from Tang et al. (2021). Copyright 2021, American Chemical Society

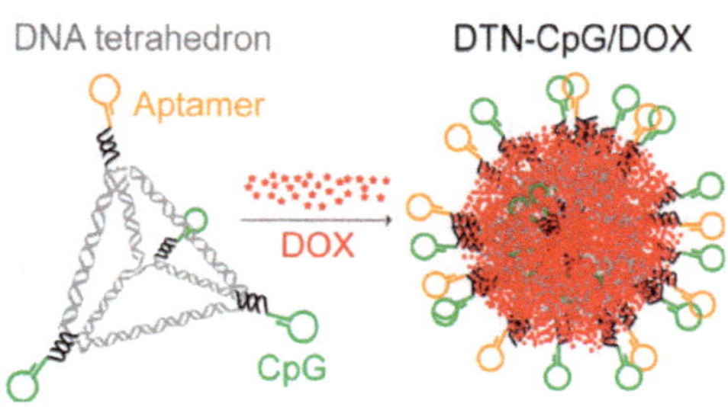

Fig. 1.39 Schematic representations of DNA systems for multimodal therapy. Co-delivery of chemotherapeutic drugs and immune adjuvants using nanoscale DNA tetrahedrons for synergistic cancer therapy. Reprinted with the permission from Wang et al. (2022a). Copyright 2022, American Chemical Society

and CpG for the treatment of malignant melanoma. Lu et al. (2022) also used DNA tetrahedron to simultaneously load Dox and CpG to facilitate ICD-related chemoimmunotherapy.

As a highly efficient nanocore, gold nanoparticles (AuNPs) have many excellent characteristics suitable for multimodal therapy, including high loading capacity, unique plasma properties, high fluorescence quenching efficiency, fast luminescence, strong cell uptake capacity, and good biocompatibility (Cheng et al. 2017). Many composite materials that combine gold nanoparticles with self-assembled DNA nanostructures show great potential in tumor multimodal therapy. Yu et al. (2021) developed an aptamer-tethered deoxyribonucleic acid-gold particle (Apt-DNA-Au) nanomachine for in situ imaging and targeted multimodal synergistic therapy of mammary carcinoma. The Apt-DNA-Au nanomachine is specifically internalized into MCF-7 cells via aptamer mediation; it can be activated by tumor-associated TK1 mRNA via a DNA strand displacement cascade, releasing a fluorophore for in situ imaging, antisense DNA for survivin suppression, and executing photothermal therapy based on AuNP aggregation. In addition, the nanomachine also facilitates controlled drug release for chemotherapy, while the loaded photosensitizers enable photodynamic therapy. This strategy integrates multiple functions such as dynamic monitoring, targeted drug delivery, and multimodal therapy, providing a promising nanoplatform for in situ image-guided combined anticancer therapy.

DNA nanostructures have also been employed to investigate the synergies between various therapies, including chemo-photodynamic therapy (Zhang et al. 2022b; Wang et al. 2022c), gene-photodynamic therapy (Guo et al. 2022), and gene-immunotherapy (Shapouri-Moghaddam et al. 2018; Qian et al. 2023). In summary, multimodal therapy leverages the complementarity and synergistic effects of multiple treatment strategies to enhance treatment outcomes and offer potential solutions for tumor types that are challenging to address with single-agent therapies.

In addition to the conventional cancer therapies mentioned above, there are some promising methods that are expected to become powerful tools in cancer treatment in the future, such as biological interference. Specifically, using DNA nanotechnology, DNA aggregates are formed locally in living cells in response to intracellular triggers, inducing bio-interference. Yang et al. proposed a series of strategies for bio-interference, such as a K^+-mediating dynamic assembly of DNA tetrahedrons for mitochondrial interference (Li et al. 2022b), proton-driven dynamic assembly of DNA nanoframework for lysosome interference (Dong et al. 2022), controlled sequential assembly of DNA nanoparticles for mitochondrial interference (Yuan et al. 2024), and telomerase-mediated self-assembly of DNA network for mitochondrial interference (Guo et al. 2022). These studies have important implications for the development of nanobiologic drugs, especially in the treatment of cancer.

1.5 Conclusions and Perspectives

In this chapter, we have summarized remarkable advances in the development of self-assembling DNA nanostructures and nanomaterials based on them. The application of these nanomaterials in biomedicine is extensive, with a particular focus on cancer diagnosis and treatment. DNA nanomaterials have shown great potential in cancer diagnosis, being widely utilized for biosensing and bioimaging of tumor-related biomarkers, which is advantageous for early detection and monitoring. In cancer treatment, we have explored the use of various DNA nanomaterials as delivery carriers for chemotherapy drugs, nucleic acid therapies, photosensitizers, and immune agents, as well as multifunctional nanoplatforms for multimodal cancer therapy. Although nucleic acid nanomaterials offer diverse functionalities and the ability to respond to a range of external stimuli, most stimuli typically elicit a single signal output, making it challenging to achieve an accurate diagnosis of the complex tumor microenvironment. Additionally, it is crucial to design certain nucleic acid functional units to minimize potential immunogenicity. Advances in DNA nanotechnology must focus on enhancing sensitivity to develop more effective detectors for cancer prevention, early detection, treatment, and prognosis. Several challenges need to be addressed before DNA nanomaterials can be translated into clinical practice. One key issue is the structural heterogeneity of DNA nanocarriers, which may lead to unintended side effects. Evidence suggests that partially assembled and fully intact DNA nanostructures exhibit different organ distribution preferences, indicating that denaturation during circulation could impact their targeting efficacy (Jiang et al. 2020, 2021). Secondly, alterations in the surface chemical properties of DNA nanomaterials can influence their biological distribution. It is essential to characterize these effects in vivo, including assessing circulation time, organ distribution, and toxicity. Additionally, the long-term biosafety of DNA-based diagnostic and therapeutic systems must be thoroughly evaluated. We believe that self-assembled DNA nanomaterials hold great promise and have the potential to become a powerful tool in the fight against cancer in the future.

References

Alberts B (2002) Molecular biology of the cell, vol 18, 4th edn. Garland Science, New York, p 289

Alexopoulou L, Holt AC, Medzhitov R et al (2001) Recognition of double-stranded RNA and activation of NF-κB by toll-like receptor 3. Nature 413(6857):732–738

Alix Panabières C, Pantel K (2013) Circulating tumor cells: liquid biopsy of cancer. Clin Chem 59(1):110–118

Andersen ES, Dong M, Nielsen MM et al (2009) Self-assembly of a nanoscale DNA box with a controllable lid. Nature 459(7243):73–76

Azad RN, Zafiropoulos D, Ober D et al (2018) Experimental maps of DNA structure at nucleotide resolution distinguish intrinsic from protein-induced DNA deformations. Nucleic Acids Res 46(5):2636–2647

Bao J, Qiu X, Wang D et al (2021) Carbon nanomaze for biomolecular detection with zeptomolar sensitivity. Adv Funct Mater 31(14):2006521

Bath J, Green S, Turberfield A (2005) A free-running DNA motor powered by a nicking enzyme. Angew Chem Int Ed 44(28):4358–4361

Bell CJ, Manfredi G, Griffiths EJ et al (2007) Luciferase expression for ATP imaging: application to cardiac myocytes. Methods Cell Biol 80:341–352

Bertoli G, Cava C, Castiglioni I (2015) MicroRNAs: new biomarkers for diagnosis, prognosis, therapy prediction and therapeutic tools for breast cancer. Theranostics 5(10):1122

Biswas S, Kinbara K, Niwa T et al (2013) Biomolecular robotics for chemomechanically driven guest delivery fuelled by intracellular ATP. Nat Chem 5(7):613–620

Bordanaba-Florit G, Royo F, Kruglik SG et al (2021) Using single-vesicle technologies to unravel the heterogeneity of extracellular vesicles. Nat Protoc 16(7):3163–3185
Cai Z, Wang A, Wang Y et al (2022) Smart programmable scalable dual-mode diagnostic logic nanoflare strategy for dual-tumor marker detection. Anal Chem 94(27):9715–9723
Cai S, Chen X, Chen H et al (2023) A fluorescent aptasensor for ATP based on functional DNAzyme-walker and terminal deoxynucleotidyl transferase-assisted formation of DNA-AgNCs. Analyst 148(4):799–805
Calin GA, Croce CM (2006) MicroRNA signatures in human cancers. Nat Rev Cancer 6(11):857–866
Chai X, Fan Z, Yu MM et al (2021) A redox-activatable DNA nanodevice for spatially-selective, AND-gated imaging of ATP and glutathione in mitochondria. Nano Lett 21(23):10047–10053
Chandrasekaran AR, Wady H, Subramanian HKK (2016) Nucleic acid nanostructures for chemical and biological sensing. Small 12(20):2689–2700
Chao Q, Zhang Y, Li Q et al (2023) Compute-and-release logic-gated DNA cascade circuit for accurate cancer cell imaging. Anal Chem 95(19):7723–7734
Chen F, Bai M, Cao K et al (2017) Programming enzyme-initiated autonomous DNAzyme nanodevices in living cells. ACS Nano 11(12):11908–11914
Chen Y, Tyagi D, Lyu M et al (2019a) Regenerative nanooctopus based on multivalent-aptamer-functionalized magnetic microparticles for effective cell capture in whole blood. Anal Chem 91(6):4017–4022
Chen J, Yang HH, Yin W et al (2019b) Metastable dumbbell probe-based hybridization chain reaction for sensitive and accurate imaging of intracellular-specific microRNAs in situ in living cells. Anal Chem 91(7):4625–4631
Chen C, Tong Y, Zheng Y et al (2021) Cytosolic delivery of thiolated Mn-cGAMP nanovaccine to enhance the antitumor immune responses. Small 17(17):2006970
Chen J, Fu S, Zhang C et al (2022a) DNA logic circuits for cancer theranostics. Small 18(20):2108008
Chen H, Huang C, Wu Y et al (2022b) Exosome metabolic patterns on aptamer-coupled polymorphic carbon for precise detection of early gastric cancer. ACS Nano 16(8):12952–12963
Chen Q, Wang X, Chen J et al (2022c) Multiple-aptamer-integrated DNA-origami-based chemical nose sensors for accurate identification of cancer cells. Anal Chem 94(28):10192–10197
Chen J, Xie M, Shi M et al (2022d) Spatial confinement-derived double-accelerated DNA cascade reaction for ultrafast and highly sensitive in situ monitoring of exosomal miRNA and exosome tracing. Anal Chem 94(4):2227–2235
Chen Y, Lin M, Ye D et al (2024) Functionalized tetrahedral DNA frameworks for the capture of circulating tumor cells. Nat Protoc 19(4):985–1014
Cheng et al (2017) Light-triggered assembly of gold nanoparticles for photothermal therapy and photoacoustic imaging of tumors in vivo. Adv Mater 29(6):1604894
Ciccia A, Elledge SJ (2010) The DNA damage response: making it safe to play with knives. Mol Cell 40(2):179–204
Cohen L, Cui N, Cai Y et al (2020) Single molecule protein detection with attomolar sensitivity using droplet digital enzyme-linked immunosorbent assay. ACS Nano 14(8):9491–9501
Curtin NJ (2012) DNA repair dysregulation from cancer driver to therapeutic target. Nat Rev Cancer 12(12):801–817
Deng J, Walther A et al (2020) ATP-responsive and ATP-fueled self-assembling systems and materials. Adv Mater 32(42):2002629
Deng J, Liu C, Sun J (2024) DNA-based nanomaterials for analysis of extracellular vesicles. Adv Mater 36(22):2303092
Ding P, Wang Z, Wu Z et al (2020) Aptamer-based nanostructured interfaces for the detection and release of circulating tumor cells. J Mater Chem B 8(16):3408–3422
Ding TL, Yang J, Wang J et al (2022) Shaped DNA origami carrier nanopore translocation influenced by aptamer based surface modification. Biosens Bioelectron 195:113658
Dong C, Song C, Chao J et al (2022) Multi-armed tetrahedral DNA probes for visualizing the whole-course of cell apoptosis by simultaneously fluorescence imaging intracellular cytochrome c and telomerase. Biosens Bioelectron 205:114059

Dong Y, Wang G, Afshan N et al (2024) Protease-responsive bitext editing biopolymer for breast cancer imaging in vivo. ACS Mater Lett 6(8):3815–3823
Douglas SM, Dietz H, Liedl T et al (2009) Self-assembly of DNA into nanoscale three-dimensional shapes. Nature 459(7245):414–418
Douglas SM, Bachelet I, Church GM (2012) A logic-gated nanorobot for targeted transport of molecular payloads. Science 335(6070):831–834
Fang et al (2021) Spatial regulation of T-cell signaling by programmed death-ligand 1 on wireframe DNA origami flat sheets. ACS Nano 15(2):3441–3452
Février B, Raposo G (2004) Exosomes: endosomal-derived vesicles shipping extracellular messages. Curr Opin Cell Biol 16(4):415–421
Fidler IJ (2003) The pathogenesis of cancer metastasis: the 'seed and soil' hypothesis revisited. Nat Rev Cancer 3(6):453–458
Fu TJ, Seeman NC (1993) DNA double-crossover molecules. Biochemistry 32(13):3211–3220
Gao R, Xu L, Hao C et al (2019) Circular polarized light activated chiral satellite nanoprobes for the imaging and analysis of multiple metal ions in living cells. Angew Chem 131(12):3953–3957
Gao X, Li X, Sun X et al (2020a) DNA tetrahedra-cross-linked hydrogel functionalized paper for onsite analysis of DNA methyltransferase activity using a personal glucose meter. Anal Chem 92(6):4592–4599
Gao C, Yu H, Zhang L et al (2020b) Ultrasensitive paper-based photoelectrochemical sensing platform enabled by the polar charge carriers-created electric field. Anal Chem 92(4):2902–2906
Gao J, Zhang H, Wang Z (2020c) A DNA tetrahedron nanoprobe-based fluorescence resonance energy transfer sensing platform for intracellular tumor-related miRNA detection. Analyst 145(10):3535–3542
Gao Y, Zhang S, Wu C et al (2021) Self-protected DNAzyme walker with a circular bulging DNA shield for amplified imaging of miRNAs in living cells and mice. ACS Nano 15(12):19211–19224
Ge Z, Fan C, Yan H (2014) Research progress and applications of self-assembled DNA nanostructures. Chin Sci Bull 59:146
Goodman RP, Heilemann M, Doose S et al (2008) Reconfigurable, braced, three-dimensional DNA nanostructures. Nat Nanotechnol 3(2):93–96
Gottesman MM, Fojo T, Bates SE (2002) Multidrug resistance in cancer: role of ATP-dependent transporters. Nat Rev Cancer 2(1):48–58
Gourine AV, Llaudet E, Dale N et al (2005) ATP is a mediator of chemosensory transduction in the central nervous system. Nature 436(7047):108–111
Guo X, Wu X, Sun M et al (2019) Tetrahedron probes for ultrasensitive in situ detection of telomerase and surface glycoprotein activity in living cells. Anal Chem 92(2):2310–2315
Guo L, Zhang Y, Wang Y et al (2022) Directing multivalent aptamer-receptor binding on the cell surface with programmable atom-like nanoparticles. Angew Chem Int Ed 134(18):e202117168
György B, Szabó TG, Pásztói M et al (2011) Membrane vesicles, current state-of-the-art: emerging role of extracellular vesicles. Cell Mol Life Sci 68:2667–2688
Han DR, Pal S, Nangreave J et al (2011) DNA origami with complex curvatures in three-dimensional space. Science 332(6027):342–346
Han D, Pal S, Yang Y et al (2013) DNA gridiron nanostructures based on four-arm junctions. Science 339(6126):1412–1415
Han Q, Wang C, Liu P et al (2021) Functionalized europium-porphyrin coordination polymer: rational design of high performance electrochemiluminescence emitter for mucin 1 sensing. Biosens Bioelectron 191:113422
Han Y, Wang Y, Zhou F et al (2022) Molecular beacon-based DNA tetrahedrons for APE 1 activity detection in living cells. Sens Actuators B Chem 355:131258
He Y, Chen Y, Liu H et al (2005) Self-assembly of hexagonal DNA two-dimensional (2D) arrays. J Am Chem Soc 127(35):12202–12203
He Y, Tian Y, Ribbe AE et al (2006) Highly connected two-dimensional crystals of DNA six-point-stars. J Am Chem Soc 128(50):15978–15979

He L, Lu DQ, Liang H et al (2017) Fluorescence resonance energy transfer-based DNA tetrahedron nanotweezer for highly reliable detection of tumor-related mRNA in living cells. ACS Nano 11(4):4060–4066

He L, Lu D, Liang H et al (2018) mRNA-initiated, three-dimensional DNA amplifier able to function inside living cells. J Am Chem Soc 140(1):258–263

He M, He M, Zhang J et al (2020) A spatial-confinement hairpin cascade reaction-based DNA tetrahedral amplifier for mRNA imaging in live cells. Talanta 207:120287

He J, Luo S, Deng H et al (2023) Fluorescent features and applicable biosensing of a core-shell Ag nanocluster shielded by a DNA tetrahedral nanocage. Anal Chem 95(39):14805–14815

Herrmann IK, Wood MJA, Fuhrmann G (2021) Extracellular vesicles as a next-generation drug delivery platform. Nat Nanotechnol 16(7):748–759

Hong F, Zhang F, Liu Y et al (2017) DNA origami: scaffolds for creating higher order structures. Chem Rev 117(20):12584–12640

Hou T, Xu N, Wang W et al (2019) Label-free and immobilization-free photoelectrochemical biosensing strategy using methylene blue in homogeneous solution as signal probe for facile DNA methyltransferase activity assay. Biosens Bioelectron 141:111395

Hu J, Liu MH, Zhang CY (2019) Construction of tetrahedral DNA-quantum dot nanostructure with the integration of multistep forster resonance energy transfer for multiplex enzymes assay. ACS Nano 13(6):7191–7201

Hu Y, Wang Y, Yan J et al (2020) Dynamic DNA assemblies in biomedical applications. Adv Sci 7(14):2000557

Hu W, Chang Y, Huang J et al (2022) Tetrahedral DNA nanostructure with multiple target-recognition domains for ultrasensitive electrochemical detection of mucin 1. Anal Chem 94(18):6860–6865

Huang H, Tan Y, Shi J et al (2010) DNA aptasensor for the detection of ATP based on quantum dots electrochemiluminescence. Nanoscale 2(4):606–612

Huang R, He N, Li Z (2018) Recent progresses in DNA nanostructure-based biosensors for detection of tumor markers. Biosens Bioelectron 109:27–34

Huang W, Hu GB, Yao LY et al (2020) Matrix coordination-induced electrochemiluminescence enhancement of tetraphenylethylene-based hafnium metal-organic framework: an electrochemiluminescence chromophore for ultrasensitive electrochemiluminescence sensor construction. Anal Chem 92(4):3380–3387

Huang B, Liang B, Zhang R et al (2022a) Molecule fluorescent probes for adenosine triphosphate imaging in cancer cells and in vivo. Coord Chem Rev 452:214302

Huang X, Zhan Y, Chen J et al (2022b) Dual-locked DNAzyme platform for in vitro and in vivo discrimination of cancer cells. Anal Chem 94(35):12221–12230

Huang G, Xie Q, Chi J et al (2024) Magneto-fluorescent nanobiosensor with tetrahedral framework nucleic acid-confined aptamer for trace adenosine triphosphate detection. Sens Actuators B Chem 417:136129

Idzko M, Ferrari D, Eltzschig HK (2014) Nucleotide signalling during inflammation. Nature 509(7500):310–317

Iinuma R, Ke Y, Jungmann R et al (2014) Polyhedra self-assembled from DNA tripods and characterized with 3D DNA-PAINT. Science 344(6179):65–69

Imamura H, Huynh Nhat KP, Togawa H et al (2009) Visualization of ATP levels inside single living cells with fluorescence resonance energy transfer-based genetically encoded indicators. Proc Natl Acad Sci 106(37):15651–15656

Jiang D, England CG, Cai W (2016) DNA nanomaterials for preclinical imaging and drug delivery. J Control Release 239:27–38

Jiang Y, Kong X, Jiang Y et al (2020) DNA nanodevices for base excision repair regulates ATP in situ imaging and tumor therapy. ACS Appl Bio Mater 3(12):8507–8514

Jiang S, Ge Z, Mou S et al (2021) Designer DNA nanostructures for therapeutics. Chem 7(5):1156–1179

Jiang Y, Zhao W, Zhou H et al (2022a) ATP-triggered intracellular in situ aggregation of a gold-nanoparticle-equipped triple-helix molecular switch for fluorescence imaging and photothermal tumor therapy. Langmuir 38(12):3755–3764

Jiang Y, Zhou H, Zhao W et al (2022b) ATP-triggered drug release of self-assembled 3D DNA nanostructures for fluorescence imaging and tumor therapy. Anal Chem 94(18):6771–6780

Jiao K, Zhu B, Guo L et al (2020) Programming switchable transcription of topologically constrained DNA. J Am Chem Soc 142(24):10739–10746

Jinek M, Chylinski K, Fonfara I et al (2012) A programmable dual-RNA-guided DNA endonuclease in adaptive bacterial immunity. Science 337(6096):816–821

Jolma A, Yan J, Whitington T et al (2013) DNA-binding specificities of human transcription factors. Cell 152(1):327–339

Jung C, Allen PB, Ellington AD et al (2016) A stochastic DNA walker that traverses a microparticle surface. Nat Nanotechnol 11(2):157–163

Kallenbach NR, Ma RI, Seeman NC et al (1983) An immobile nucleic acid junction constructed from oligonucleotides. Nature 305(5937):829–831

Kalluri R, LeBleu VS (2020) The biology, function, and biomedical applications of exosomes. Science 367(6478):6977

Kaur V, Kaur C, Sen T (2023) Single-molecule fluorescence enhancement based detection of ATP using DNA origami-assembled au@ag nanostar optical antennas. J Phys Chem C 127(15):7308–7318

Kern et al (2021) Tight nanoscale clustering of Fcγ receptors using DNA origami promotes phagocytosis. Elife 10:e68311

Kishi JY, Lapan SW, Beliveau BJ et al (2019) SABER amplifies FISH: enhanced multiplexed imaging of RNA and DNA in cells and tissues. Nat Methods 16(6):533–544

Knowles JR (1980) Enzyme-catalyzed phosphoryl transfer reactions. Annu Rev Biochem 49(1):877–919

Kooijmans SA, Vader P, van Dommelen SM et al (2012) Exosome mimetics: a novel class of drug delivery systems. Int J Nanomedicine 7:1525–1541

Krebs MG, Metcalf RL, Carter L et al (2014) Molecular analysis of circulating tumour cells-biology and biomarkers. Nat Rev Clin Oncol 11(3):129–144

Krieg AM (2006) Therapeutic potential of toll-like receptor 9 activation. Nat Rev Drug Discov 5(6):471–484

Labean T, Yan H, Kopatsch J et al (2000) Construction, analysis, ligation, and self-assembly of DNA triple crossover complexes. J Am Chem Soc 122:1848–1860

Lee DS, Qian H, Tay CY et al (2016) Cellular processing and destinies of artificial DNA nanostructures. Chem Soc Rev 45(15):4199–4225

Lee JU, Kim WH, Lee HS et al (2019) Quantitative and specific detection of exosomal miRNAs for accurate diagnosis of breast cancer using a surface-enhanced Raman scattering sensor based on plasmonic head-flocked gold nanopillars. Small 15(17):1804968

Lei Y, Fei X, Ding Y et al (2023) Simultaneous subset tracing and miRNA profiling of tumor-derived exosomes via dual-surface-protein orthogonal barcoding. Sci Adv 9(40):eadi1556

Li J, Pei H, Zhu B et al (2011) Self-assembled multivalent DNA nanostructures for noninvasive intracellular delivery of immunostimulatory CpG oligonucleotides. ACS Nano 5(11):8783–8789

Li W, Liu X, Hou T et al (2015) Ultrasensitive homogeneous electrochemical strategy for DNA methyltransferase activity assay based on autonomous exonuclease III-assisted isothermal cycling signal amplification. Biosens Bioelectron 70:304–309

Li L, Zhang Y, Zhang L et al (2016) Based device for colorimetric and photoelectrochemical quantification of the flux of H2O2 releasing from MCF-7 cancer cells. Anal Chem 88(10):5369–5377

Li C, Xue C, Wang J et al (2019) Oriented tetrahedron-mediated protection of catalytic DNA molecular-scale detector against in vivo degradation for intracellular miRNA detection. Anal Chem 91(18):11529–11536

Li C, Luo M, Wang J et al (2020a) Rigidified DNA triangle-protected molecular beacon from endogenous nuclease digestion for monitoring microRNA expression in living cells. ACS Sens 5(8):2378–2387

Li J, Wang J, Liu S et al (2020b) Amplified FRET nanoflares:an endogenous mRNA-powered nanomachine for intracellular microRNA imaging. Angew Chem 132(45):20279–20286

Li Y, Deng J, Han Z et al (2021) Molecular identification of tumor-derived extracellular vesicles using thermophoresis-mediated DNA computation. J Am Chem Soc 143(3):1290–1295
Li CH, Lv WY, Yang FF et al (2022a) Simultaneous imaging of dual microRNAs in cancer cells through catalytic hairpin assembly on a DNA tetrahedron. ACS Appl Mater Interfaces 14(10):12059–12067
Li F, Song N, Dong Y et al (2022b) A proton-activatable DNA-based nanosystem enables co-delivery of CRISPR/Cas9 and dnazyme for combined gene therapy. Angew Chem Int Ed 61(9):e202116569
Li C, Zhang J, Gao Y et al (2022c) Nonenzymatic autonomous assembly of cross-linked network structures from only two palindromic DNA components for intracellular fluorescence imaging of miRNAs. ACS Sens 7(2):601–611
Li L, Li S, Wang J et al (2023a) Extracellular ATP-activated hybridization chain reaction for accurate and sensitive detection of cancer cells. Chin Chem Lett 34(12):108399
Li T, Sun M, Zhou Q et al (2023b) Endogenous enzyme-powered DNA nanomotor operating in living cells for microRNA imaging. Anal Chem 95(40):15025–15032
Li L, Wang J, Jiang H et al (2023c) DNA tetrahedron-based split aptamer probes for reliable imaging of ATP in living cells. Chin Chem Lett 34(3):107506
Li S, Leng M, Li Z et al (2024a) Confined DNA tetrahedral molecular sieve for size-selective electrochemiluminescence sensing. Anal Chim Acta 1304:342561
Li Z, Shen Q, Wang WX et al (2024b) Shu membrane protein guided and ATP-gated dual module DNA nanodevice for amplified fluorescence labeling of cancer cells. Sens Actuators B Chem 415:136000
Liang L, Li J, Li Q et al (2014) Single-particle tracking and modulation of cell entry pathways of a tetrahedral DNA nanostructure in live cells. Angew Chem Int Ed 53(30):7745–7750
Liang L, Shuai X, Cao L et al (2023) Endogenous adenosine triphosphate-assisted three-dimensional DNA walker assembled on soft nanoparticles for intracellular microRNA imaging. Anal Chem 95(28):10721–10727
Liao R, Zhao F, Hamada S et al (2020) DNA-based engineering system for improving human and environmental health: identification, detection, and treatment. Nano Today 35:100958
Lin M, Guo JT (2019) New insights into protein-DNA binding specificity from hydrogen bond based comparative study. Nucleic Acids Res 47(21):11103–11113
Lin Y, Yang Z, Lake RJ (2019) Enzyme-mediated endogenous and bioorthogonal control of a DNAzyme fluorescent sensor for imaging metal ions in living cells. Angew Chem Int Ed 58(47):17061–17067
Liu D, Wang M, Deng Z et al (2004) Tensegrity: construction of rigid DNA triangles with flexible four-arm DNA junctions. J Am Chem Soc 126(8):2324–2325
Liu W, Halverson J, Tian Y et al (2016) Self-organized architectures from assorted DNA-framed nanoparticles. Nat Chem 8(9):867–873
Liu M, Ma W, Li Q et al (2019a) Aptamer-targeted DNA nanostructures with doxorubicin to treat protein tyrosine kinase 7-positive tumours. Cell Prolif 52(1):e12511
Liu J, Wu T, Lu X et al (2019b) A self-assembled platform based on branched DNA for sgRNA/Cas9/antisense delivery. J Am Chem Soc 141(48):19032–19037
Liu R, Zhang S, Zheng TT et al (2020) Intracellular nonenzymatic in situ growth of three-dimensional DNA nanostructures for imaging specific biomolecules in living cells. ACS Nano 14(8):9572–9584
Liu F, Jiang Q, Liu Q et al (2021a) Logic-gated plasmonic nanodevices based on DNA-templated assembly. CCS Chem 3(3):985–993
Liu S, Jiang Q, Zhao X et al (2021b) A DNA nanodevice-based vaccine for cancer immunotherapy. Nat Mater 20(3):421–430
Liu HY, Kumar R, Zhong C et al (2021c) Rapid capture of cancer extracellular vesicles by lipid patch microarrays. Adv Mater 33(35):2008493

Liu W, Wang Y, Jiang P et al (2024) DNAzyme and controllable cholesterol stacking DNA machine integrates dual-target recognition CTCs enable homogeneous liquid biopsy of breast cancer. Biosens Bioelectron 261:116493
Llaudet E, Hatz S, Droniou M et al (2005) Microelectrode biosensor for real-time measurement of ATP in biological tissue. Anal Chem 77(10):3267–3273
Lu CH, Willner B, Willner I (2013) DNA nanotechnology: from sensing and DNA machines to drug-delivery systems. ACS Nano 7(10):8320–8332
Lu L, Li B, Ding S et al (2020) NIR-II bioluminescence for in vivo high contrast imaging and in situ ATP-mediated metastases tracing. Nat Commun 11(1):4192
Lu Z, Shi Y, Ma Y et al (2022) Fast and specific enrichment and quantification of cancer-related exosomes by DNA-nanoweight-assisted centrifugation. Anal Chem 94(26):9466–9471
Lund K, Manzo AJ, Dabby N et al (2010) Molecular robots guided by prescriptive landscapes. Nature 465(7295):206–210
Ma C, Yang X, Wang K et al (2008) A novel kinase-based ATP assay using molecular beacon. Anal Biochem 372(1):131–133
Ma PQ, Liang CP, Zhang HH et al (2018) A highly integrated DNA nanomachine operating in living cells powered by an endogenous stimulus. Chem Sci 9(13):3299–3304
Mao C, Sun W, Shen Z et al (1999) A nanomechanical device based on the B–Z transition of DNA. Nature 397(6715):144–146
Mao X, Liu M, Yan L et al (2020) Programming biomimetically confined aptamers with DNA frameworks. ACS Nano 14(7):8776–8783
Marar C, Starich B, Wirtz D (2021) Extracellular vesicles in immunomodulation and tumor progression. Nat Immunol 22(5):560–570
Mattson MP, Gleichmann M, Cheng A (2008) Mitochondria in neuroplasticity and neurological disorders. Neuron 60(5):748–766
Meng F, Chai H, Ma X et al (2019a) FRET investigation toward DNA tetrahedron-based ratiometric analysis of intracellular telomerase activity. J Mater Chem B 7(11):1926–1932
Meng L, Ma W, Shi LS et al (2019b) Tetrahedral DNA nanostructure-delivered DNAzyme for gene silencing to suppress cell growth. ACS Appl Mater Interfaces 11(7):6850–6857
Meng L, Ma W, Zhang M et al (2021a) Aptamer-guided DNA tetrahedrons as a photo-responsive drug delivery system for mucin 1-expressing breast cancer cells. Appl Mater Today 23:101010
Meng X, Yang F, Dong H et al (2021b) Recent advances in optical imaging of biomarkers in vivo. Nano Today 38:101156
Micalizzi DS, Maheswaran S, Haber DA (2017) A conduit to metastasis: circulating tumor cell biology. Genes Dev 31(18):1827–1840
Mittal S, Kaur H, Gautam N et al (2017) Biosensors for breast cancer diagnosis: a review of bioreceptors, biotransducers and signal amplification strategies. Biosens Bioelectron 88:217–231
Mo L, Liang D, He W et al (2021) Ratiometric and amplified fluorescence nanosensor based on a DNA tetrahedron for miRNA imaging in living cells. J Mater Chem B 9(39):8341–8347
Möller A, Lobb RJ (2020) The evolving translational potential of small extracellular vesicles in cancer. Nat Rev Cancer 20(12):697–709
Moro AJ, Cywinski PJ, Körsten S et al (2010) An ATP fluorescent chemosensor based on a Zn (II)-complexed dipicolylamine receptor coupled with a naphthalimide chromophore. Chem Commun 46(7):1085–1087
Nadal R, Lorente JA, Rosell R et al (2013) Relevance of molecular characterization of circulating tumor cells in breast cancer in the era of targeted therapies. Expert Rev Mol Diagn 13(3):295–307
Nelson DL, Cox MM (2000) Lehninger principle of biochemistry, vol 16, 3rd edn. Worth Publisher, New York, pp 567–597
Nishikawa M, Matono M, Rattanakiat et al (2008) Enhanced immunostimulatory activity of oligodeoxynucleotides by Y-shape formation. Immunology 124(2):247–255
Nishikawa M, Mizuno Y, Mohri K et al (2011) Biodegradable CpG DNA hydrogels for sustained delivery of doxorubicin and immunostimulatory signals in tumor-bearing mice. Biomaterials 32(2):488–494

Nishikawa M, Tan M, Liao W et al (2019) Nanostructured DNA for the delivery of therapeutic agents. Adv Drug Del Rev 147:29–36

Ochmann SE, Vietz C, Trofymchuk K et al (2017) Optical nanoantenna for single molecule-based detection of zika virus nucleic acids without molecular multiplication. Anal Chem 89(23):13000–13007

Okamoto A, Tanaka K, Saito I (2004) DNA logic gates. J Am Chem Soc 126(30):9458–9463

Ouyang C, Zhang S, Xue C et al (2020) Precision-guided missile-like DNA nanostructure containing warhead and guidance control for aptamer-based targeted drug delivery into cancer cells in vitro and in vivo. J Am Chem Soc 142(3):1265–1277

Pan M, Jiang Q, Sun J et al (2020) Programming DNA nanoassembly for enhanced photodynamic therapy. Angew Chem Int Ed 132(5):1913–1192

Pan W, Niu H, Luo S et al (2024) Intelligent reconfiguration-promoted cellular internalization of core-shell DNA nanoprobe equipped with successive dual stimuli-responsive protective satellites for amplification fluorescence imaging of tumor cells. Small 20(28):2311388

Pantel K, Brakenhoff RH (2004) Dissecting the metastatic cascade. Nat Rev Cancer 4(6):448–456

Park SH, Pistol C, Ahn SJ et al (2006) Finite-size, fully addressable DNA tile lattices formed by hierarchical assembly procedures. Angew Chem Int Ed 45(5):735–739

Pei H, Zuo X, Zhu D et al (2014) Functional DNA nanostructures for theranostic applications. Acc Chem Res 47(2):550–559

Penders J, Nagelkerke A, Cunnan EM et al (2021) Single particle automated Raman trapping analysis of breast cancer cell-derived extracellular vesicles as cancer biomarkers. ACS Nano 15(11):18192–18205

Peng R, Zheng X, Lyu Y et al (2018) Engineering a 3D DNA-logic gate nanomachine for bispecific recognition and computing on target cell surfaces. J Am Chem Soc 140(31):9793–9796

Peng Y, Gao Z, Qiao B et al (2023) Size-controlled DNA tile self-assembly nanostructures through caveolae-mediated endocytosis for signal-amplified imaging of microRNAs in living cells. Adv Sci 10(21):2300614

Pisitkun T, Shen RF, Knepper MA (2004) Identification and proteomic profiling of exosomes in human urine. Proc Natl Acad Sci USA 101(36):13368–13373

Puchkova A, Vietz C, Pibiri E et al (2015) DNA origami nanoantennas with over 5000-fold fluorescence enhancement and single-molecule detection at 25 μM. Nano Lett 15(12):8354–8359

Qi X, Liu X, Matiski L et al (2020) RNA origami nanostructures for potent and safe anticancer immunotherapy. ACS Nano 14(4):4727–4740

Qian L, Wang Y, Zhang Z et al (2006) Analogic China map constructed by DNA. Chin Sci Bull 51(24):2973–2976

Qian JJ, Yang YX, Gong F et al (2023) Ratiometric fluorescence biosensing of silver nanocluster beacons for ATP detection based on ligation- triggered rolling cycle amplification. Microchem J 190:108663

Quyang X, De Stefano M, Krissanaprasit A, Kodal ALB et al (2017) Docking of antibodies into cavities in DNA origami. Angew Chem Int Ed 56(46):14423–14427

Rajendran A, Endo M, Sugiyama H (2012) Single-molecule analysis using DNA origami. Angew Chem Int Ed 51(4):874–890

Rattanakiat S, Nishikawa M, Funabashi H et al (2009) The assembly of a short linear natural cytosine-phosphate-guanine DNA into dendritic structures and its effect on immunostimulatory activity. Biomaterials 30(29):5701–5706

Reishus D, Shaw B, Brun Y et al (2005) Self-assembly of DNA double-double crossover complexes into high-density, doubly connected, planar structures. J Am Chem Soc 127(50):17590–17591

Ren LJ, Xu P, Zhang P et al (2021) Label-free fluorescence aptasensor based on AuNPs and CQDs for the detection of ATP. AIP Adv 11(1):015316

Riethdorf S, Müller V, Zhang L et al (2010) Detection and HER2 expression of circulating tumor cells: prospective monitoring in breast cancer patients treated in the neoadjuvant GeparQuattro trial. Clin Cancer Res 16(9):2634–2645

Roh YH, Ruiz RCH, Peng SM et al (2011) Engineering DNA-based functional materials. Chem Soc Rev 40(12):5730–5744

Rothemund PWK (2006) Folding DNA to create nanoscale shapes and patterns. Nature 440(7082):297–302

Sabaté del Río J, Woo HK, Park J et al (2022) SEEDING to enable sensitive electrochemical detection of biomarkers in undiluted biological samples. Adv Mater 34(24):2200981

Salas-Benito et al (2021) Paradigms on immunotherapy combinations with chemotherapy. Cancer Discov 11(6):1353–1367

Salta SP, Nunes S, Fontes-Sousa M et al (2018) A DNA methylation-based test for breast cancer detection in circulating cell-free DNA. J Clin Med 7(11):420

Schuller VJ, Heidegger S, Sandholzer N et al (2011) Cellular immunostimulation by CpG-sequence-coated DNA origami structures. ACS Nano 5(12):9696–9702

Schwarzenbach H, Hoon DS, Pantel K (2011) Cell-free nucleic acids as biomarkers in cancer patients. Nat Rev Cancer 11(6):426–437

Seeman NC (1982) Nucleic acid junctions and lattices. J Theor Biol 99(2):237–247

Sena LA, Chandel NS (2012) Physiological roles of mitochondrial reactive oxygen species. Mol Cell 48(2):158–167

Shan L, Chen Y, Tan X et al (2023) Tetrahedral DNA nanostructure-engineered paper-based sensor with an enhanced antifouling ability for photoelectrochemical sensing. Anal Chem 95(10):4760–4767

Shang Y, Chen Y, Wang Q et al (2022) Construction of an endogenously activated catalytic DNA circuit for highly robust in vivo microRNA imaging. Nano Today 45:101553

Shapouri-Moghaddam et al (2018) Macrophage plasticity, polarization, and function in health and disease. J Cell Physiol 233(9):6425–6440

Shay JW, Wright WE (2019) Telomeres and telomerase: three decades of progress. Nat Rev Genet 20(5):299–309

Shen B, Li L, Guo M et al (2024) Advances in DNA walking nanomachine-based biosensors. Interdiscip Med 2(1):e20230046

Sherman WB, Seeman NC (2004) A precisely controlled DNA biped walking device. Nano Lett 4(7):1203–1207

Shi X, Meng HM, Geng X et al (2020) DNAzyme-metal-organic framework two-photon nanoprobe for in situ monitoring of apoptosis-associated Zn^{2+} in living cells and tissues. ACS Sens 5(10):3150–3157

Shin JS, Pierce NA (2004) A synthetic DNA walker for molecular transport. J Am Chem Soc 126(35):10834–10835

Singh SK, Szulik MW, Ganguly M et al (2011) Characterization of DNA with an 8-oxoguanine modification. Nucleic Acids Res 39(15):6789–6801

Slattery M, Zhou T, Yang L et al (2014) Absence of a simple code: how transcription factors read the genome. Trends Biochem Sci 39(9):381–399

Song TQ, Eshra A, Shah SL, Bui H et al (2019) (2019) fast and compact DNA logic circuits based on single-stranded gates using strand-displacing polymerase. Nat Nanotechnol 14:1075–1081

Spari D, Beldi G (2020) Extracellular ATP as an inter-kingdom signaling molecule: release mechanisms by bacteria and its implication on the host. Int J Mol Sci 21(15):5590

Sullivan LB, Chandel NS (2014) Mitochondrial reactive oxygen species and cancer. Cancer Metab 2:1–12

Sun S, Yang S, Hu X et al (2020) Combination of immunomagnetic separation with aptamer-mediated double rolling circle amplification for highly sensitive circulating tumor cell detection. ACS Sens 5(12):3870–3878

Sun Y, Sun J, Xiao M et al (2022a) DNA origami-based artificial antigen-presenting cells for adoptive T cell therapy. Sci Adv 8(48):eadd1106

Sun S, Xu H, Yang Y et al (2022b) Intracellular in situ assembly of palindromic DNA hydrogel for predicting malignant invasion and preventing tumorigenesis. Chem Eng J 428:131150

Sun SS, Yang HQ, Wu ZY et al (2024) CRISPR/Cas systems combined with DNA nanostructures for biomedical applications. Chem Commun 60(23):3098–3117

Tang et al (2021) An aptamer-modified DNA tetrahedron-based nanogel for combined chemo/gene therapy of multidrug-resistant tumors. ACS Appl Bio Mater 4(10):7701–7707

Tang J, Jia X, Li Q et al (2023a) A DNA-based hydrogel for exosome separation and biomedical applications. Proc Natl Acad Sci USA 120(28):e2303822120
Tang W, Tong T, Wang H et al (2023b) A DNA origami-based gene editing system for efficient gene therapy in vivo. Angew Chem Int Ed 62(51):e202315093
Tian Y, He Y, Chen Y et al (2005) A DNAzyme that walks processively and autonomously along a one-dimensional track. Angew Chem Int Ed 44(28):4355–4358
Tian T, Li J, Xie C et al (2018) Targeted imaging of brain tumors with a framework nucleic acid probe. ACS Appl Mater Interfaces 10(4):3414–3420
Tian R, Shang Y, Wang Y et al (2023) DNA nanomaterials-based platforms for cancer immunotherapy. Small Methods 7(5):2201518
Tinland B, Pluen A, Sturm J et al (1997) Persistence length of single-stranded DNA. Macromolecules 30(19):5763–5765
Tsuboi T, Lippiat JD, Ashcroft FM et al (2004) ATP-dependent interaction of the cytosolic domains of the inwardly rectifying K^+ channel Kir6.2 revealed by fluorescence resonance energy transfer. Proc Natl Acad Sci 101(1):76–81
Tyagi S (2009) Imaging intracellular RNA distribution and dynamics in living cells. Nat Methods 6(5):331–338
Valadi H, Ekström K, Bossios A et al (2007) Exosome-mediated transfer of mRNAs and microRNAs is a novel mechanism of genetic exchange between cells. Nat Cell Biol 9(6):654–659
Vander Heiden MG, Cantley LC, Thompson CB (2009) Understanding the Warburg effect: the metabolic requirements of cell proliferation. Science 324(5930):1029–1033
Veneziano R, Ratanalert S, Zhang K et al (2016) Designer nanoscale DNA assemblies programmed from the top down. Science 352(6293):1534–1534
Vlaskovska M, Kasakov L, Rong W et al (2001) P_2X_3 knock-out mice reveal a major sensory role for urothelially released ATP. J Neurosci 21(15):5670–5677
Wang D, Xiang DB, Yang XQ et al (2009) APE1 overexpression is associated with cisplatin resistance in non-small cell lung cancer and targeted inhibition of APE1 enhances the activity of cisplatin in A549 cells. Lung Cancer 66(3):298–304
Wang F, Lu CH, Willner I (2014a) From cascaded catalytic nucleic acids to enzyme-DNA nanostructures: controlling reactivity, sensing, logic operations, and assembly of complex structures. Chem Rev 114(5):2881–2941
Wang Y, Tang L, Li Z et al (2014b) In situ simultaneous monitoring of ATP and GTP using a graphene oxide nanosheet–based sensing platform in living cells. Nat Protoc 9(8):1944–1955
Wang P, Gaitanaros S, Lee S et al (2016) Programming self-assembly of DNA origami honeycomb two-dimensional lattices and plasmonic metamaterials. J Am Chem Soc 138(24):7733–7740
Wang H, Li C, Liu X et al (2018) Construction of an enzyme-free concatenated DNA circuit for signal amplification and intracellular imaging. Chem Sci 9(26):5842–5849
Wang DX, Wang J, Cui YX et al (2019a) Nanolantern-based DNA probe and signal amplifier for tumor-related biomarker detection in living cells. Anal Chem 91(20):13165–13173
Wang J, Wang DX, Ma JY et al (2019b) Three-dimensional DNA nanostructures to improve the hyperbranched hybridization chain reaction. Chem Sci 10(42):9758–9767
Wang J, Wang H, Wang H et al (2019c) Nonviolent self-catabolic DNAzyme nanosponges for smart anticancer drug delivery. ACS Nano 13(5):5852–5863
Wang H, He D, Wan K et al (2020a) In situ multiplex detection of serum exosomal microRNAs using an all-in-one biosensor for breast cancer diagnosis. Analyst 145(9):3289–3296
Wang H, Peng P, Wang Q (2020b) Environment-recognizing DNA-computation circuits for the intracellular transport of molecular payloads for mRNA imaging. Angew Chem Int Ed 59(15):6099–6107
Wang S, Liu Z, Tong Y et al (2021a) Improved cancer phototheranostic efficacy of hydrophobic IR780 via parenteral route by association with tetrahedral nanostructured DNA. J Control Release 330:483–492
Wang YX, Wang DX, Ma JY et al (2021b) DNA nanolantern-based split aptamer probes for in situ ATP imaging in living cells and lighting up mitochondria. Analyst 146(8):2600–2608

Wang DX, Wang J, Wang YX et al (2021c) DNA nanostructure-based nucleic acid probes: construction and biological applications. Chem Sci 12(22):7602–7622
Wang Y, Bai Y, Cao LP et al (2022a) Catalytic hairpin assembled polymeric tetrahedral DNA frameworks for MicroRNA imaging in live cells. Biosens Bioelectron 197:113783
Wang W, Gao Y, Chen Y et al (2022b) Outward movement of targeting ligands from a built-in reserve pool in nuclease-resistant 3D hierarchical DNA nanocluster for in vivo high-precision cancer therapy. Adv Sci 9(33):2203698
Wang XJ, Mao ZF, Chen RP et al (2022c) Self-assembled DNA origami-based duplexed aptasensors combined with centrifugal filters for efficient and rechargeable ATP detection. Biosens Bioelectron 211:114336
Wei J, Wang H, Wu Q et al (2020) A smart, autocatalytic, DNAzyme biocircuit for in vivo, amplified, microRNA imaging. Angew Chem 132(15):6021–6027
Wen Y, Li L, Li J et al (2020) DNA framework-mediated electrochemical biosensing platform for amplification-free microRNA analysis. Anal Chem 92(6):4498–4503
Wen X, Yang M, Li L et al (2024) Enzymatically controlled DNA tetrahedron nanoprobes for specific imaging of ATP in tumor. Chin Chem Lett 35(8):109291
Wile BM, Ban K, Yoon YS et al (2014) Molecular beacon-enabled purification of living cells by targeting cell type-specific mRNAs. Nat Protoc 9(10):2411–2424
Wu Q, Wang H, Gong K et al (2019) Construction of an autonomous nonlinear hybridization chain reaction for extracellular vesicles-associated MicroRNAs discrimination. Anal Chem 91(15):10172–10179
Wu H, Chen TT, Wang XN et al (2020) RNA imaging in living mice enabled by an in vivo hybridization chain reaction circuit with a tripartite DNA probe. Chem Sci 11(1):62–69
Wu Y, Li Z, Shi M et al (2021) Programmable DNAzyme cmputing for specific in vivo imaging: intracellular stimulus-unlocked target sensing and signal amplification. Anal Chem 93(36):12456–12463
Wu Y, Wu M, Liu M et al (2022) Closing-upon-repair DNA tetrahedron nanoswitch for FRET imaging the repair activity of 8-oxoguanine DNA glycosylase in living cells. Anal Chim Acta 1196:339481
Wu Y, He Y, Han M et al (2023a) Dual-miRNA-propelled three-dimensional DNA walker for highly specific and rapid discrimination of breast cancer cell subtypes in clinical tissue samples. CCS Chem 5(7):1561–1573
Wu X, Yang C, Wang H et al (2023b) Genetically encoded DNA origami for gene therapy in vivo. J Am Chem Soc 145(16):9343–9353
Wu J, Zheng X, Lin W et al (2024) Persistent targeting DNA nanocarrier made of 3D structural unit assembled from only one basic multi-palindromic oligonucleotide for precise gene cancer therapy. Adv Healthc Mater 13(11):2303865
Xia Z, Wang P, Liu X et al (2016) Tumor-penetrating peptide-modified DNA tetrahedron for targeting drug delivery. Biochemistry 55(9):1326–1331
Xia K, Kong H, Cui Y et al (2018) Systematic study in mammalian cells showing no adverse response to tetrahedral DNA nanostructure. ACS Appl Mater Interfaces 10(18):15442–15448
Xie N, Huang J, Yang X et al (2016) Competition-mediated FRET-switching DNA tetrahedron molecular beacon for intracellular molecular detection. ACS Sens 1(12):1445–1452
Xie M, Wang F, Yang J et al (2022) DNA zipper mediated membrane fusion for rapid exosomal miRNA detection. Anal Chem 94(38):13043–13051
Xie Z, Chen Z, Li A et al (2024) Specific small-molecule detection using designed nucleic acid nanostructure carriers and nanopores. Anal Chem 96(21):8528–8533
Xiong M, Yang Z, Lake RJ et al (2020) DNAzyme-mediated genetically encoded sensors for ratiometric imaging of metal ions in living cells. Angew Chem Int Ed 59(5):1891–1896
Xu C, Zhang F, Wang K et al (2019) A three-dimensional DNAzyme motor for sensitive imaging of telomerase activity in living cells. Sens Actuators B Chem 298:126930
Xu Y, Li X, Niu C et al (2022) Janus wireframe DNA cube-based 3D nanomachine for rapid and stable fluorescence detection of exosomal microRNA. Biosens Bioelectron 212:114405

Xu H, Zheng Y, Xie J et al (2024) Imaging mRNA in vitro and in vivo with nanofirecracker probes via intramolecular hybridization chain reaction. Biosens Bioelectron 248:115973
Xue H, Ding F, Zhang J et al (2019a) DNA tetrahedron-based nanogels for siRNA delivery and gene silencing. Chem Commun 55(29):4222–4225
Xue C, Zhang S, Li C et al (2019b) Y-shaped backbone-rigidified triangular DNA scaffold-directed stepwise movement of a DNAzyme walker for sensitive microRNA imaging within living cells. Anal Chem 91(24):15678–15685
Xue C, Zhang S, Yu X et al (2020) Periodically ordered, nuclease-resistant DNA nanowires decorated with cell-specific aptamers as selective theranostic agents. Angew Chem Int Ed 132(40):17693–17700
Xue C, Hu SY, Gao ZH et al (2021a) Programmably tiling rigidified DNA brick on gold nanoparticle as multi-functional shell for cancer-targeted delivery of siRNAs. Nat Commun 12(1):2928
Xue C, Wang L, Huang H et al (2021b) Stimuli-induced upgrade of nuclease-resistant DNA nanostructure composed of a single molecular beacon for detecting mutant genes. ACS Sens 6(11):4029–4037
Xue C, Huang H, Wang L et al (2022) Swelling of serum-stable DNA nanoparticles upon target-induced conformational rearrangement of sensing probes for the signal-on detection of cancer-related genes. Anal Chem 94(6):2749–2756
Yan H, Zhang XP, Shen ZY et al (2002) A robust DNA mechanical device controlled by hybridization topology. Nature 415(6867):62–65
Yan H, Park SH, Finkelstein G et al (2003) DNA-templated self-assembly of protein arrays and highly conductive nanowires. Science 301(5641):1882–1884
Yang Y, Huang J, Yang X et al (2015) FRET nanoflares for intracellular mRNA detection: avoiding false positive signals and minimizing effects of system fluctuations. J Am Chem Soc 137(26):8340–8343
Yang XJ, Zhang K, Zhang TT et al (2017) Reliable forster resonance energy transfer probe based on structure-switching DNA for ratiometric sensing of telomerase in living cells. Anal Chem 89(7):4216–4222
Yang B, Chen Y, Shi J (2019) Exosome biochemistry and advanced nanotechnology for next-generation theranostic platforms. Adv Mater 31(2):1802896
Yang F, Cheng Y, Zhang Y et al (2020a) Bioinspired framework nucleic acid capture sensitively and rapidly resolving microRNAs biomarkers in living cells. Anal Chem 92(6):4411–4418
Yang Y, Sun X, Xu J et al (2020b) Circular bispecific aptamer-mediated artificial intercellular recognition for targeted T cell immunotherapy. ACS Nano 14(8):9562–9571
Yang H, Weng B, Liu S et al (2022) Acid-improved DNAzyme-based chemiluminescence miRNA assay coupled with enzyme-free concatenated DNA circuit. Biosens Bioelectron 204:114060
Yang C, Shi Y, Zhang Y et al (2023a) Modular DNA tetrahedron nanomachine-guided dual-responsive hybridization chain reactions for discernible bivariate assay and cell imaging. Anal Chem 95(27):10337–10345
Yang S, Zhao Z, Wang B et al (2023b) Modular engineering of a DNA tetrahedron-based nanomachine for ultrasensitive detection of intracellular bioactive small molecules. ACS Appl Mater Interfaces 15(19):23662–23670
Yao W, Wang L, Wang H et al (2009) An aptamer-based electrochemiluminescent biosensor for ATP detection. Biosens Bioelectron 24(11):3269–3274
Yin P, Choi HMT, Calvert CR et al (2008a) Programming biomolecular self-assembly pathways. Nature 451(7176):318–322
Yin P, Hariadi RF, Sahu S et al (2008b) Programming DNA tube circumferences. Science 321(5890):824–826
Yu M, Yao Y, Cui B et al (2018) Metal-enhanced near infrared fluorescence-based sensor with highly improved sensitivity for adenosine triphosphate. ACS Appl Nano Mater 2(1):48–57
Yu W, Sun J, Liu F et al (2020) Enhanced immunostimulatory activity of a cytosine-phosphate-guanosine immunomodulator by the assembly of polymer DNA wires and spheres. ACS Appl Mater Interfaces 12(15):17167–17176

Yu Y, Guo Q, Jiang W et al (2021) Dual-aptamer-assisted and logic gate for cyclic enzymatic signal amplification electrochemical detection of tumor-derived small extracellular vesicles. Anal Chem 93(32):11298–11304

Yu L, Yang S, Liu Z et al (2022) Programming a DNA tetrahedral nanomachine as an integrative tool for intracellular microRNA biosensing and stimulus-unlocked target regulation. Mater Today Bio 15:100276

Yuan K, Meng HM, Wu Y et al (2022) Extracellular milieu and membrane receptor dual-driven DNA nanorobot for accurate in vivo tumor imaging. CCS Chem 4(5):1597–1609

Yuan CQ, Zhou F, Xu ZH et al (2024) Functionalized DNA origami-enabled detection of biomarkers. Chembiochem 25(13):e202400227

Yue X, Qiao Y, Gu D et al (2020) Reliable FRET-ON imaging of telomerase in living cells by a tetrahedral DNA nanoprobe integrated with structure-switchable molecular beacon. Sens Actuators B Chem 312:127943

Yue S, Qiao Z, Wang X et al (2022) Enzyme-free catalyzed self-assembly of three-dimensional hyperbranched DNA structures for in situ SERS imaging and molecular logic operations. Chem Eng J 446:136838

Yurke B, Turberfield AJ, Mills AP et al (2000) A DNA-fuelled molecular machine made of DNA. Nature 406(6796):605–608

Zeng WJ, Li XR, Liu W et al (2024) AND logic gate-regulated DNAzyme nanoflower for monitoring the activity of multiple DNA repair enzymes. Anal Chem 96(5):2117–2123

Zhan Y, Ma W, Zhang Y et al (2019) DNA-based nanomedicine with targeting and enhancement of therapeutic efficacy of breast cancer cells. ACS Appl Mater Interfaces 11(17):15354–15365

Zhang DY, Seelig G (2011) Dynamic DNA nanotechnology using strand-displacement reactions. Nat Chem 3(2):103–113

Zhang C, Su M, He Y et al (2008) Conformational flexibility facilitates self-assembly of complex DNA nanostructures. Proc Natl Acad Sci 105(31):10665–10669

Zhang Y, Jiang S, Zhang D et al (2017a) DNA-affibody nanoparticles for inhibiting breast cancer cells overexpressing HER2. Chem Commun 53(3):573–576

Zhang G, Zhang Z, Yang J (2017b) DNA tetrahedron delivery enhances doxorubicin-induced apoptosis of HT-29 colon cancer cells. Nanoscale Res Lett 12:1–7

Zhang YZ, Tu J, Wang DQ et al (2018) Programmable and multifunctional DNA-based materials for biomedical applications. Adv Mater 30(24):1703658

Zhang L, Abdullah R, Hu X et al (2019a) Engineering of bioinspired, size-controllable, self-degradable cancer-targeting DNA nanoflowers via the incorporation of an artificial sandwich base. J Am Chem Soc 141(10):4282–4290

Zhang Y, Deng Y, Wang C et al (2019b) Probing and regulating the activity of cellular enzymes by using DNA tetrahedron nanostructures. Chem Sci 10(23):5959–5966

Zhang J, He M, Nie C et al (2019c) Biomineralized metal-organic framework nanoparticles enable enzymatic rolling circle amplification in living cells for ultrasensitive MicroRNA imaging. Anal Chem 91(14):9049–9057

Zhang X, Liu C, Pei Y et al (2019d) Preparation of a novel Raman probe and its application in the detection of circulating tumor cells and exosomes. ACS Appl Mater Interfaces 11(32):28671–28680

Zhang H, Wang Z, Wang F et al (2020) In situ formation of gold nanoparticles decorated Ti_3C_2 MXenes nanoprobe for highly sensitive electrogenerated chemiluminescence detection of exosomes and their surface proteins. Anal Chem 92(7):5546–5553

Zhang Y, Chen W, Fang Y et al (2021a) Activating a DNA nanomachine via computation across cancer cell membranes for precise therapy of solid tumors. J Am Chem Soc 143(37):15233–15242

Zhang P, Ouyang Y, Sohn YS et al (2021b) pH-and miRNA-responsive DNA-tetrahedra/metal-organic framework conjugates: functional sense-and-treat carriers. ACS Nano 15(4):6645–6657

Zhang Y, Chen J, Yang H et al (2022a) Light-controlled recruitable hybridization chain reaction on exosome vehicles for highly sensitive microRNA imaging in living cells. Anal Chem 94(27):9665–9673

Zhang B, Tian T, Xiao D et al (2022b) Facilitating in situ tumor imaging with a tetrahedral DNA framework-enhanced hybridization chain reaction probe. Adv Funct Mater 32(16):2109728
Zhang R, Zhang R, Zhao C et al (2022c) A DNA tetrahedron docking assembly for imaging telomerase activity in cancerous cells. Anal Chim Acta 1193:339395
Zhang X, Zhu X, Li Y et al (2023) A colorimetric and photothermal dual-mode biosensing platform based on nanozyme-functionalized flower-like DNA structures for tumor-derived exosome detection. Talanta 258:124456
Zhang Y, Sun M, Xie J et al (2024a) Dual-signal amplification strategy based on catalytic hairpin assembly and APE1-assisted amplification for high-contrast miRNA imaging in living cells. Anal Chem 96(2):910–916
Zhang M, Zhang Y, Wang Q et al (2024b) An APE1 gated signal amplified biosensor driven by catalytic hairpin assembly for the specific imaging of microRNA in situ. Int J Biol Macromol 262:129902
Zhang M, Zhang Y, Zhang X et al (2024c) An enzymatically activated self-powered and self-recycled strategy for in situ tumor cell-specific molecular imaging. Chem Eng J 489:151226
Zhao CQ, Ding SN (2019) Perspective on signal amplification strategies and sensing protocols in photoelectrochemical immunoassay. Coord Chem Rev 391:1–14
Zhao J, Liu C, Li Y et al (2020b) Thermophoretic detection of exosomal microRNAs by nanoflares. J Am Chem Soc 142(11):4996–5001
Zhao H, Wang M, Xiong X et al (2020b) Simultaneous fluorescent detection of multiplexed miRNA of liver cancer based on DNA tetrahedron nanotags. Talanta 210:120677
Zhao Y, Zhao J, Zhang J et al (2022) Enzymatically controlled nanoflares for specific molecular recognition and biosensing. Anal Chem 94(25):8883–8889
Zhao S, Zhang S, Hu H et al (2023) Selective in situ analysis of mature microRNAs in extracellular vesicles using a DNA cage-based thermophoretic assay. Angew Chem Int Ed 135(24):e202303121
Zhou W, Li D, Xiong C et al (2016) Multicolor-encoded reconfigurable DNA nanostructures enable multiplexed sensing of intracellular microRNAs in living cells. ACS Appl Mater Interfaces 8(21):13303–13308
Zhou T, Luo R, Li Y et al (2020a) Activity assay and intracellular imaging of APE1 assisted with tetrahedral DNA nanostructure modified-dnazyme and molecular beacon. Sens Actuators B Chem 317:128203
Zhou Z, Sohn YS, Nechushtai R et al (2020b) DNA tetrahedra modules as versatile optical sensing platforms for multiplexed analysis of miRNAs, endonucleases, and aptamer-ligand complexes. ACS Nano 14(7):9021–9031
Zhou Z, Wang J, Willner I (2020c) Dictated emergence of nucleic acid-based constitutional dynamic networks by DNA replication machineries. J Am Chem Soc 143(1):241–251
Zhou Y, Yang Q, Wang F et al (2021a) Self-assembled DNA nanostructure as a carrier for targeted siRNA delivery in glioma cells. Int J Nanomed 16:1805–1817
Zhou Y, Zou L, Li G et al (2021b) A cooperatively activatable DNA nanoprobe for cancer cell-selective imaging of ATP. Anal Chem 93(41):13960–13966
Zhu C, Yang J, Zheng J et al (2019) Triplex-functionalized DNA tetrahedral nanoprobe for imaging of intracellular pH and tumor-related messenger RNA. Anal Chem 91(24):15599–15607
Zhu D, Wei Y, Sun T et al (2021a) Encoding DNA frameworks for amplified multiplexed imaging of intracellular microRNAs. Anal Chem 93(4):2226–2234
Zhu L, Xu Y, Wei X et al (2021b) Coupling aptamer-based protein tagging with metabolic glycan labeling for in situ visualization and biological function study of exosomal protein-specific glycosylation. Angew Chem Int Ed 133(33):18259–18263

Chapter 2
Application of Small Molecule Fluorescent Probe in Tumor Diagnosis and Treatment

Lei Yang

Abstract Currently, cancer poses a significant global health challenge. The precise identification and management of this disease are crucial for mitigating its progression and enhancing patient survival. Numerous theranostic agents that integrate diagnostic and therapeutic functions have emerged as vital tools in the realm of precision oncology. Notably, fluorescent theranostic probes have gained prominence in cutting-edge medical research, owing to their capacity for real-time monitoring, minimal toxicity, and superior imaging quality. These probes typically consist of a fluorescent molecule coupled with a latent therapeutic agent, which is activated by specific triggers, either internal or external. Upon activation, the release of the drug is indicated by a visible fluorescence. Crucially, the selection of suitable fluorophores allows for the tracking of drug distribution over time and space, and supports applications in photodynamic and photothermal therapies. This chapter will delve into recent advancements in fluorescent theranostic probes, with a special emphasis on those designed to respond to the unique conditions of the tumor microenvironment.

Keywords Fluorescent probe · Theranostic · Chemotherapy · Photodynamic therapy · Photothermal therapy

2.1 Introduction

Cancer, an ailment characterized by uncontrolled cell proliferation that is strongly linked to aging, has emerged as a leading cause of mortality globally and is likely to persist as a major health threat (Bray et al. 2020). In tandem with the rapid

L. Yang (✉)
Shandong Provincial Key Laboratory of Tumor Imaging Equipment Development and Tumor Diagnosis and Treatment Integration Technology, School of Medicine, Linyi University, Linyi, Shandong, People's Republic of China
e-mail: yanglei@lyu.edu.cn

© The Author(s), under exclusive license to Springer Nature Singapore Pte Ltd. 2026

P. Shi et al. (eds.), *Functional Organic Materials for Tumor Diagnostics and Therapy*, https://doi.org/10.1007/978-981-96-9478-5_2

advancement of conventional medical technologies for tumor treatment, theranostic agents, which integrate diagnostic and therapeutic functions, are gaining prominence (Porret et al. 2020; Shetty et al. 2019). These "theranostic probes" are molecular or nanoscale entities that offer both imaging and therapeutic capabilities on a single platform (Bardhan et al. 2011). The theranostics approach enables precise therapy through visual imaging and targeted delivery (Du et al. 2019; Newton et al. 2018). Unlike traditional diagnostic techniques such as MRI, X-ray, and photoinduced electron transfer (PET), fluorescence imaging provides benefits like minimal invasiveness, reduced radiation, low toxicity, real-time responsiveness, and superior spatial resolution (Allison 2016; De Boer et al. 2015; Nagaya et al. 2017; Upchurch et al. 2017). Consequently, numerous fluorescence-based methods have been developed for cancer detection and treatment, including targeted probes for surgical guidance, responsive probes for early cancer detection, and near-infrared probes for vascular imaging (Nagaya et al. 2017; Li et al. 2019, 2020a). Additionally, light-activated therapies such as photothermal therapy (PTT) and PDT have become integral in the design of multifunctional fluorescent theranostic probes (Denkova et al. 2018). The evolution of these probes is closely tied to the advancements in fluorescent material research.

To date, numerous organic fluorescent materials have been employed in the creation of fluorescence theranostic systems. Organic fluorescent dyes have consistently been a key component in the design of these probes, serving as the signaling element. Fluorescent prodrug probes, which integrate fluorescence imaging with chemotherapy, show significant potential for clinical applications. These prodrugs are typically engineered to respond to the tumor microenvironment, releasing therapeutic agents that selectively target and eliminate cancer cells, thereby minimizing harm to healthy tissues. As the drugs are released, the fluorophores are also liberated, enabling real-time monitoring and precise visualization of the tumor. Furthermore, through strategic design, many small molecule dyes can function as both photosensitizers and photothermal agents, enhancing the tumor-targeting capabilities of photothermal and photodynamic probes (Almeida-Marrero et al. 2018; Liu et al. 2020a; Tsolekile et al. 2019; Yan et al. 2019).

In this chapter, we will delve into the contemporary knowledge of tumor biology as it pertains to targeted drug delivery mechanisms and intelligently crafted cancer treatments. We will particularly highlight fluorogenic theranostic agents that are activated in specific tumor conditions and their role in enhancing chemotherapy, photodynamic therapy (PDT), photothermal therapy (PTT), and combined therapeutic approaches.

2.2 Theranostic Fluorescence Probes in Chemotherapy

The idea of using cytotoxic drugs and agents to eliminate cancer cells, known as chemotherapy, emerged following a 1947 report that mustard gas could damage lymphatic tissues. Subsequent studies in mice confirmed the efficacy of nitrogen mustards in suppressing lymphoma tissue (Needham et al. 1947; Chen et al. 2018). Chemotherapeutic drugs generally function by slowing down the growth of cancer cells and alleviating tumor-related stress. Given that cancer cells grow and multiply more rapidly than normal cells, these drugs tend to have a greater impact on cancer cells. Over the years, numerous anticancer agents have been developed, each with distinct mechanisms of action. Some disrupt cellular metabolism, while others target essential enzymes. Most of these agents interfere with critical cellular processes, such as DNA repair, DNA replication, immune responses, and apoptosis regulation (Fig. 2.1) (Dieci et al. 2013; Petrova et al. 2018). Despite their high activity, these drugs often lack specificity, affecting both cancerous and normal tissues. This lack of selectivity can cause off-target toxicities, a major challenge in treatment. To address this, researchers have modified the structure of these agents to create theranostic probes that specifically target cancer cells or the tumor microenvironment (TME) and release an active payload. By carefully designing masking and

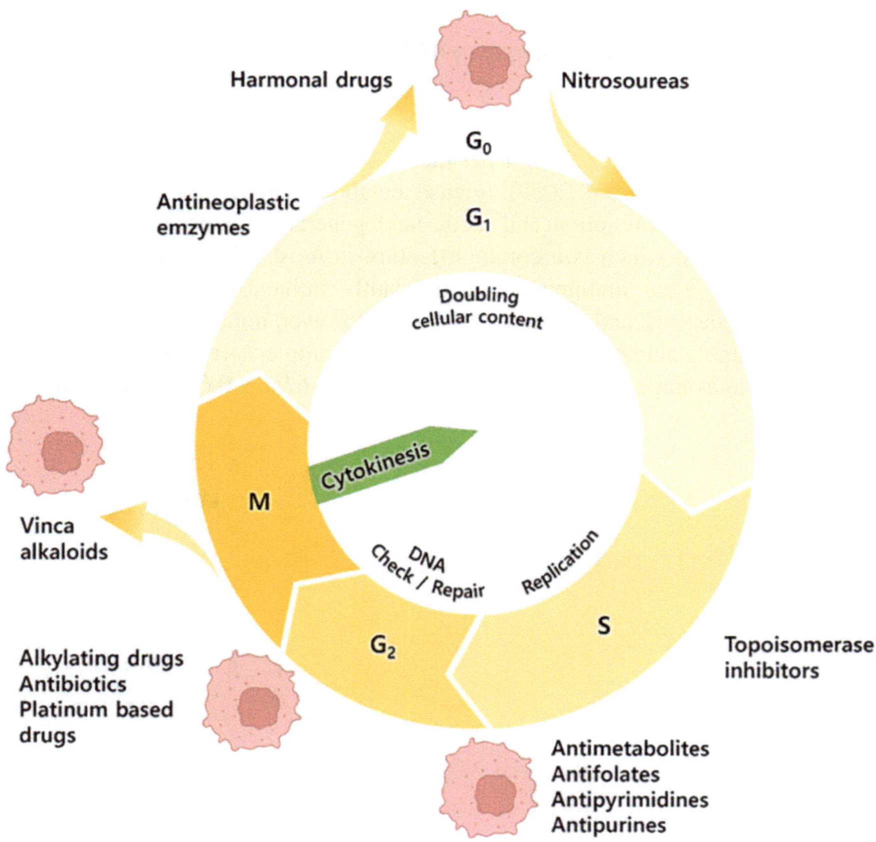

Fig. 2.1 Different classes of chemotherapeutic agents in clinical practices. *Go* resting phase; *G1* growth; *S* DNA synthesis and replication; *G2* growth and preparation for mitosis; *M* mitosis (reproduced from Sharma et al. 2019)

demasking steps, these probes can selectively deliver nonselective chemotoxins to cancer cells (Kratz et al. 2008; Pollock and Parker Hannah 2022). Similar strategies have been employed to develop activatable theranostic fluorescent probes. In this section, we will examine various types of fluorescent theranostic probes, categorized by their activation methods.

The microenvironment of healthy tissues and cancerous tumors exhibits notable differences. Particular biological markers, including a lower pH, increased reactive oxygen species (ROS), elevated intracellular glutathione, heightened enzyme activity, and a reductive or oxygen-deprived environment, set cancerous tissues apart from their adjacent normal counterparts and organs (Fig. 2.2).

In this section, we have categorized the theranostic agents utilized in different therapeutic approaches (such as chemotherapy, photodynamic therapy, and photothermal therapy) into distinct segments based on their activation mechanisms within the tumor microenvironment.

2.2.1 *pH-Responsive Fluorescent Probes*

The pH levels in healthy tissues generally hover around 7.4, whereas the tumor microenvironment often exhibits a pH that is 0.5 to 1.0 units lower (Liberti and Locasale 2016; Vander et al. 2009). Intense metabolic activity in cancer cells leads to rapid glucose consumption and lactic acid generation, especially under low-oxygen conditions, which is a common feature in most solid tumors. The acidic environment of these malignancies significantly influences tumor progression, recurrence, metastasis, and resistance to drugs. However, not all tumors are characterized by a highly acidic microenvironment. For example, soft tissue sarcomas and adenocarcinomas have been observed with a pHe of 6.94 ± 0.08, while malignant

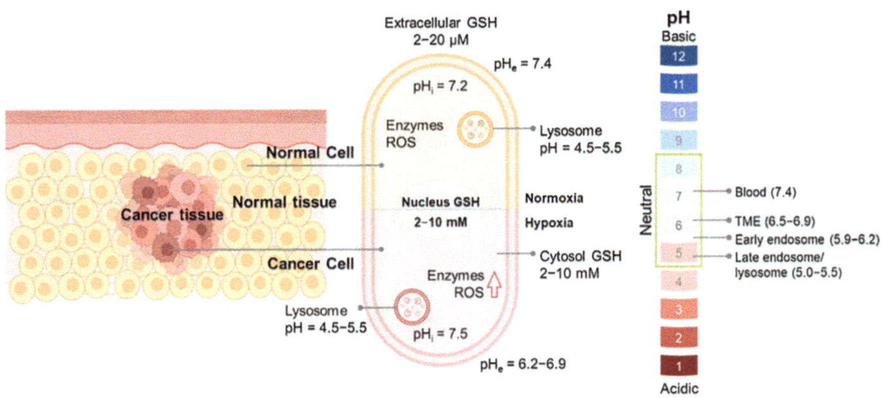

Fig. 2.2 Schematic diagram showing different microenvironments of normal and tumor tissues [pH_i intracellular pH; pH_e extracellular pH, *ROS* reactive oxygen species] (Reproduced from Sharma et al. 2019)

melanomas and squamous cell carcinomas show a pHe of 7.20 ± 0.07 (Engin et al. 1995). Despite this variability, an abnormal pH is still considered a consistent indicator of solid tumors, regardless of their type or stage. Consequently, pH is frequently utilized as a trigger for targeted drug delivery systems (DDS) in cancer therapy. To date, pH-sensitive DDS have been primarily studied in MCF-7 (Ahn et al. 2018; Guo et al. 2018), HeLa (Jin et al. 2015a), and BxPC-3 (Wang et al. 2013) cancer models.

The primary aim of developing DDS is to boost the therapeutic effectiveness of the base drugs or cytotoxic agents and to reduce harmful side effects (Giang et al. 2014; Mahato et al. 2011). To meet these goals, the system must be robust enough for delivery and circulation while effectively releasing the drug at the intended site. In pH-triggered DDS, the increased proton concentration in lysosomes, endosomes, cells, or tumor tissues acts as the trigger for drug release. Over the past few decades, several pH-sensitive theranostic systems have been developed that utilize this approach. These systems typically involve drugs linked to acid-sensitive chemical bonds such as acetals, hydrazines, oximes, and imines. These bonds are advantageous because they remain stable at the physiological pH of 7.4 but rapidly break down in the acidic environment of endosomes, ensuring targeted release at acidic tumor sites (Frangioni 2008). However, this method is limited to drugs that have free aldehyde or ketone groups capable of forming these labile linkages.

The acid-mediated hydrolysis of hydrazone linkers has been extensively studied for creating pH-responsive drug delivery systems (DDS), especially in the design of theranostics based on small molecules, nanocarriers, and polymer-based platforms (Kanamala et al. 2016; Sonawane et al. 2017). These systems are used to transport cytotoxic agents to tumor locations. Kalia and colleagues conducted comprehensive NMR analyses in deuterated buffer, which showed that the key step in the hydrolysis of hydrazone- and oxime-linked drugs is the nucleophilic attack by a water molecule on the imine carbon (Kalia and Raines 2008). Following this, the imine nitrogen is protonated, leading to further hydrolysis and the release of the free drug (Fig. 2.3). However, substrates with electron-withdrawing groups have a lower tendency for their imine nitrogen to be protonated, thus reducing the overall hydrolysis rate. The stability hierarchy of different linkers at pH 7.4 is as follows: trialkylhydrazonium ≫ oxime ≫ acyl hydrazone > primary-hydrazone ≫ sec-hydrazone > imine. Since quaternary ammonium hydrazones are highly stable at pH 5, they are not ideal for pH-sensitive drug delivery. In contrast, acyl hydrazone linkers are particularly appealing because of their high stability at neutral pH and their susceptibility to hydrolysis in acidic environments (pH 5).

The anticancer medication doxorubicin (Dox), an anthracycline, is a cornerstone treatment for various cancers, such as those of the breast, lung, ovary, and stomach, as well as multiple myeloma and pediatric cancers, including Hodgkin's and non-Hodgkin's lymphomas. However, its use is frequently constrained by adverse effects like fatigue, liver damage, and heart toxicity (Cortés-Funes and Cyntia 2007; Thorn et al. 2011; Weiss 1992). To mitigate these side effects, several approaches have been suggested. One approach involves enhancing the delivery of Dox to tumor sites, with various efforts made in this direction. For example, Dox has been conjugated via an acid-sensitive hydrazone linker to a cancer-targeting integrin, GRDS-oligopeptide, and a coumarin component, resulting in the theranostic agent **1** (Fig. 2.3) (Li et al. 2014). Free Dox is naturally fluorescent, emitting light at 595 nm when excited at 470 nm, making it a popular choice for imaging in cancer biology (Chen et al. 2012; Euhus et al. 1986; Lankelma et al. 1999; Motlagh et al. 2016). In agent **1**, the fluorescence of Dox is quenched before release due to a contact-mediated process involving the Dox and coumarin components. At a pH of 5, 94% of Dox is released from agent 1 over 11 h, compared to only 41% at a neutral pH of 7.4. Agent **1** demonstrated dose-dependent toxicity (IC_{50} = 0.19 μg mL^{-1}) in integrin-positive human glioblastoma U87 cells. Additionally, the distinct fluorescence of free Dox (red) and coumarin (blue) allows for real-time tracking of prodrug activation and drug distribution.

Integrating markers of cell apoptosis into scaffold design enables the noninvasive evaluation of cytotoxin activation and localization, as well as the precise adjustment of therapeutic dosage. This approach is exemplified by the dual Forster resonance energy transfer (FRET) agent **2** (Fig. 2.3) (Li et al. 2015a, b). In this setup, the

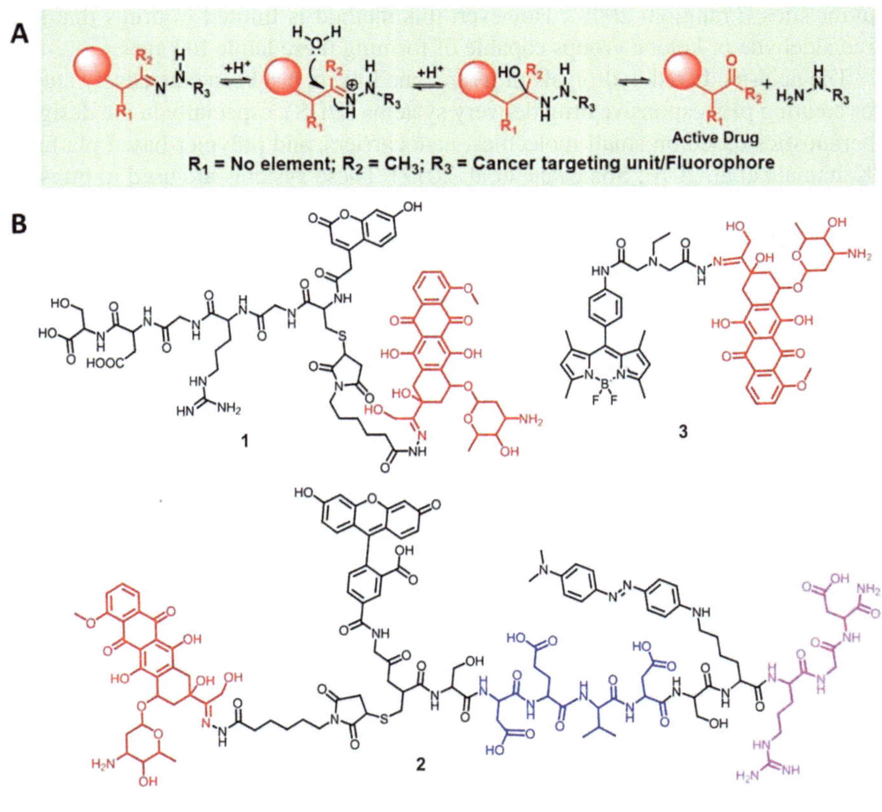

Fig. 2.3 (**a**) Mode of activation of hydrazone-based pH-responsive theranostic agents. (**b**) Chemical structures of acid-sensitive theranostic agents (**1**–**3**)

anticancer drug (Dox) is attached via an acid-sensitive hydrazone linker to a strong fluorescence quencher, 4-(dimethylamino azo)benzene-4-carboxylic acid (Dabcyl), and a peptide sequence (Asp-Glu-Val-Asp, DEVD) that reacts to the apoptosis marker, caspase-3. To enable real-time tracking of drug activation at the cellular level, a 5(6)-carboxylfluorescein (FAM) unit was included in the design. The construct was also labeled with an integrin-specific sequence (Arg-Gly-Asp, RGD) to ensure cancer-specific targeting and uptake. Initial solution studies showed that agent **2** released 90% of Dox in its active form in acidic conditions (pH = 5.0), compared to only 19% at pH 7.4. In U87 cancer cells (integrin positive), agent **2** exhibited time-dependent fluorescence enhancement due to DEVD peptide cleavage, resulting in green fluorescence (indicating the release of the FAM group from its quenched state). The associated Dox release led to significant therapeutic effects ($IC_{50} = 4.3 \times 10^{-6}$ M) and simultaneous caspase-3 activation.

Significant endeavors have been made to create simpler constructs than **2**, which enable cancer-specific activation and real-time tracking of drug release at the cellular level. For instance, a theranostic agent, designated as **3**, was engineered by linking Dox to a BODIPY fluorophore (4,4-difluoro-4-bora-3a,4a-diaza-s-indacene) via a hydrazone bond (Fig. 2.3) (Fernandez et al. 2017). This construct exhibited minimal fluorescence under normal physiological pH conditions. However, in the acidic environment of phagosomes (pH 6.5–4.5), the hydrazone linker was easily cleaved, leading to a noticeable fluorescence from the liberated BODIPY. Consequently, agent **3** was evaluated for its potential in treating immune-related disorders. RAW264.7 macrophages were employed as a model, and these cells were stimulated with lipopolysaccharide (LPS) to induce proinflammatory M1 macrophages, causing phagosome acidification. When treated with **3**, LPS-activated M1 macrophages showed dose-dependent toxicity and a fluorescence "Turn-On" effect. In contrast, no such effects were observed in anti-inflammatory M2 macrophages induced by IL-4. A comparable outcome was noted in zebrafish exposed to LPS, where apoptotic macrophages emitted red fluorescence, indicative of Dox, while adjacent areas displayed green fluorescence, corresponding to BODIPY. The theranostic agent **3** was further tested in a zebrafish regeneration model to observe the activity of phagocytic M1 macrophages. Treatment with **3** and LPS led to enhanced M1 macrophage polarization, which correlated with improved tissue regeneration.

Theranostic agent **4**, created by Kim and colleagues, was evaluated for its capacity to enhance the cancer cell-specific uptake of Dox compared to normal cells (Fig. 2.4) (Park et al. 2015). In this study, the inherent fluorescence of Dox was harnessed to track prodrug activation and intracellular distribution. A nitrobenzene group served as a fluorescence quencher through photoinduced electron transfer (PET) and was connected to Dox via a hydrazone linkage. Additionally, a biotin component was integrated into the structure to target cancer cells selectively. The agent maintained its stability under physiological conditions for up to 7 h. Under acidic conditions, however, the hydrazone bond was hydrolyzed, causing a fluorescence "Turn-On" effect due to the release of free Dox. In vitro experiments showed

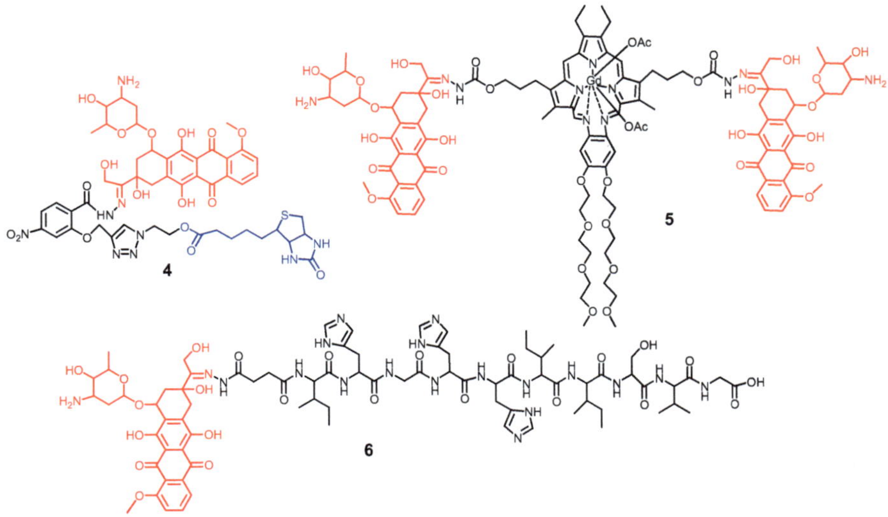

Fig. 2.4 Chemical structures of acid-sensitive theranostic agents (**4**–**6**)

that agent **4** exhibited high selectivity for biotin-positive cancer cells (HepG2) over biotin-negative normal WI38 cells.

The research team also devised a multimodal approach to evaluate the cellular intake and activation of Dox through fluorescence imaging and magnetic resonance imaging (MRI) (Lee et al. 2016). This method, represented by compound **5** (Fig. 2.4), integrates two Dox (fluorescent) components with a single paramagnetic motexafin gadolinium (MGd) core via a hydrazone bond. The MGd core is a unique water-soluble texaphyrin. The reduced fluorescence of the intact compound **5** could be restored in acidic environments, a behavior also observed in cancer cells (A549, CT26). Notably, no significant toxicity was detected in normal NIH3T3 fibroblast cells even at a 100 μM concentration of compound **5**. Compound **5** exhibited T1-contrast relaxivities of approximately 20.1 ± 0.4 mM^{-1} s^{-1} (at 60 MHz) and 6.1 ± 0.2 mM^{-1} s^{-1} (at 200 MHz) in PBS, surpassing those of a conventional Gd^{3+} contrast agent. Additionally, at a low concentration of 4 μM, cell pellet phantoms treated with compound **5** achieved saturation at a high T1 relaxivity value (cf. Fig. 2.4). Similarly, a tumor-targeting theranostic agent **6** was created, linking a targeting sequence AP2H (IHGHHIISVG) to Dox via a hydrazone bond (Jin et al. 2015b). This system, designed to break down under mildly acidic conditions, showed dose-dependent toxicity in LAPTM4B-positive lung cancer A549 cells (IC_{50} = 1.14 μM) and liver cancer HepG2 cells (IC_{50} = 4.0 μM). Fluorescence studies confirmed that agent 6 was activated in endosomes and lysosomes, followed by translocation to the nucleus in these cancer cell lines (Fig. 2.4), while minimal activation was observed in normal HEK293 cells.

Although hydrazone-based linkers are prevalent in small molecule systems, oxime-based linkers have also garnered interest, albeit with fewer reports, primarily in polymeric contexts. This review focuses on discrete molecule-based theranostic systems. Jin and colleagues utilized a terephthaldehyde group to create an oxime linker in a PEG-Dox micelle system, which exhibited slow hydrolysis at pH 5.0 ($t_{1/2}$ = 15 h) (Jin et al. 2011). Conversely, at neutral pH, the drug release was more rapid, reaching 20% within 12 h. These findings highlight the lower stability of oxime linkages compared to acyl hydrazone-based linkers. Xu et al. developed a dextran-Dox conjugate (**7**) for acid-sensitive Dox delivery into HepG2 cancer cells (Fig. 2.5). Fluorescence studies indicated that at pH 7.4, approximately 25.9% of Dox was released over 72 h. Under acidic conditions, the release increased significantly: 40.4% at pH 6.8, 64.7% at pH 6.0, and 87% at pH 5.0 over 72 h (Xu et al. 2015). A reduced version of **7**, with the oxime linker modified, did not show acid sensitivity. Confocal laser scanning microscopy confirmed endocytosis-mediated uptake and subsequent hydrolysis in acidic environments in HepG2 cells. The cytotoxicity of **7**, after 72 h of incubation, was comparable to free Dox, with IC_{50} values of 0.73 μg mL^{-1} and 0.62 μg mL^{-1}, respectively. Tumor inhibition studies in H22-xenograft mice showed that 7 (administered intravenously) significantly inhibited tumor growth (71.0%) and improved survival rates with fewer side effects, outperforming other controls (Dox 45.4%, reduced analog of **7** 19.2%).

The aforementioned research underscores the promising role of acid-labile chemical linkers. Nevertheless, it also emphasizes a critical principle that is often taken for granted: the necessity of comprehending the fundamental chemistry before selecting a specific linker for a given drug delivery system.

Yang and colleagues introduced the N-ethoxybenzylimidazole (NEBI) component as a versatile, acid-sensitive chemical linker for selective delivery of the anti-cancer drug indenoisoquinoline to folate-positive cancer cells (Cao and Jerry 2014). The acid-induced hydrolysis in this system is mediated by the "aminol" ether functionality. In Construct **8**, an imidazole ring nitrogen atom forms part of the aminol ether (Fig. 2.5). Under acidic conditions, the protonation of the imidazole ring promotes the spontaneous release of the active drug. The rate of hydrolysis and subsequent drug release can be adjusted by modifying the groups attached to the phenyl ring. For instance, a nitro group, which withdraws electrons, slows down the

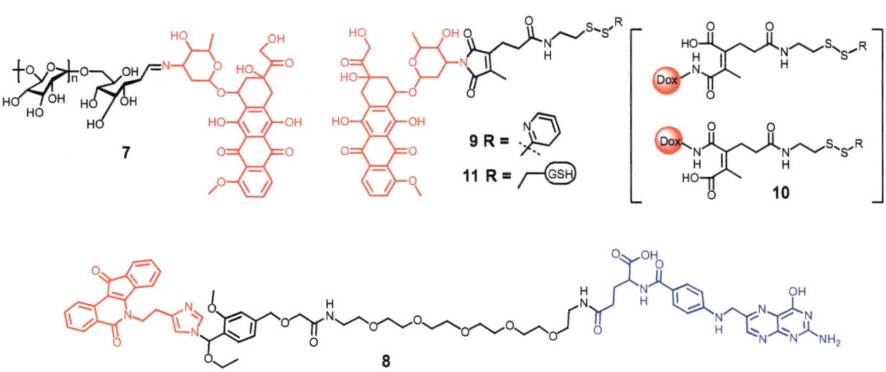

Fig. 2.5 Chemical structure of acid-sensitive theranostic agents (**7–11**)

hydrolysis ($t_{1/2}$ = 6900 h at pH 5.5), while a methoxy group, which donates electrons, accelerates the release ($t_{1/2}$ = 0.6 h at pH 5.5). Regardless of the substituent, the prodrug's half-life at pH 7.4 was ten times longer than at pH 5.5. In vitro experiments with folate-positive KB cells demonstrated targeted delivery and release of the active drug, with an IC_{50} of 60 μM. However, reducing the expression of the folate receptor diminished the drug's effectiveness (IC_{50} = 655 μM). Further, incorporating PEG linkers into the prodrugs reduced their activity in folate-positive KB cells (IC_{50} = 250 μM). The inherent fluorescence of the parent drug enabled the monitoring of cellular uptake and activation of prodrug **8** via fluorescence microscopy. This research underscores the potential of the NEBI linker for drug delivery through folate receptor-mediated endocytosis.

Derivatives of maleic acid have been employed as pH-sensitive linkers in drug delivery systems. The mechanism for drug release is based on the intramolecular cyclization of the malonyl amine, which occurs at a pH lower than the pKa of the free carboxylic acid group. Initial studies in this field concentrated on the use of *cis*-aconitic anhydride, enabling the formation of the final conjugate. However, these *cis*-aconitic acid-based conjugates frequently experienced decarboxylation and *trans*-isomerization (Dinand et al. 2002). A specific example of this method is the prodrug **9**, derived from Dox (Fig. 2.5) (Zhang et al. 2017a). At a pH below the pKa of the carboxylic acid, intermediate **10** becomes the predominant form. Prodrug **9** can also be connected to GSH to create theranostic agent **11**. A significant release of Dox was observed when the pH was lowered, with approximately 70% of Dox released at pH 6.0. In contrast, only 10% of Dox was released at pH 7.0, even after 5 h. Dox, with its free amino group (pKa = 8.2), is highly responsive to acidic tumor conditions, affecting both its uptake and toxicity. In ES-2 (ovarian cancer cells), Dox shows greater toxicity at an extracellular pH of 6.7 ($IC_{50} = 1.6 \pm 0.1$ μM) compared to a neutral pH of 7.4 ($IC_{50} = 0.7 \pm 0.1$ μM). When using maleamic acid-based linkers, this pH sensitivity is altered. Conjugate 10 demonstrated similar efficacy at pH 6.7 ($IC_{50} = 2.1 \pm 0.3$ μM) and pH 7.4 ($IC_{50} = 3.2 \pm 0.4$ μM). Notably, although conjugate **11** was less potent than the original drug, it reduced the undesirable pH-dependent toxicity of Dox. Additionally, the intrinsic fluorescence of Dox allowed for the monitoring of drug activation and cellular localization, making **11** a suitable acid-sensitive theranostic agent.

In conclusion, the ongoing advancement of acid-labile chemical linkers is driving the evolution of small molecule drug delivery systems. In acidic conditions, such as those found in tumor microenvironments, these pH-responsive linkers facilitate the release of drugs through an acid-catalyzed process. The selection of acid-sensitive linkers can be tailored to the specific target site. For instance, maleamic acid-based linkers are particularly effective for delivering cytotoxic agents to solid tumors (pH 6.0–7.0) due to their sensitivity to minor pH variations. Conversely, acyl hydrazone-based linkers are more appropriate for targeting lysosomes and endosomes (pH 4.5–6.0). Despite these advancements, there remains a significant need for further development of chemical linkers, which could lead to enhanced cancer-targeted drug delivery.

2.2.2 GSH Responsive Fluorescent Probes

Various redox reactions take place in diverse intracellular and extracellular settings, as well as within tissues. Common redox pairs include nicotinamide adenine dinucleotide phosphate (NADP+/NADPH), oxygen/superoxide ($O_2/O_2{\bullet}^-$), thioredoxin (TrxSS/Trx$(SH)_2$), and glutathione (l-γ-glutamyl-l-cysteinyl-glycine) (GSH/GSSG). The GSH/GSSG pair is particularly significant in theranostic and drug delivery applications because of the relatively high GSH levels inside cells (1–10 mM), which are crucial for maintaining cellular integrity, differentiation, metabolism, and apoptosis. GSH also acts as a key antioxidant, protecting against ROS-induced damage (Sies 1999; Smith et al. 1996; Zheng et al. 2005). Conversely, GSH concentrations are much lower in the extracellular matrix, blood, and cell surfaces (2–20 μM) due to the higher protein content that stabilizes disulfides. The intracellular environment remains reductive, primarily through the actions of NADPH and GSH reductase, ensuring GSH is predominantly in its reduced form (Cheng et al. 2011; Meng et al. 2009). A notable GSH concentration gradient exists within intracellular compartments (Montero et al. 2013). Tumor tissues have approximately four times the GSH concentration of healthy tissues, attributed to the rapid proliferation of cancer cells (Estrela et al. 2006; Saito et al. 2003; Schafer and Buettner 2001). This elevated GSH can reduce the effectiveness of certain cancer treatments, such as those for ovarian, breast, lung, liver, bone marrow, and colon cancers (Parodi 2007). However, these high GSH levels in tumors can be leveraged to develop redox-responsive theranostic agents that release active drugs or fluorophores specifically within tumor cells. Since the extracellular GSH levels are lower, prodrugs can be designed to remain stable until they are internalized by cancer cells, where they release cytotoxins. The disulfide (S–S) bridge is a widely used chemical linker for this purpose, as it undergoes thiol–disulfide exchange in the presence of free bithiols (Jiang et al. 2015; Jones et al. 2006). GSH can act as a potent bioactivator, triggering the release of drugs from S–S containing prodrugs.

The disulfide approach has been widely investigated in the creation of prodrugs, utilizing a range of carriers such as antibodies, peptides, and small molecules. In the realm of small molecule-based theranostics, numerous drug candidates and fluorophores have been examined, including Dox, camptothecin (Dongbang et al. 2014; Hou et al. 2017), paclitaxel (Han et al. 2016; Lim et al. 2009), gemcitabine (Dasari et al. 2013; Maiti et al. 2013; Yang et al. 2013), naphthalimide (Lee et al. 2012), hemicyanine (Kong et al. 2016), dicyano-methylene-4H-pyran (Kong et al. 2017; Liu et al. 2017; Wu et al. 2014a), and fluorescein (Cui et al. 2012). These drug molecules and fluorophores are often modified to incorporate the necessary disulfide linkage, which is crucial for GSH-mediated release and the generation of a free thiol group. Currently, three primary strategies are employed: (1) linking chemotherapeutic drugs to a fluorophore via a cleavable chemical linker, (2) connecting intrinsically fluorescent chemotherapeutic drugs to a targeting ligand through chemical linkers, and (3) integrating chemotherapeutic drugs with fluorophores and a targeting ligand using a multi-component strategy (Fig. 2.6).

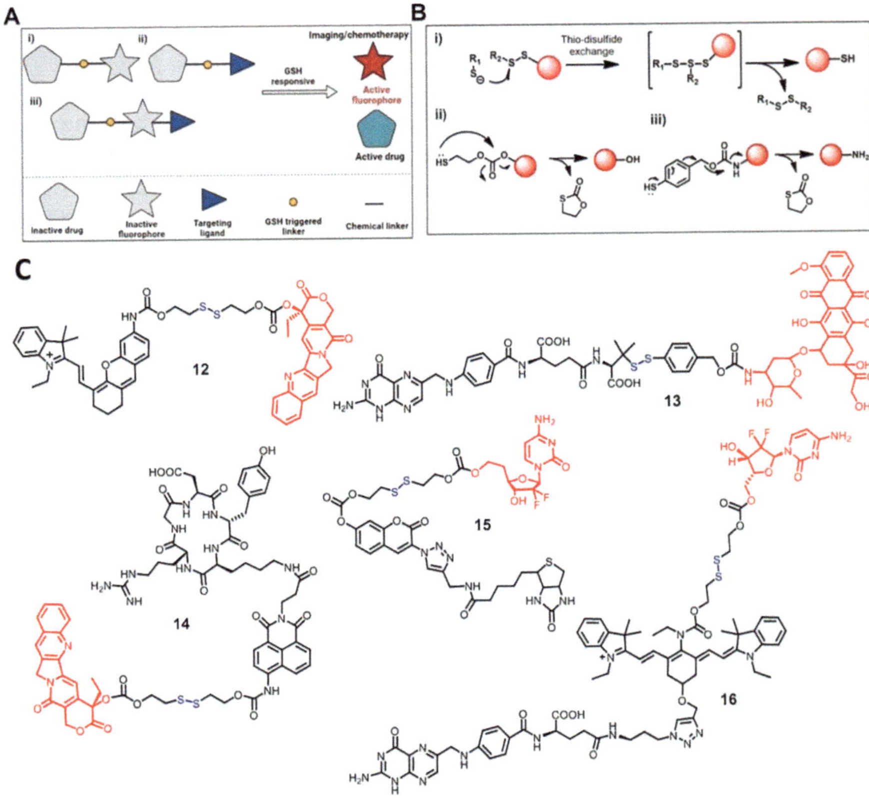

Fig. 2.6 (**a**) Schematic illustration of GSH-responsive activatable molecular theranostic probes. (**b**) Disulfide cleavage and self-immolative drug release mechanisms. (**c**) Chemical structures of selected GSH-responsive theranostic probes (**12–16**)

For the first category, various methods have been explored. For example, drugs or fluorophores with carbonate or carbamate esters of free hydroxyl or amine groups can be connected through a disulfide-containing linker. This is exemplified by theranostic agent **12** (Fig. 2.6), designed to monitor GSH-responsive activation of camptothecin (CPT) in an H22 tumor mouse model (Kong et al. 2016). In this agent, CPT was linked to a hemicyanine-based fluorophore via an S–S bond. Theranostic **12** exhibited low fluorescence, likely due to the blocking of the fluorophore's free amino group by the linker. Upon GSH exposure, the disulfide bond is reduced, leading to enhanced fluorescence (at 702 nm) from the free hemicyanine fluorophore and the concurrent release of CPT. Cell studies confirmed that theranostic **12** showed significant cytotoxicity in the HepG2 cancer cell line compared to the HL-7702 normal liver cell line (Fig. 2.6), as indicated by a strong fluorescence signal (at 702 nm) in the HepG2 cells, attributable to higher GSH levels. The increased emission also demonstrated the preferential tumor localization of theranostic **12**, verified through IVIS luminal in vivo imaging (intravenous administration). Similar approaches have been used by many researchers to observe GSH-mediated activation of theranostic agents in different tumor models (Dasari et al. 2013; Kong et al. 2016, 2017; Liu et al. 2016a, 2017; Wu et al. 2014b; Dutta et al. 2016; Hu and Fang 2017; Tian et al. 2020; Ye et al. 2016; Zhang et al. 2017b; Zhang and Zhijie 2018; Zheng et al. 2019).

Under strategy (2), anticancer agents that emit fluorescence are often attached to cancer-specific ligands via a disulfide bond that can be cleaved. Strategy (1) is applicable for drugs with hydroxyl or amine groups. A notable example is the GSH-responsive folate-doxorubicin theranostic **13**, which includes an α,α-dimethyl-substituted p-thiophenyl urethane-based disulfide and a carbamate ester (Zheng et al. 2019). The inclusion of two methyl groups enhances the stability of aromatic thiols by a factor of two compared to standard aliphatic disulfide linkers. Additionally, the lower pKa value of aromatic thiols (approximately 6) compared to aliphatic thiols (pKa 8–9) was anticipated to enhance the release kinetics. This theranostic probe **13** also features a folate targeting moiety. Folate receptors (FR) are overexpressed in many types of cancer, indicating that conjugate **13** (Fig. 2.6) would likely be internalized through recycling endosomal pathways (Sabharanjak and Satyajit 2004). Initial studies in solution demonstrated that theranostic **13** remains stable in 0.2 mM GSH in a phosphate-buffered solution, simulating the redox conditions of blood circulation. However, when exposed to 10 mM GSH, mimicking the intracellular environment of cancer cells, a steady drug release was observed, with a half-life of about 1.3 h. These findings suggest that **13** could be suitable for clinical application, with an expected half-life of around 30 min in the circulatory system (Li et al. 2009). Besides **13**, multiple other theranostic agents following strategy (2) have been developed and evaluated in various cancer models, both in vitro and in vivo (Liu et al. 2021a; Santra et al. 2011).

Strategy (3) presents an opportunity to enhance the precision of therapeutic payloads in theranostics, significantly lowering systemic toxicity by attaching cancer-targeting elements to the drug components. For nonfluorescent drug candidates, this method also facilitates the visualization of the drug delivery system through additional fluorescence tagging. Kim and colleagues illustrated the potential of this technique by creating a RGD peptide-embellished naphthalimide pro-camptothecin (pro-CPT) theranostic agent **14** (Fig. 2.6) for image-guided chemotherapy (Lee et al. 2012). The multifunctional agent **14** comprises an RGD cyclic peptide for cancer targeting, a naphthalimide component as a fluorophore, and the topoisomerase I inhibitor CPT as the anticancer drug, connected via a GSH-responsive disulfide linker. Agent **14** was engineered to be internalized into U87 cancer cells due to the interaction between the RGD moiety and the overexpressed avβ3 integrin receptor. The elevated GSH levels in these cancer cells were anticipated to cause the disulfide bond to break, releasing the active cytotoxin CPT and the naphthalimide fluorophore. When U87 cancer cells were incubated with **14**, a strong red-shifted emission at 535 nm was observed in the endoplasmic reticulum. Simultaneously, the CPT moved to the cell nucleus, inducing an antiproliferative effect. Further studies were conducted to confirm the broad applicability of this approach for other non-fluorescent drug candidates. For instance, theranostic agent **15** (Fig. 2.6) was developed to monitor the GSH-selective delivery of gemcitabine, another anticancer drug (Maiti et al. 2013). System **15** includes biotin as the cancer-targeting element, coumarin as the fluorophore, and gemcitabine as the model anticancer drug. By tracking the coumarin fluorescence, it was found that the theranostic is retained in the cell lysosomes, with drug activation occurring in the cytoplasm. Due to coumarin's inherent two-photon absorption (Cao et al. 2019), the therapeutic effects of **15** could be evaluated at the subcellular level.

Various theranostic compounds have been developed and evaluated to enhance treatment outcomes (Bhuniya et al. 2014). One notable instance is theranostic **16** (Fig. 2.6), which combines cyanine-7 (Cy7) with gemcitabine via a disulfide bond and includes folate for cancer targeting, attached to the cyclohexene ring of the fluorophore (Yang et al. 2013). Cyanine dyes are characterized by two nitrogen atoms connected to a π-conjugated polymethine chain (Shindy 2012). These dyes exhibit high molar absorptivity, narrow absorption and emission spectra, and adjustable spectroscopic properties in the UV-Vis and NIR ranges (Gui et al. 2018; Hameed et al. 2019; Liu et al. 2019a). They also show low systemic toxicity and excellent biocompatibility. Due to their minimal autofluorescence, deep tissue penetration, reduced light scattering, and low phototoxicity, cyanines (>650 nm) are frequently used in in vivo studies (Choi et al. 2020; Escobedo et al. 2010; Luo et al. 2011). When tested with theranostic **16**, folate-positive KB cancer cells showed a strong fluorescence signal within 20 min, whereas no such signal was detected in folate-negative A549 cancer cells. This indicated that theranostic **16** was selectively internalized by folate-positive KB cells through a folate-mediated endocytosis process. Furthermore, colocalization studies using fluorescence imaging demonstrated that the drug was activated in the endoplasmic reticulum, leading to apoptosis.

In addition to disulfide linkers, various other chemical linkers have been explored for GSH-responsive drug delivery, with selenium being a notable focus. Selenium and sulfur, both belonging to the same elemental group, share similar chemical characteristics (Maroney and Hondal 2018). The Se–Se bond has a lower redox potential of 172 kJ mol^{-1}, compared to the C–Se bond at 244 kJ mol^{-1} and the S–S bond at 268 kJ mol^{-1} (Li et al. 2020b; Ma et al. 2010; Wei et al. 2020). This makes Se–Se linkers more sensitive to GSH than their S–S counterparts. Fang et al. developed two seleno-based CPT theranostic agents, **17** and **18** (Fig. 2.7), which, upon incubation with GSH, exhibit enhanced fluorescence at 430 nm (excitation = 365 nm), as confirmed by HPLC analysis (Zhao et al. 2021a). In a Tris-EDTA buffer, these agents showed almost complete drug release within 5 min in the presence of 1 mM GSH. The released selenol intermediates are thought to boost the efficacy of **17** and **18** by reducing intracellular GSH levels and increasing ROS production. In a HepG2 xenograft mouse model, **17** and **18** reduced tumor weight by approximately 75%, while CPT therapy alone achieved only a 53% reduction. Despite these encouraging outcomes, diselenide linkers are less commonly used than disulfide linkers, likely due to synthetic challenges and their dual sensitivity to both GSH and ROS, which can lead to premature activation and unintended drug release (Deepagan et al. 2016).

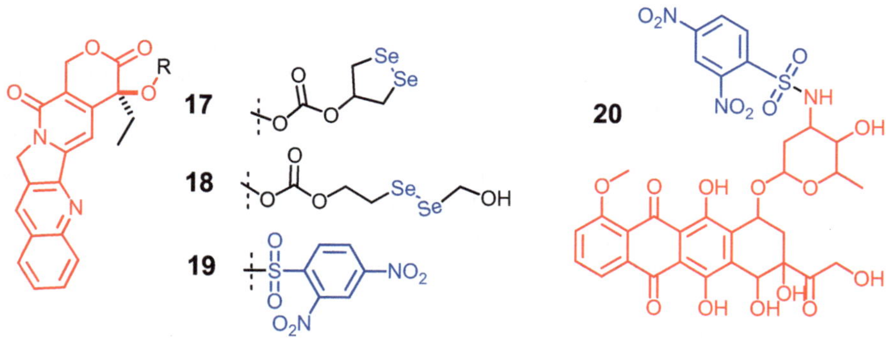

Fig. 2.7 Chemical structures of GSH-responsive theranostic probes **17–20**

Other biothiols, such as cysteine (Cys) and homocysteine (Hcy), are crucial for maintaining the redox equilibrium in cellular microenvironments (Deepagan et al. 2016). A lack of Cys can lead to various health issues, including cardiovascular diseases, leukopenia, liver damage, neurological disorders, cancer, and hematopoietic dysfunction (Rehman et al. 2020; Zhang et al. 2004). Many linkers that are sensitive to glutathione (GSH) also react with Cys, causing unintended drug activation in healthy tissues where both Cys and GSH are present at specific levels. On the other hand, systems that respond exclusively to Cys can offer greater selectivity. Recently, 2,4-dinitrobenzenesulfonyl (DBS) has been recognized as a Cys-sensitive linker for drug delivery and is increasingly being used with drug candidates that have free hydroxyl (OH) or amine (NH_2) groups. In separate studies by Jo and Johansson et al., theranostic agents **19** (derived from SN-38, an active metabolite of irinotecan) (Whang et al. 2018) and **20** (using DOX) (van Gisbergen et al. 2016) were developed by directly attaching the DBS group to the respective parent drugs (Fig. 2.7). This DBS modification leads to fluorescence quenching in both theranostic agents.

2.2.3 *Hydrogen Sulfide Responsive Theranostic Probes*

Hydrogen sulfide (H_2S) is a toxic, flammable gas with a pungent, rotten egg odor and corrosive properties (Reiffenstein et al. 1992). Alongside carbon monoxide (CO) and nitric oxide (NO), endogenous H_2S is a key gasotransmitter (Paul and Snyder 2015; Szabo 2016). It is crucial in regulating various physiological processes such as cell differentiation, proliferation, survival/apoptosis, and metabolism (Sen 2017; Zhang et al. 2017c). Abnormal levels of H_2S have been associated with diseases like cancer, Alzheimer's, and diabetes (Wallace and Rui 2015). H_2S shields neurons from oxidative stress by increasing GSH levels, which indirectly neutralizes reactive oxygen species (ROS). (Zhong et al. 2020). The significance of H_2S has driven the creation of H_2S-responsive theranostic agents for potential cancer treatments. H_2S not only acts as a nucleophilic agent, especially in its deprotonated forms, but also exhibits reductive properties, enabling it to selectively convert azides into amines. The strong electron-donating capability of amines can be harnessed to trigger the release of fluorophores or drugs. Qian et al. designed an H_2S-responsive theranostic agent, **21** (Fig. 2.8), which includes amonafide, an anticancer drug targeted for glioblastoma (Ge et al. 2021a, b). The natural fluorescence of amonafide is suppressed due to a photoinduced electron transfer (PET) effect. When exposed to H_2S, the azide is reduced to an amine, leading to the breakdown of a self-immolative linker and the release of the active drug through a 1,6-elimination reaction. Fluorescence studies showed that the drug accumulates in the lysosomes of U87MG cells and then moves to the nucleus, where it causes cytotoxicity by inducing DNA damage and mitochondrial dysfunction. In a 3D spheroid model of U87MG cells, this theranostic agent exhibited significant toxicity at a low dose (30 μM, 2 days treatment) and was more effective than cisplatin (40 μM, 2 days treatment), as indicated by the destruction of tumor spheroid integrity. A similar approach was used to create theranostic probe **22** (Fig. 2.8) for H_2S-mediated delivery of SN-38 in colon cancer cells (Bobba et al. 2019).

Leveraging the nucleophilic properties of H_2S, Li and colleagues engineered the theranostic compound **23** (Fig. 2.8), which integrates the anticancer medication SN-38 with a dinitrophenyl (DNP) ether group that withdraws electrons (Huang et al. 2022). Initially, this agent does not emit light due to the suppression of potential intramolecular charge transfer (ICT). Upon encountering H_2S, the ether bond is severed, leading to increased fluorescence and the release of the active drug. In vitro

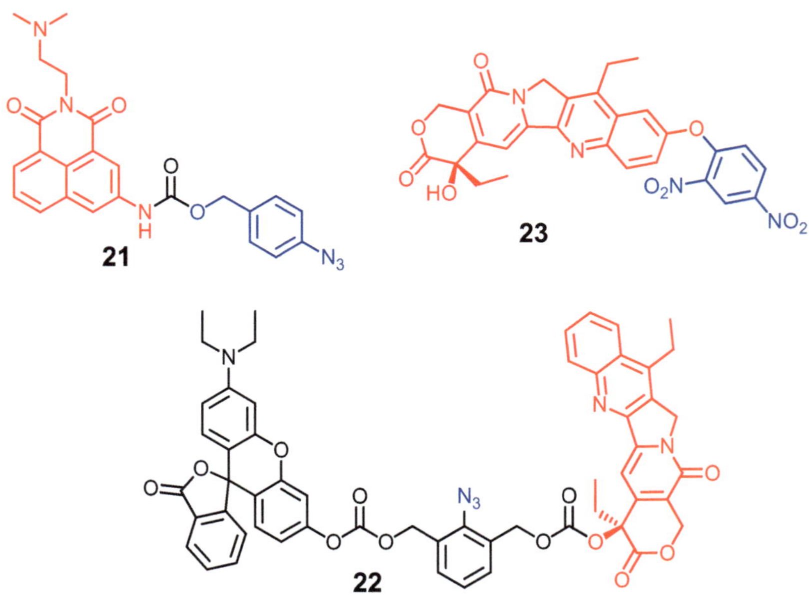

Fig. 2.8 Chemical structures of hydrogen sulfide-responsive theranostic probes **21–23**

experiments demonstrated that compound **23** was selectively absorbed and activated in H_2S-enriched HCT116 and 4T1 cancer cells, exhibiting cytotoxic effects. The inherent fluorescence of the SN-38 component allows for real-time tracking of drug activation and therapeutic efficacy. Although the use of H_2S's nucleophilicity for prodrug activation is promising, additional research is required to assess its selectivity in the presence of other nucleophiles. Despite this encouraging finding, no other theranostic agents that utilize the unique characteristics of H_2S are currently known to the authors.

2.2.4 Hydrogen Peroxide Responsive Theranostic Probes

"ROS" denotes unstable molecules and free radicals that arise from the reduction of oxygen (O_2), encompassing hydrogen peroxide (H_2O_2), superoxide (O_2^-), singlet oxygen ($^1O_2$), and hydroxyl radicals (•OH) (Bayr 2005). These reactive species are mainly produced endogenously via mitochondrial processes and reactions catalyzed by NADPH-dependent enzymes (Circu and Yee 2010; Lassègue et al. 2012). External sources, such as UV radiation and foreign substances, can also generate ROS. Under regulated conditions, ROS are essential for maintaining cellular balance and serve as signaling molecules in cell growth, division, movement, and programmed cell death (Bandyopadhyay et al. 1999). Their capacity to modify protein functions makes them crucial in regulating blood vessels, sensing oxygen, supporting immune responses, and activating genes (Alfadda and Sallam Reem 2012; Di et al. 2016; Pourova et al. 2010). However, excessive ROS can induce oxidative stress, causing damage to nucleic acids, lipids, and proteins. Imbalanced ROS levels have been associated with various diseases, including aging (Haigis and Yankner 2010), cancer (Trachootham et al. 2009), heart diseases (Moris et al. 2017), diabetes (Hoffmann and Griffiths 2018), and neurodegenerative conditions (Ahn et al. 2018; Blesa et al. 2015).

Within the mitochondria, hydrogen peroxide (H_2O_2) is generated from superoxide anions via processes mediated by superoxide dismutase (SOD). Cells that are inflamed or cancerous produce significantly higher levels of H_2O_2 (0.5 nmol/10^4 cells/h) compared to their normal counterparts (0.050 ± 0.004 nmol/10^4 cells/h) (Wang et al. 2017a). H_2O_2 can serve as a precursor for other highly reactive species, including peroxynitrite, hypochlorite, and hydroxyl radicals. Given its relatively long half-life ($t_{1/2}$ = 1 ms) compared to other reactive oxygen species (ROS), which typically have a half-life of less than 1 μs, and its tendency to accumulate during oxidative stress, H_2O_2 has become a focal point in the development of ROS-responsive drug delivery systems (Yang et al. 2020). In the realm of cancer-specific theranostic probes, various H_2O_2-sensitive chemical linkers, such as arylboronic esters, thioethers/thioketals, amino acrylates, selenium/tellurium compounds, and polyproline, have been investigated. This section reviews theranostic probes that are activated by oxidative stress, specifically H_2O_2.

Kim and colleagues have introduced a probe activated by H_2O_2, creating a molecular theranostic agent (Fig. 2.9). This compound integrates two anticancer drug segments, 5′-deoxy -5-fluorouridine (which becomes the active toxin 5-fluorouracil), connected via a self-immolative linker with two phenyl boronic acid groups, and an ethidium fluorophore that marks mitochondrial apoptosis (Kumar et al. 2014). In biotin-positive A549 lung cancer cells, the theranostic agent is selectively absorbed and accumulates in mitochondria, driven by the positive charge of the ethidium. The increased H_2O_2 concentration in these cancer cells initiates the release of 5′-deoxy -5-fluorouridine, which thymidine phosphorylase then converts to 5-fluorouridine (5′-FU). The resulting fluorescence enhancement from the free ethidium and its subsequent intercalation into DNA serves as a real-time indicator of cellular apoptosis. The theranostic probe was also tested in an A549 xenograft mouse model. Upon intravenous administration, the probe generated a fluorescent signal when activated by endogenous H_2O_2 in tumor tissues. Additionally, a more pronounced fluorescence was detected in the group treated with lipopolysaccharide (LPS, 10 μL, injected directly into the tumor), aligning with the oxidative stress-induced activation. The theranostic probe demonstrated selective accumulation in tumors and facilitated greater tumor reduction compared to the control (PBS) and 5′-FU groups.

Shabat and colleagues utilized a near-infrared fluorophore (Cy7) to track the activation and distribution of a drug with a precision suitable for in vivo studies (Redy-Keisar et al. 2015). For this purpose, they developed the theranostic compound **25** (Fig. 2.9), which integrates a boronic ester linked to a cyanine-based fluorophore and camptothecin (CPT). Upon exposure to H_2O_2, the boronate group was

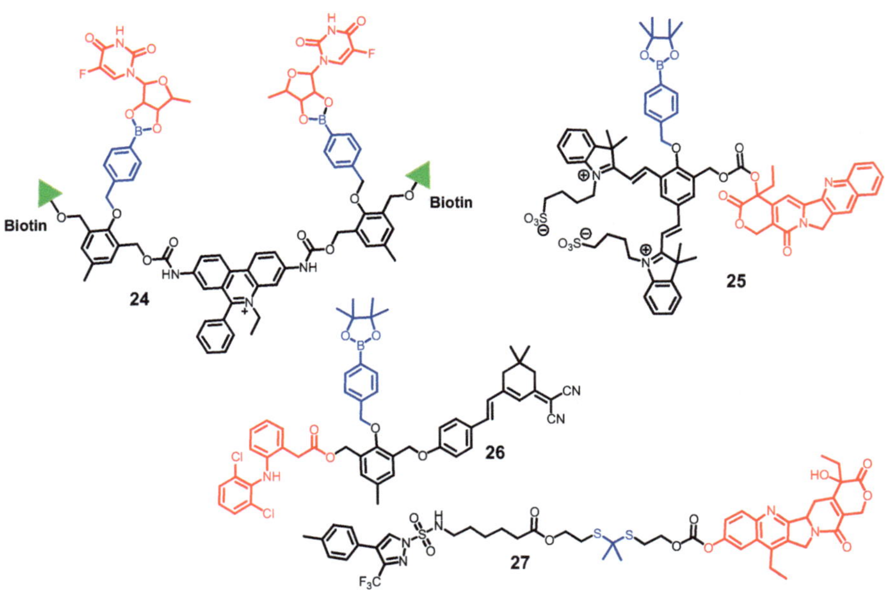

Fig. 2.9 Chemical structures of H_2O_2-responsive theranostic probes (**24–27**)

cleaved, initiating a sequence of reactions that released active CPT over about 90 min, accompanied by the liberation of free Cy7 and an increased fluorescence signal peaking at 720 nm. In human glioblastoma multiforme (GBM) U87 cells, compound **25** exhibited significant cytotoxicity when treated with H_2O_2 (IC_{50} = 40 nM), but minimal toxicity without H_2O_2 (IC_{50} = 250 nM). The researchers observed that this theranostic agent was less toxic than the parent drug (CPT, IC_{50} = 20 nM); however, its reduced off-target effects make it a promising candidate for various therapeutic approaches, such as personalized medicine. Further animal studies demonstrated that the theranostic probe **25**, whether administered intratumorally or intravenously, significantly enhanced fluorescence (Cy7), indicating the release of CPT. Another related NIR-based theranostic probe, **26**, has been designed for H_2O_2-mediated release and monitoring of diclofenac (a nonsteroidal anti-inflammatory drug, NSAID) in inflammation-induced macrophages (Sufian et al. 2022).

As previously mentioned, the primary site of H_2O_2 production is within the mitochondria. Consequently, a theranostic agent that does not efficiently concentrate in the mitochondria of cancer cells may not elicit a strong anticancer effect. Landfester and colleagues have proposed that generating H_2O_2 directly within cancer cells could address this issue. They created a multifunctional theranostic probe **27** (Fig. 2.9) that includes celecoxib-labeled SN-38 connected via an H_2O_2-sensitive thioketal linker. This probe was encapsulated in silica nanoparticles coated with folate, which were then further modified with GOX enzymes through covalent bonding to form the nanoreactor F-GOX@NR (Jo et al. 2022). Upon entering folate-positive cancer cells via receptor-mediated endocytosis, the nanoreactor F-GOX@NR utilizes glucose to produce H_2O_2 through GOX catalysis. The inactive form of celecoxib-labeled SN-38 binds to COX-2, promoting its retention in the cytoplasm. The overproduction of H_2O_2 by the nanoreactor F-GOX@NR + **27** causes the thioketal linker to break, releasing active SN-38, which inhibits topoisomerase and ultimately induces cell death. Comparative studies demonstrated that the nanoreactor system F-GOX@NR + **27**, when activated by intracellular H_2O_2, synergistically releases the active drug, showing significantly greater efficacy in MCF-7 cancer cells that are positive for both folate and COX-2 (IC_{50} = 0.14 μM, 24 h) compared to SN-38 alone (IC_{50} = 1.2 μM, 24 h) and celecoxib-SN-38 conjugates (IC_{50} = 2.8 μM, 24 h). Additionally, the ratiometric fluorescence change of SN-38 (shifting from 452 nm to 560 nm in the presence of H_2O_2) enables the monitoring of drug activation and localization.

2.2.5 *Other ROS-Responsive Theranostic Probes*

Apart from H_2O_2, other reactive oxygen species (ROS) such as $^1O_2$ and •OH have garnered significant interest due to their high reactivity. The hydroxyl radical •OH can interact with a wide range of biomolecules, including amino acids, carbohydrates, lipids, and nucleotides, but these interactions often occur without specificity (Gutteridge 1984). Singlet oxygen ($^1O_2$), an excited form of O_2, possesses a higher oxidative potential compared to the •OH radical (Barrios et al. 2021; Halliwell 2006). In biological systems, •OH is generated through the Fenton reaction, which involves the decomposition of H_2O_2 catalyzed by copper(I) or iron(II) (Lyngsie et al. 2018). On the other hand, $^1O_2$ is mainly produced by photosensitizers (PSs) (Li et al. 2016). Both $^1O_2$ and •OH have shorter lifespans than H_2O_2 in physiological conditions.

Zhang and colleagues engineered a theranostic probe, designated as **28** (Fig. 2.10), which integrates gemcitabine with a red light-activatable fluorescent photosensitizer (meso-tetraphenylporphyrin; TPP) via a reactive oxygen species (ROS)-sensitive thioketal linker for fluorescence-guided therapy (Liu et al. 2016b). In this study, the 5′-OH group of gemcitabine was altered to inhibit its anticancer activity. Given that singlet oxygen ($^1O_2$) has a brief half-life and limited diffusion in aqueous environments (approximately 40 ns and 20–200 nm, respectively), the $^1O_2$ levels in cells without illumination are insufficient to cause cytotoxicity (Sugimoto et al. 2005). However, when exposed to light (658 nm, 280 mW cm^{-2}), TPP generates $^1O_2$, leading to the cleavage of the thioketal bond and the subsequent release of the drug. The liberated drug can then spread to neighboring cells, inducing toxicity in areas not directly illuminated, a phenomenon known as the bystander effect. Compared to control groups (TPP, TPP-UCL-GEM; UCL denotes an ROS-insensitive alkyl linker), the theranostic agent **28** demonstrated concentration-dependent toxicity in HeLa cells upon illumination (IC_{50} = 0.25 μM). Tumor suppression experiments using subcutaneous H22-bearing mice showed that intravenous administration of PEG2000-PLA2000-modified theranostic **28** resulted in selective tumor accumulation and, under low-power light, inhibited tumor growth. The inherent fluorescence of TPP was also utilized to track the in vivo distribution of the theranostic agent. Wang and associates used a comparable approach to deliver Dox to breast cancer sites in mouse models with minimal adverse effects (Pei et al. 2019).

Several theranostic agents that are activated by $^1O_2$ have been developed. One notable approach involves the use of aminoacrylate-based chemical linkers, which act as bridges to covalently connect a photosensitizer (PS) with a therapeutic drug.

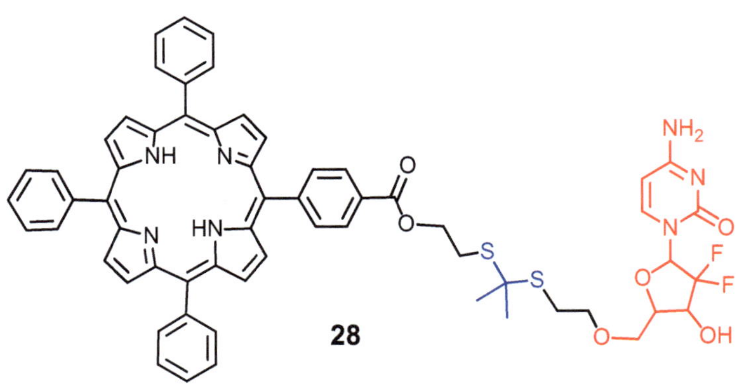

Fig. 2.10 Chemical structures of ROS-responsive theranostic probes (**28**)

For drugs containing hydroxyl groups, these can be easily linked to the carboxylic acid of an aminoacrylate linker, creating a straightforward platform for drug loading. On the other hand, the amino group of the acrylate linker can be attached to the PS, leading to the formation of effective theranostic agents. This method has enabled researchers to create various delivery systems incorporating drugs such as SN-38, combretastatin, paclitaxel, and NSAIDs like ibuprofen and naproxen as the active components.

2.2.6 *Enzyme Responsive Theranostic Probes*

Enzymes, primarily composed of proteins, expedite chemical reactions within and outside cells (Robinson 2015). Their precise substrate selectivity, strong reactivity, and specific cellular localization render them ideal for initiating chemical processes (Liu and Tadhg 2020). The malfunction of critical enzymes is a common feature in various diseases, such as inflammatory conditions (Mariaule et al. 2021; Zhao et al. 2021b), malignancies (Benjamin et al. 2013; Takahashi 2018), and neurodegenerative disorders (Cleland et al. 2021; Quinones et al. 2020). By strategically selecting enzyme-specific substrates, it is feasible to conceal an anticancer agent and develop theranostic probes with targeted activation, enhanced stability, and superior therapeutic outcomes. A review of these advancements is provided below.

Quinone is a prevalent component in numerous natural and synthetic compounds, such as anticancer drugs, antimicrobial agents, dyes, vitamin K, and enzyme cofactors (Ross and David 2021). Its significance lies in its ability to participate in redox reactions. NQO1, or DT-diaphorase, is a two-electron reductase primarily found in the cytosol, with smaller amounts in the endoplasmic reticulum and mitochondria (Haefeli et al. 2011; Siegel et al. 2012). This enzyme is crucial for detoxification and is implicated in early stages of carcinogenesis. Elevated levels of DT-diaphorase are observed in various cancers, including ovarian, thyroid, breast, colon, and pancreatic tumors (Cui et al. 2015). Consequently, the increased presence of DT-diaphorase in cancerous tissues compared to normal tissues has been leveraged for targeted drug delivery in tumor treatment.

Indolequinone is acknowledged as a substrate for NQO1. Leveraging this, Nishimoto and colleagues engineered and synthesized the theranostic compound **29a** (Fig. 2.11) to transport the cytotoxic agent SN-38 to malignant cells (Zhang et al. 2005). The activation of SN-38 was detected through fluorescence variations following DT-diaphorase-mediated reduction. To enhance its tumor-specificity, the SN-38 component in the theranostic compound **29b** was coupled with indolequinone and an integrin-targeting peptide (Huang et al. 2010). This modified agent **29b** exhibited selective uptake by $\alpha v\beta 3$ integrin-expressing cancer cells, where the release of active SN-38 via DT-diaphorase activity led to a 50–70% inhibition of human cervical carcinoma (KB) cell proliferation. Research has indicated that the alkenyliminium intermediate-generated postdrug release is potent in DNA

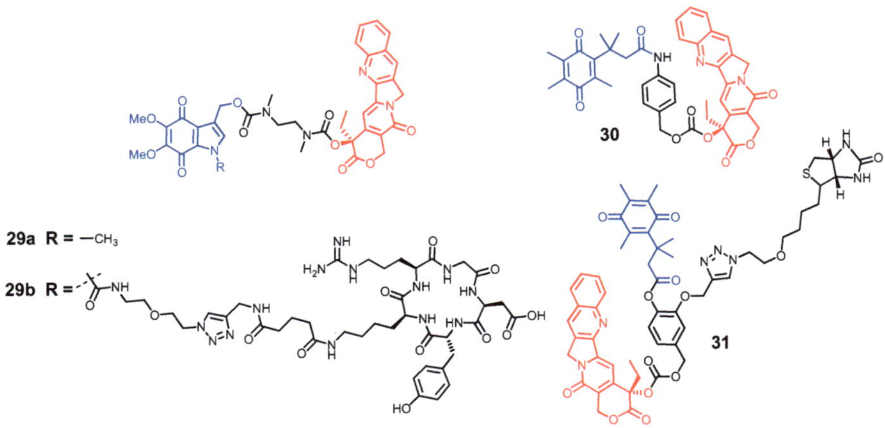

Fig. 2.11 Chemical structures of DT-diaphorase responsive theranostic probes (**29–31**)

alkylation, a mechanism believed to underlie the observed cytotoxic effects (Tanabe et al. 2009).

Numerous investigators have explored quinone-based prodrugs to enhance their release characteristics. Notably, a quinone derivative with a "trialkyl lock" developed by Wang and colleagues has attracted significant interest (Wang et al. 1996). Under conditions of DT-diaphorase-mediated reduction, this quinone transforms into hydroquinone, enabling a hydroxyl group to form a six-membered ring with a carbonyl group through lactonization. This process triggers the detachment of a drug linked to the carbonyl group of hydroquinone. This approach was exemplified in the theranostic agent **30**, created by Wu et al., designed to deliver SN-38 to cancer cells (Liu et al. 2015). The inherent fluorescence of SN-38 enabled real-time monitoring of both drug release and its therapeutic impact (Fig. 2.11). Subsequently, Kim et al. incorporated an additional functional group to attach biotin as a targeting element in theranostic **31** (Shin et al. 2016), which exhibits enhanced uptake by cancer cells and improved therapeutic outcomes following drug activation.

Azoreductases, which are flavin-reliant enzymes, exist in both eukaryotic and bacterial species. These enzymes, predominantly found in the cytosol, are crucial for maintaining cellular equilibrium. They facilitate the reduction of various substrates with the aid of an electron donor, usually NADH or NADPH. Elevated levels of azoreductases have been observed in several types of cancer, including lung (Li et al. 2015a), breast (Yang et al. 2014), and pancreatic cancers (Lewis et al. 2005). Consequently, there has been a focus on creating theranostic agents that target these enzymes.

Kim and colleagues engineered a theranostic compound, designated as **32** (Fig. 2.12), aimed at delivering a chemotherapeutic agent specifically to the mitochondria of cancer cells (Verwilst et al. 2017). This system integrates a rhodamine 123/B derivative linked to N,N′-bis(2-chloroethyl)-1,4-benzenediamine, a nitrogen mustard analog, via an azo bond. A lipophilic triphenyl phosphonium group is also

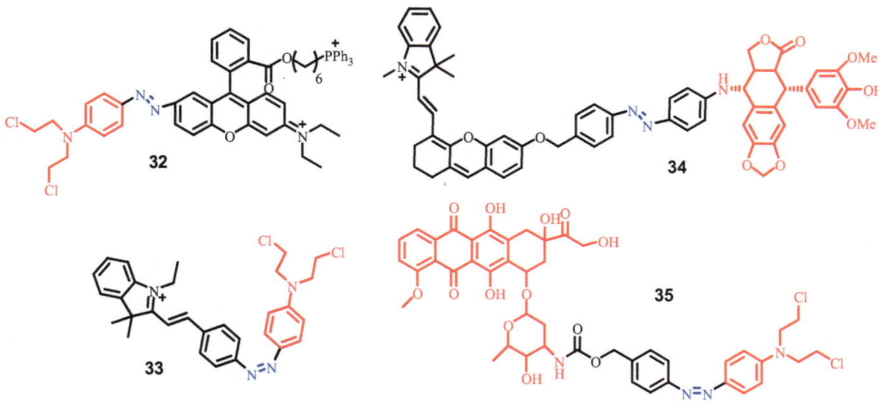

Fig. 2.12 Chemical structure of azoreductase-responsive theranostic probes (**32**–**35**)

incorporated to enhance mitochondrial targeting. When the azo bond is reduced, it breaks, releasing both the active drug and a fluorescent marker, enabling the monitoring of drug activation and localization in the hypoxic environment of cancer cells. Nonetheless, the intricate multistep synthesis and limited solubility of compound **32** hindered its practical application. To overcome these challenges, Xie created theranostic **33** (Fig. 2.12), which features a positively charged fluorophore for mitochondrial targeting and a reduced hydrophobic surface area to improve water solubility (Hu et al. 2018a).

Shi and his team employed a similar strategy to develop the theranostic agent **34** (Fig. 2.12), which incorporates Cy as a near-infrared fluorescent tag and has proven effective for in vivo drug monitoring (Zhao et al. 2020a). Similarly, Yu and his group designed a theranostic system 3**5** (Fig. 2.12), combining nitrogen mustard and Dox via an azo bond (Li et al. 2018a, b). Under reductive conditions, both components are released, causing an increase in fluorescence due to the liberation of free Dox. This enhanced fluorescence can be used to monitor drug activation and cellular uptake. Studies using 4T1 cell models, both in vitro and in vivo, revealed that theranostic agent **35** displays higher cytotoxicity and reduced side effects compared to the control groups (i.e., PBS and free Dox).

Nitroreductases (NTRs), which contain flavin mononucleotide (FMN), are overproduced in various types of cancer. The expression levels of NTRs are strongly associated with hypoxic conditions within solid tumors (Bonnet and Dick 1997). Type 1 NTRs, predominantly found in bacteria, are not affected by oxygen and use FMN as their active site. In contrast, type 2 NTRs are sensitive to oxygen and can utilize either FMN or flavin adenine dinucleotide (Wilkinson et al. 2002). Hypoxic areas in tumors often result from inadequate blood vessel networks or insufficient support for rapid tumor growth, leading to restricted blood and oxygen flow. It is estimated that 50–60% of solid tumors, especially in their central regions, experience hypoxia (Vaupel and Arnulf 2007). Extensive research has been conducted on hypoxia-responsive drug delivery systems (DDS), as evidenced by a plethora of published studies (Kumari et al. 2020; Sharma et al. 2019; Thambi et al. 2016; Yang et al. 2021). This section provides an overview of the latest developments in hypoxia-responsive DDS for molecular theranostics.

One of the major hurdles in creating hypoxia-sensitive therapeutic agents is the tumor's ability to generate new blood vessels, which increases its oxygen levels and diminishes the effectiveness of treatments. Consequently, integrating strategies that inhibit angiogenesis with hypoxia-activated drug delivery mechanisms could significantly improve the performance of these therapies.

Nitroaromatic compounds act as ideal substrates for NTRs, leading to the generation of hydroxylamines and amine derivatives. In a self-immolative setup, the ensuing electron rearrangement in the aromatic ring triggers the release of attached therapeutic agents. Building on this principle, Kim and colleagues created theranostic compound **36** (Fig. 2.13), which integrates an NSAID (indomethacin) with SN-38 via a nitrobenzyl alcohol-based connector. The inclusion of indomethacin in the theranostic construct serves dual roles: it targets tumors by exploiting COX-2 overexpression and suppresses COX-2-induced angiogenesis. Theranostic agent **36** demonstrated toxicity in COX-2-positive A549 and HeLa cells that increased with concentration under low oxygen conditions (1% O_2). Additionally, probe **36** showed intense fluorescence indicative of active SN-38 in multicellular tumor spheroids (A549) of different sizes (110, 235, 300 μm). This observation highlights the compound's ability to penetrate deep tissues and activate the prodrug, facilitated by indomethacin's antiangiogenic properties.

The research team created a theranostic duo, **37a** and **37b**, for the visualization and treatment of cancer stem cells (CSCs) (Fig. 2.13) (Kim et al. 2021). In this investigation, they utilized dimethyl nitrothiophene as the hypoxia-sensitive trigger, replacing the p-nitrobenzyl group, due to its lower reduction potential, which was anticipated to enhance drug release. For therapeutic purposes, a curcumin derivative, 3,4-difluorobenzylidene curcumin, was selected. To target CSCs, acetazolamide, a recognized carbonic anhydrase (CAIX) inhibitor, was employed (Lock et al. 2013). For imaging, a naphthylamide fluorophore was used in place of the drug. The key distinction in this theranostic pair, designed for separate imaging and

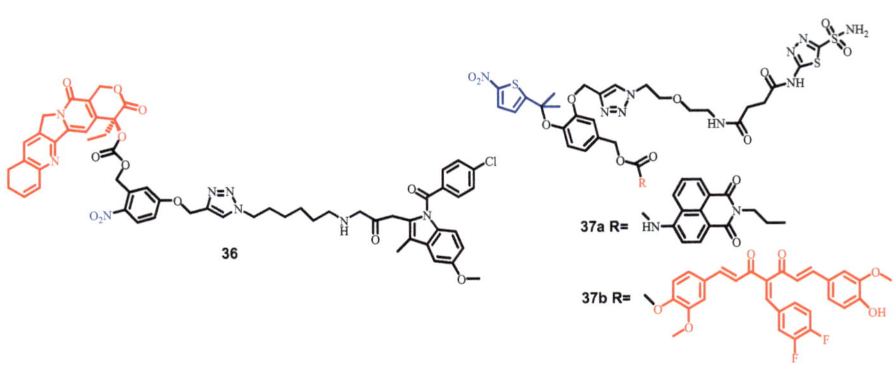

Fig. 2.13 Chemical structures of nitroreductase-responsive theranostic probes (**36** and **37**)

treatment, lies in the presence of either the drug or the phthalimide unit, while the targeting moiety and hypoxia-responsive element remain consistent. This structural similarity led to the assumption that the cellular uptake and activation patterns of these compounds would be similar. Further cell-based studies using $CD^{+}133$ MDA-MB-231 breast cancer cells are required to validate this hypothesis. Both **37a** and **37b** showed hypoxia-dependent activation, enabling both imaging and anticancer effects. Treatment with **37b**, administered via tail vein injection, slowed tumor growth in $CD^{+}133$ MDA-MB-231 xenograft mice.

Carboxylesterases (CESs) are a prevalent group of hydrolases that facilitate the breakdown of ester, amide, and carbamate bonds (Fukami and Tsuyoshi 2012; Kohnz et al. 2015). These enzymes serve as crucial detoxifiers for foreign substances. CES1 and CES2, two well-studied carboxylesterases, have been the focus of extensive research. Notably, CES2 has been found to be more abundant in diseased tissues than in healthy ones (Aranda et al. 2014; Wells and Rubin 2003).The increased presence of CES in cancerous cells is believed to enhance their ability to invade, migrate, survive, and proliferate (Nomura et al. 2010; Qin and Zhi-Hua 2014). In colorectal cancer, CES activity is observed to be two to four times higher (0.45 ± 0.25 U/L male; 0.45 ± 0.35 U/L female) compared to normal cells (0.17 ± 0.09 U/L male; 0.12 ± 0.07 U/L female) (NIU et al. 2012). Consequently, various theranostic agents responsive to esterase activity have been developed to target tumors and enhance treatment efficacy. For instance, Kunimoto and colleagues created a CEs-responsive variant of SN-38, known as agent **38** (Fig. 2.14) (Kunimoto et al. 1987). They attached a water-soluble δ-lactone ring to the SN-38 drug via a

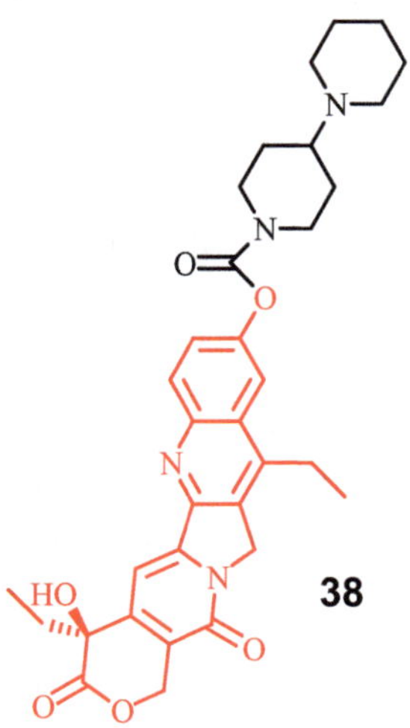

38

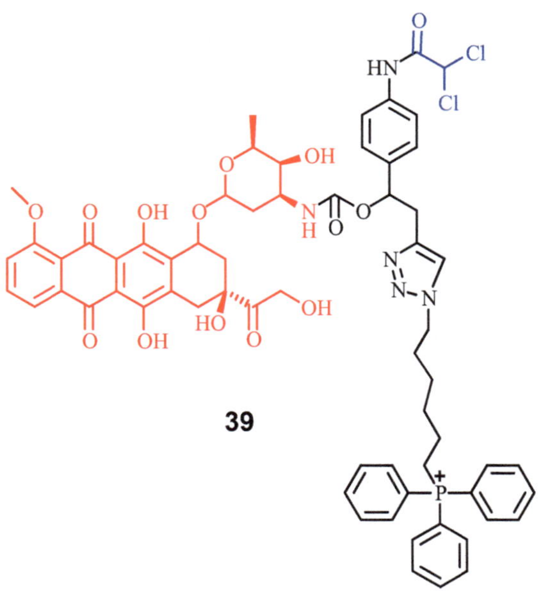

39

Fig. 2.14 Chemical structures of esterase-responsive theranostic probes **38** and **39**

carbamate bond. When exposed to CES, the carbamate bond was cleaved, releasing the active SN-38.

One of the most significant hurdles in cancer chemotherapy is the emergence of multidrug resistance (MDR), a condition where cancer cells persist despite exposure to different chemotherapeutic agents (Zahreddine and Katherine 2013). Statistically, MDR accounts for approximately 90% of cancer-related deaths (Bukowski et al. 2020). MDR can be attributed to various factors, including decreased drug uptake, increased drug efflux, enhanced DNA repair, elevated xenobiotic metabolism, and genetic changes such as gene amplifications, mutations, and epigenetic modifications (Dallavalle et al. 2020; Luqmani 2008; Wang et al. 2017b; Wu et al. 2014a). These factors collectively diminish the effectiveness of administered drugs. To tackle this issue, Kim and colleagues created theranostic agent **39** (Fig. 2.14), which consists of a dichloroacetic acid (DCA) component connected via an amide bond to a self-immolative linker, Dox, and a lipophilic TPP mitochondrial targeting unit (Sharma et al. 2018a). The design aims to sensitize cancer cells by redirecting their abnormal glycolytic metabolism to mitochondrial phosphorylation. The rationale also includes the potential for improved therapeutic outcomes through the delayed release of active Dox. DCA, known for inhibiting pyruvate dehydrogenase (PDK), can shift glycolysis to glucose oxidation, thereby reducing lactate and intracellular ATP levels and causing mitochondrial dysfunction (Nayak et al. 2018). Additionally, agent **39** initially localizes in the mitochondria, evading early drug efflux through ATP-dependent ABC transporters (Choi and Ai-Ming 2014). Over time, Dox migrates to the nucleus to exert its anticancer effects. Studies have shown that the cytotoxicity of agent **39** is CEs-dependent and that it is preferentially taken up by A549 and HepG2 cancer cells compared to normal cells (NHDF, IMR90). Agent **39** outperforms both Dox alone and the combined administration of DCA and Dox at a 1:1 ratio. Furthermore, agent **39** demonstrates efficacy in both drug-sensitive and drug-resistant MCF breast cancer cells, as well as in MCF xenograft mouse models.

Proteases, a diverse group of enzymes, facilitate the breaking of peptide bonds (López-Otín and Bond 2008). These enzymes are essential for various aspects of protein metabolism, such as catabolism, digestion, and cellular communication. Increased levels of proteases have been linked to numerous health conditions, including inflammatory disorders (Banerjee and Bond 2008), malignancies (Matters et al. 2005; Su et al. 2018), cardiovascular issues (Hua and Sreejayan 2015), and neurodegenerative ailments (Seo and Mikyoung 2020). Extensive research has been conducted to create inhibitors for these enzymes, as well as diagnostic and therapeutic tools that respond to their activity. Among the proteases implicated in inflammation, cancer, and apoptosis, caspases stand out. Caspase-3, a specific cysteine-aspartic acid protease, can be triggered by both internal (apoptosis) and external (radiation or chemotherapy) stimuli (Duclos et al. 2017; Fujita et al. 2001; Huang et al. 2011).

Byun and colleagues engineered a caspase-3-sensitive theranostic compound (Fig. 2.15) that integrates a tumor-targeting integrin (Arg-Gly-Asp) tripeptide, a

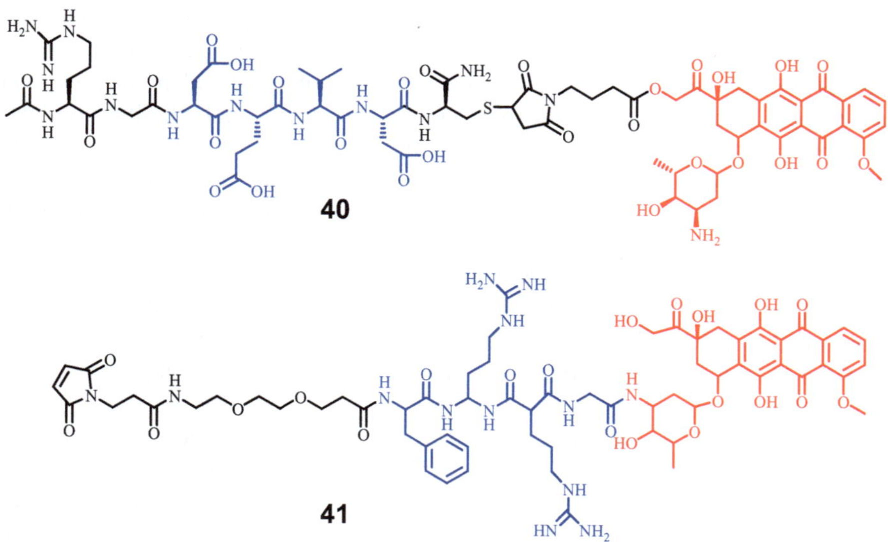

Fig. 2.15 Chemical structures of theranostic probes **40** and **41**

caspase-3-cleavable DEVD (Asp-Glu-Val-Asp) tetrapeptide, a cellular ester-responsive bond, and Dox as the anticancer component (Chung et al. 2018). Dox is recognized for its ability to trigger caspase-3, a key marker in apoptosis (Ueno et al. 2006). The hypothesis was that caspase-3, once activated by Dox, would facilitate the release of Dox through ester hydrolysis, thereby further activating caspase-3. This feedback loop was expected to enhance the therapeutic impact. According to this theory, the cleavage of the DEVD linker by caspase-3, followed by the breakdown of the ester bond, led to the discharge of the active drug, along with an increase in fluorescence. The theranostic agent **40** was more readily absorbed by integrin-positive human glioma U-87MG cells via integrin-mediated endocytosis, showing greater toxicity to these cells than to integrin-negative HT-29 colon cancer cells. Cellular experiments demonstrated that this self-amplifying enzyme apoptosis theranostic agent could boost caspase-3 activity by approximately 154 times. In vivo tests on U-87 MG tumor-bearing mice also indicated that treatment with agent **40** could almost completely suppress tumor growth compared to various control groups.

Cathepsins, which are lysosomal proteolytic enzymes, mainly function in the breakdown of proteins and peptides. They are crucial for sustaining tissue equilibrium and participate in immune responses, as well as cell growth, differentiation, and programmed cell death (Reiser et al. 2010). Changes in the levels of these enzymes are linked to various diseases, including cancer, and are often associated with unfavorable outcomes (Chen et al. 2017; Gocheva and Joyce 2007).Cathepsin B is particularly significant in facilitating the degradation of the extracellular matrix (ECM), thus enhancing tumor angiogenesis, invasion, and metastasis (Tan et al.

2013). Consequently, there has been a focus on creating diagnostic and therapeutic tools that respond to cathepsin B. For instance, Kim et al. designed a theranostic agent (Fig. 2.15) that combines Dox with an albumin-binding maleimide group via a cathepsin B-sensitive peptide sequence (FRRG) (Cho et al. 2022). This modification increased the half-life of the drug from 0.25 h to 3.1 h, likely due to its binding to plasma albumin. Once the agent accumulates in the tumor through passive targeting, it is activated by cathepsin B, releasing Dox within the cancer cells. Toxicity tests showed that the theranostic agent had stronger antitumor effects in MDA-MB231 breast cancer cells (IC_{50} = 7.33 μM) compared to rat cardiomyocytes H9C2 (IC_{50} > 200 μM), attributed to the higher cathepsin B activity (24.26 ± 3.08-fold) in the cancer cells. In vivo experiments with MDA-MB231 tumor-bearing mice confirmed that the theranostic agent effectively inhibited tumor growth compared to control groups.

Galactosidases are enzymes that play a crucial role in various essential catabolic pathways (Asanuma et al. 2015; Juers et al. 2012). Specifically, β-galactosidase (β-gal) is a lysosomal hydrolase that removes the terminal galactose from glycoconjugates (Krishna et al. 1999). This enzyme is often overexpressed in different types of cancer (Chatterjee et al. 1979; Kubo et al. 2021; Wagner et al. 2015). It has been utilized for the development of cancer-targeted imaging and therapeutic compounds (Li et al. 2020c; Sakabe et al. 2013). In the field of theranostics, Kim et al. created a β-gal-responsive compound, referred to as agent **42** (Fig. 2.16), which links an anticancer drug to a galactosidase component via a self-immolative linker. (Sharma et al. 2018b). This theranostic agent remains stable and nonfluorescent in PBS. Upon exposure to β-gal, the galactosidase part undergoes hydrolysis, releasing active Dox and increasing fluorescence. Experiments with HT-29 and HepG2 cells showed that agent **42** was internalized through the overexpressed asialoglycoprotein (ASGP) receptors. HeLa cells, with lower ASGP expression, exhibited less uptake. The toxicity of probe **42** to HT-29 and HepG2 cell lines was found to be dose-dependent. In vivo studies using mice with HT-29 cancer xenografts demonstrated that agent **42** significantly inhibited tumor growth compared to treatment with Dox alone.

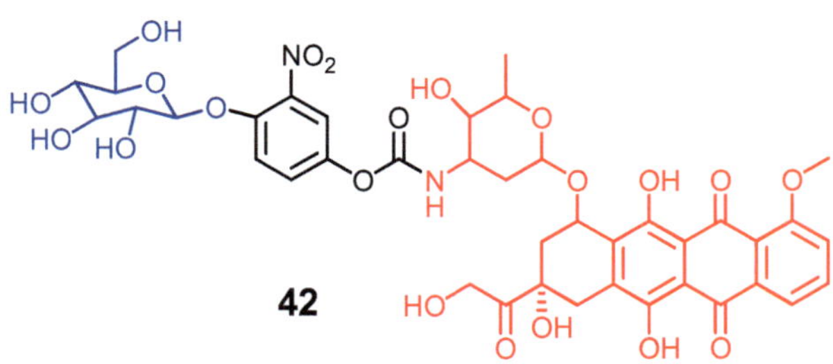

42

Fig. 2.16 Chemical structures of theranostic probe **42**

2.3 Theranostic Fluorescence Probes in PDT

2.3.1 pH-Activatable Theranostic Probes in PDT

The acidic milieu of a tumor, a hallmark of cancer, presents a unique opportunity for the creation of activatable photosensitizers (PSs). Generally, these pH-sensitive PSs incorporate elements that react to hydrogen ions, including pyridine, aniline, piperazine, morpholine, or other tertiary alkyl amine groups. These components typically suppress fluorescence and impede the production of photoinduced singlet oxygen by the PSs. Under typical physiological pH, these intelligent PSs remain dormant, but in an acidic environment, they become highly fluorescent and proficient at generating reactive oxygen species (ROS). This capability opens up novel avenues for the accurate detection and photodynamic therapy (PDT) of tumors.

For example, Siegwart and colleagues developed water-soluble, pH-sensitive near-infrared (NIR) absorbing BODIPY-based photosensitizers (PSs) for image-guided photodynamic therapy (PDT) of cancer (Fig. 2.17) (Xiong et al. 2018). Spectral analysis showed that in acidic conditions, compounds **43** and **44** exhibited robust NIR absorption and stable NIR emission, with absorption peaks at 660 and 690 nm and emission maxima at 692 and 742 nm, respectively. This characteristic makes them well-suited for noninvasive imaging of deep-seated tumors. Notably, the singlet oxygen quantum yields and fluorescence intensities of these compounds were markedly enhanced in an acidic tumor microenvironment (TME). This enhancement is attributed to the protonation of diethylaminophenyl groups, which mitigates photoinduced electron transfer (PET) quenching. Additionally, these BODIPY derivatives demonstrated selective accumulation in tumors following intravenous administration, without the need for additional targeting ligands. This underscores their superior intrinsic tumor-targeting capabilities and their potential for inducing significant tumor ablation under NIR light exposure in low pH environments.

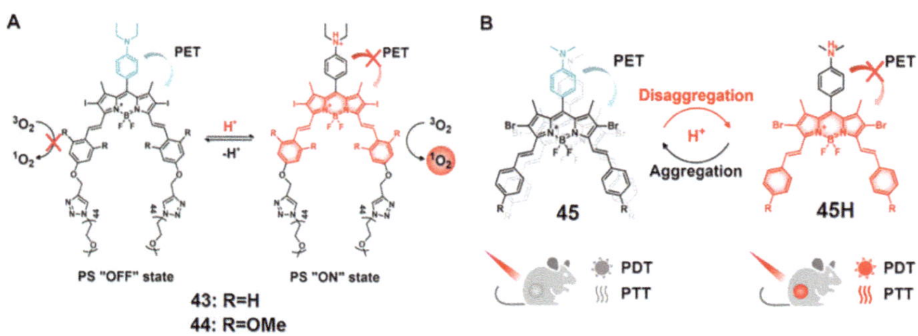

Fig. 2.17 (**a**) Structures of pH-activatable BODIPY-based **43** and **44**. (**b**) Diagram illustrating the mechanism of pH-mediated disaggregation of **45** in activating PDT and PTT (Reproduced from Sharma et al. 2019)

Tang and colleagues developed a BODIPY-based photosensitizer that displays aggregation behavior dependent on pH levels, incorporating bromine to facilitate pH-sensitive photodynamic and photothermal effects for cancer treatment (Fig. 2.17). This photosensitizer demonstrated minimal inherent toxicity and displayed weak fluorescence and $^1O_2$ generation at a neutral pH of 7.4, attributed to a dual quenching process involving a photoinduced electron transfer (PET) effect and the aggregated state (Liu et al. 2020b). In an acidic environment with a pH of 4.0, the dimethylaminophenyl group of the sensitizer became protonated, which increased its solubility and charge repulsion, effectively dispersing the aggregates. This transition enabled the activation of both photodynamic therapy (PDT) and photothermal therapy (PTT).

Cyanine dyes are renowned for their superior optical characteristics and have found extensive applications in biosensing, bioimaging, and phototherapy. Notably, cyanine-based photosensitizers (PSs) with pH-sensitive groups have been engineered for the targeted identification and elimination of cancer cells. In a research effort by Kamkaew et al., a pH-responsive heptamethine cyanine dye (**46**) was created as a theranostic agent for the treatment of HepG2 cells (Fig. 2.18) (Siriwibool et al. 2020). This probe demonstrated a high Φ_Δ value and was effectively utilized for near-infrared (NIR) imaging-guided photodynamic therapy (PDT) under acidic conditions, where the pH-sensitive N-methylpiperazine moiety became protonated. The absorption spectra of **46** showed a red shift as the pH decreased, likely due to the suppression of the intramolecular charge transfer (ICT) at lower pH levels. Additionally, the theranostic agent **46** could specifically infiltrate cancer cells, and upon exposure to 850 nm LED light for 30 min in an acidic environment, it displayed significant photocytotoxicity against HepG2 cells.

Recently, Wang and Zhou et al. utilized a pyridine group as a pH-sensitive element to develop an effective pre-photosensitizer (**47**) (Wang et al. 2022b). As illustrated in Fig. 2.18, the presence of cyano groups, known for their electron-withdrawing characteristics, enables probe **47** to demonstrate a more pronounced pH response in

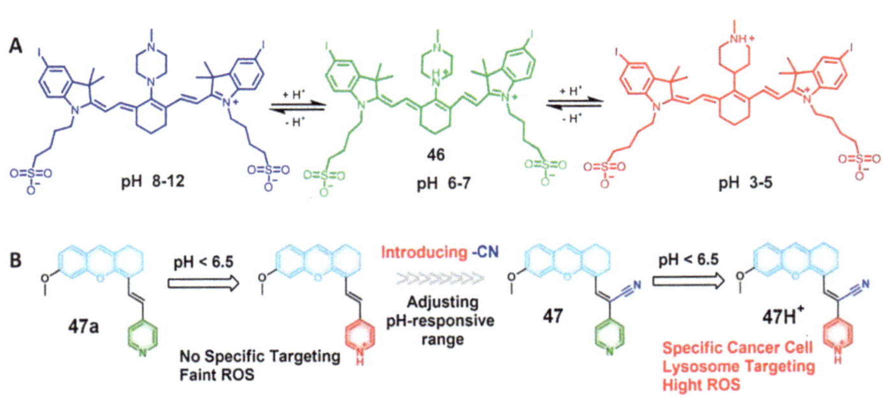

Fig. 2.18 (**a**) Chemical structure and pH-activated mechanism of **46**. (**b**) Schematic diagram showing the design concept of **47**, low pH-responsive PS

mildly acidic conditions (pH < 5.0) than the reference compound ZWZ (**47a**). Additionally, probe **47** displays a higher rate of cellular uptake and superior singlet oxygen production. This probe can be activated by intracellular protons, thereby enhancing intramolecular charge transfer (ICT) and facilitating effective photodynamic therapy (PDT) within the lysosomal milieu of cancer cells, including HepG2, HeLa, and 4T1 cells.

2.3.2 *Biothiol-Activatable Theranostic Probes in PDT*

Biothiols, including GSH, cysteine (Cys), and homocysteine (Hcy), are essential for biological processes, especially in preserving the redox equilibrium of cells (Jung et al. 2013). Cancer cells, which experience persistent oxidative stress, frequently elevate their biothiol levels as a protective response (Niu et al. 2015). For example, the GSH concentration in cancer cells can vary from 1 to 10 mM, which is more than five times higher than in normal cells (Gamcsik et al. 2012; Lee et al. 2018; Peng et al. 2023). Consequently, altered biothiol levels can be used to distinguish cancer cells from healthy ones and can also aid in the creation of biothiol-activated smart photosensitizers. These smart photosensitizers improve the efficacy of photodynamic therapy (PDT) on tumors while reducing harm to surrounding healthy tissues (Han et al. 2023; Li et al. 2020d; Lucero and Jefferson 2021; Pham et al. 2021).

The sulfhydryl (−SH) group in cysteine endows GSH and Cys with potent reductive and nucleophilic characteristics, facilitating their involvement in a wide array of chemical and biological processes. Based on the reactivity of the −SH group, biothiol-driven activation methods can be classified into several types, including the cleavage of sulfonate esters, disulfide bond interchange, nucleophilic substitution, ring formation with aldehydes or cyano groups, reduction mediated by GSH, and Michael addition reactions.

Previous research has established that 2,4-dinitrobenzenesulfonyl is a thiol-sensitive moiety. This moiety exhibits a potent electron-withdrawing effect and has been employed in the creation of GSH-activated photosensitizers (PSs). In a recent study by Sun et al., a GSH-activated near-infrared (NIR) PS (**48**) was developed by integrating iodine-substituted hemicyanine (**48**-OH) with 2,4-dinitrobenzenesulfonate (DNBS) (Fig. 2.19). The PS was designed for combined photodynamic therapy (PDT) and sulfur dioxide (SO_2) treatment (Wang et al. 2022a). SO_2, a gasotransmitter, is typically known as an air pollutant, but recent findings have highlighted its therapeutic potential, especially in enhancing cancer treatments to combat drug resistance. In this study, intracellular GSH triggered the release of SO_2 by removing the DNBS group through nucleophilic substitution. Additionally, the active PS **48-**OH was liberated, facilitating the generation of cytotoxic singlet oxygen ($^1O_2$) under red light. This PS significantly reduced cancer cell viability and inhibited tumor growth, showcasing its potent anticancer properties. The structure of **48** comprises two functional components: (1) DNBS, which acts as a GSH recognition site, an efficient quencher of the excited state, and an SO_2 producer; and (2) hemicyanine

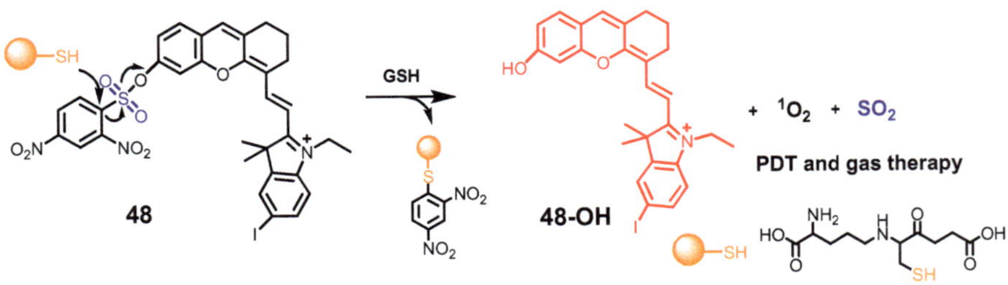

Fig. 2.19 Schematic illustration of **48** in combined PDT and SO_2 gas therapy

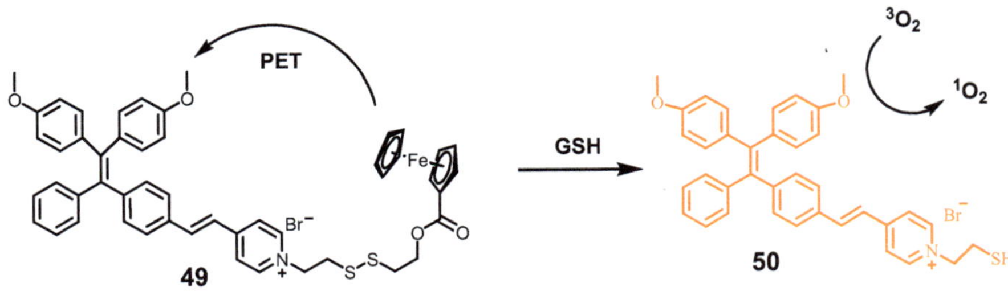

Fig. 2.20 Structure of **49** and the proposed mechanism of GSH activation of PDT

with a heavy iodine atom, serving as the core PS for generating cytotoxic reactive oxygen species (ROS) upon light exposure. In healthy tissues, **48** remains inactive due to the suppression of the intramolecular charge transfer (ICT) process, with the excited state energy dissipating through non-radiative decay. However, when taken up by tumor cells, intracellular GSH selectively removes the DNBS group via nucleophilic substitution, leading to SO_2 production. Concurrently, the active PS **48**-OH is released, producing cytotoxic $^1O_2$ and enabling red fluorescence during imaging-guided PDT. This study presents a new strategy for designing PSs that integrate PDT and SO_2 therapy, demonstrating their potential in synergistic anticancer treatments.

A disulfide bond, a thiol-responsive entity, has been extensively utilized as a connector in the development of advanced therapeutic prodrugs and reduction-sensitive drug delivery mechanisms (An et al. 2020; Ma et al. 2023; Xiao et al. 2021). Unlike traditional photosensitizers (PSs) that often suffer from aggregation-caused quenching (ACQ), the AIEgens created by Tang's team exhibit superior features, including a significant Stokes shift, intense luminescence, and robust photostability. This innovation paves the way for the creation of highly efficient PSs (Liu et al. 2021b, 2023a, b; Ouyang et al. 2022). Leveraging these benefits, Kim et al. engineered a GSH-activatable PS using AIE technology. In their design, a ferrocene component was linked to vinylpyridinium-substituted tetraphenylethylene (**49**) via a disulfide bond (Zhang et al. 2020). As depicted in Fig. 2.20, the photoinduced electron transfer (PET) between the dye and the ferrocene segments in compound **49** leads to the suppression of excited states, inhibiting fluorescence and $^1O_2$ production (Φ_Δ =

2.4%). Upon GSH-induced cleavage of the disulfide bond, the active PDT compound **50** is released, resulting in increased fluorescence and $^1O_2$ generation (Φ_Δ = 21.3%). Furthermore, following light exposure, compound **50** enables targeted imaging and apoptosis in cancer cells, a phenomenon attributed to the elevated GSH levels in these cells. This study introduces a cutting-edge intelligent PS that integrates GSH activation and AIE properties for image-guided photodynamic therapy.

A considerable body of research has demonstrated that the integration of chemotherapy and phototherapy, as a dual-mode treatment, exhibits superior anticancer efficacy compared to single-modality therapies, often leading to synergistic outcomes (Bian et al. 2021, 2022; Luo et al. 2017; Seung et al. 2022; Xie et al. 2020). However, the straightforward combination of traditional chemotherapeutic agents and continuously active photosensitizers frequently results in significant adverse effects on healthy tissues due to their nonspecific targeting. To address this issue, Sun and colleagues developed a "pro-drug-photosensitizer" compound (**51**) where the cytotoxicity of the drug (CPT) and the photodynamic activity of methylene blue (MB) are both inhibited by a disulfide linker that is activated by glutathione (GSH) (Fig. 2.21) (Yang et al. 2022). Once internalized by cancer cells, the elevated GSH levels within the cells break the S–S bond, thereby releasing CPT and MB, and concurrently activating their toxic and photosensitive properties. At a minimal laser power density (660 nm, 1 mW cm^{-2}), compound **51** induced more pronounced apoptosis and displayed greater cytotoxicity against cancer cells than either therapy alone, indicating a synergistic chemo-photodynamic effect. Furthermore, in a 4T1 tumor-bearing mouse model, compound **51** showed exceptional tumor-specific activation with minimal side effects. This study proposes an innovative approach to

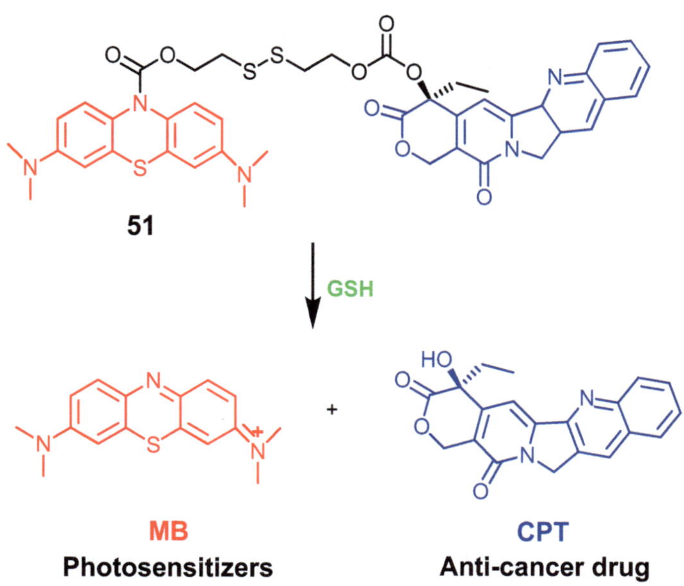

Fig. 2.21 Release mechanism of MB and CPT caused by activable **51** and GSH

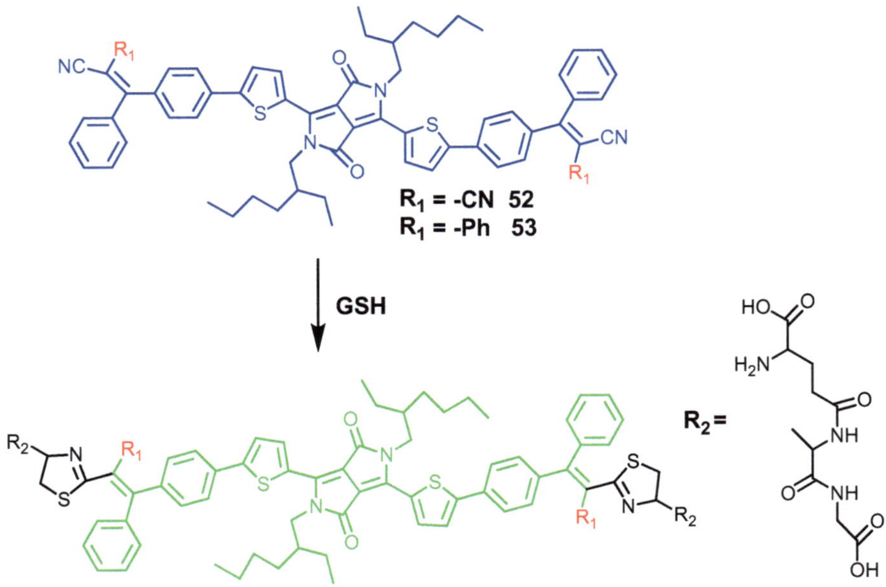

Fig. 2.22 Illustration showing the application of GSH-responsive DPP derivatives (**52**, **53**) in imaging-guided phototherapy

enhance the precision of targeted photodynamic therapy while reducing collateral damage.

The CN group can interact with the thiol group in GSH to create a thiazole ring through a Michael addition reaction. Dong and colleagues leveraged this reaction to design and synthesize two types of GSH-responsive pyrrolopyrrolidone (DPP) derivatives, each with different regio-isomers of the CN group: **52**, which contains four CN groups, and **53**, which contains two CN groups. These derivatives were then formulated into nanoparticles, **52** NPs and **53** NPs, using the nanoprecipitation method (Fig. 2.22) (Zou et al. 2019). Both types of nanoparticles can react with GSH to form a thiazole via the Michael addition reaction, enabling colorimetric detection of GSH without the issue of aggregation-induced fluorescence quenching, and improving PDT/PTT effectiveness. Notably, **52** NPs exhibit a higher $^1O_2$ quantum yield (22.3% compared to 12.5%) and photothermal conversion efficiency (45.2% compared to 34.5%) than **53** NPs. Consequently, **52** NPs, even at a low dose (0.2 mg kg^{-1}), demonstrate superior therapeutic efficacy without causing any adverse effects, in contrast to **53** NPs.

To enhance the effective accumulation and absorption of photosensitizers (PSs) by tumor cells, it is crucial to transition from a hydrophobic, inactive state to a hydrophilic, active state. Kim et al. developed a GSH-activatable photodynamic therapy (PDT) agent (**54**), which incorporates a GSH-responsive component (a ubiquinone derivative) covalently attached to a near-infrared PS (a meso-ester-2,6-iodinated-BODIPY) via an ester bond (Fig. 2.23) (Hwang et al. 2021). This compound, **54**, is highly hydrophobic and readily forms nanoparticles in aqueous

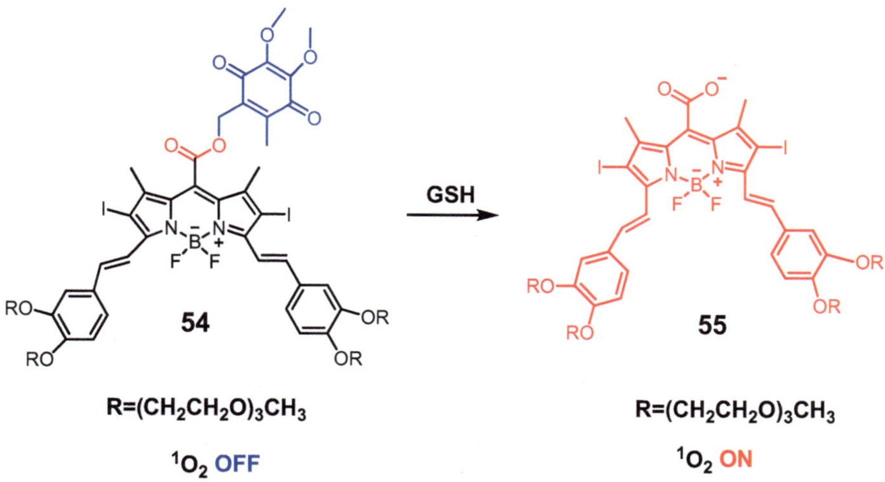

Fig. 2.23 Chemical structure of **54** conjugate and its conversion into **55** under reductive conditions

environments. This not only diminishes its photosensitivity and reduces systemic toxicity but also promotes tumor accumulation through the enhanced permeability and retention (EPR) effect. Once internalized by cancer cells, the ubiquinone part is quickly reduced by GSH, leading to the formation of ubiquinol, which then undergoes a spontaneous elimination reaction to release the anionic **55** and disassemble the **54** nanoparticles. The transformation of the aggregated, hydrophobic precursor (**54**) into the photoactive, hydrophilic **55** triggers fluorescence and singlet oxygen ($^1O_2$) generation (**55**, $\Phi_\Delta = 0.79$), enabling targeted photodynamic destruction of cancer cells and tumor eradication guided by fluorescence imaging.

Like other biothiols, Cys is essential for various physiological functions, such as maintaining redox balance and acting as an antioxidant. Increased Cys levels are linked to several diseases, including cancer and neurodegenerative conditions. The use of specific fluorescence probes to visualize intracellular Cys dynamics has gained significant interest. However, Cys-activated photosensitizers (PSs) are scarce, with only two reported instances. Kolemen et al. modified the dicyanomethylene-4H-chromene (DCM) core by incorporating a heavy iodine atom at different positions, creating two PSs (**56a**, **56b**) (Fig. 2.24) (Kilic et al. 2022). Their findings indicated that the presence of an iodine atom alone was insufficient for high $^1O_2$ generation, and its position was crucial. Specifically, **56a**, with an iodine atom on the phenolate ring, showed a much higher $^1O_2$ quantum yield (Φ_Δ = 5.2%) compared to **56b** (Φ_Δ = 0.6%), where the iodine atom was in the chromone ring. Using this insight, they created a Cys-activatable PS (**56c**) by masking the phenol of **56a** with a Cys-responsive unit (acrylic ester). This probe could be selectively activated by endogenous Cys, leading to substantial photocytotoxicity in HeLa cells (high intracellular Cys) but not in L920 cells (low intracellular Cys), along with a fluorescence "Turn-On." In the same year, Kolemen et al. also developed a Cys-responsive pro-PS based on chlorinated hemicyanine dye (Savani et al.

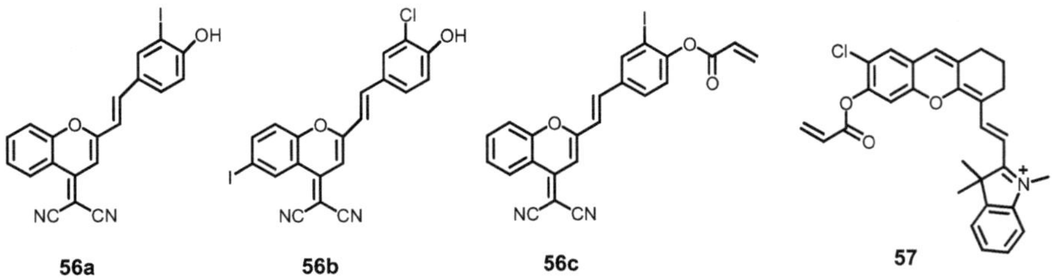

Fig. 2.24 Structures of theranostic probes **56a, 56b, 56c**, and **57**

2022). The theranostic probe **57** exhibited a significant "Turn-On" NIR fluorescence, enhanced $^1O_2$ generation (Φ_Δ = 1.8%), and improved photothermal conversion (η = 69%) after Cys activation, resulting in significant cancer cell death without harming normal cells. Overall, this research group introduced two key cysteine-activated PSs, opening a new pathway for Cys-triggered cancer-specific treatments.

2.3.3 ROS/RNS/H$_2$S-Activatable Theranostic Probes in PDT

Reactive oxygen and nitrogen species (ROS/RNS) are metabolic byproducts that naturally occur in living organisms (Cheung and Vousden 2022; Ishikawa et al. 2008; Pacher et al. 2007). These reactive molecules are essential for cellular signaling and maintaining the balance of intracellular redox states (Fang 2004; Finkel 2011; Willems et al. 2015). Given the distinct metabolic activities of cancerous and normal cells, higher concentrations of ROS/RNS are often detected in cancer cells, indicating a strong correlation between these species and the onset and progression of cancer (Kumar et al. 2014; Chen et al. 2023; Hagen et al. 2012; Li et al. 2022a, b; Zhu et al. 2014). Additionally, mounting evidence suggests that the excessive production of ROS/RNS in cancer cells can induce multiple forms of cell death (Lu et al. 2023). Consequently, investigating ROS/RNS-sensitive photosensitizers is a crucial avenue for advancing novel cancer-targeted phototheranostic strategies.

The development of stimulus-responsive photosensitizers (PSs) that employ an oxidative activation approach has attracted growing attention from the scientific community (Lu et al. 2023; Gu et al. 2022; Nguyen et al. 2022). Specifically, the higher chemical stability and significantly elevated levels of H_2O_2 (ranging from 5 μM to 1.0 mM) in cancer cells compared to normal cells (Hagen et al. 2012) have led to the creation of numerous H_2O_2-activatable phototheranostics. These are based on various PS types and exhibit a wide range of responsive mechanisms. For instance, Takahashi and Toshima et al. investigated a hypocrellin B-derived pro-photosensitizer with arylboronic ester (**58**, Fig. 2.25), designed for the targeted photodestruction of cancer cells with excessive H_2O_2 production (Kitamura et al. 2022). Notably, when the two hydroxy groups of hypocrellin B are modified with H_2O_2-cleavable phenylboronic esters, the resulting caged hypocrellin B primarily absorbs light in the UV–vis spectrum, with its photosensitizing capability significantly

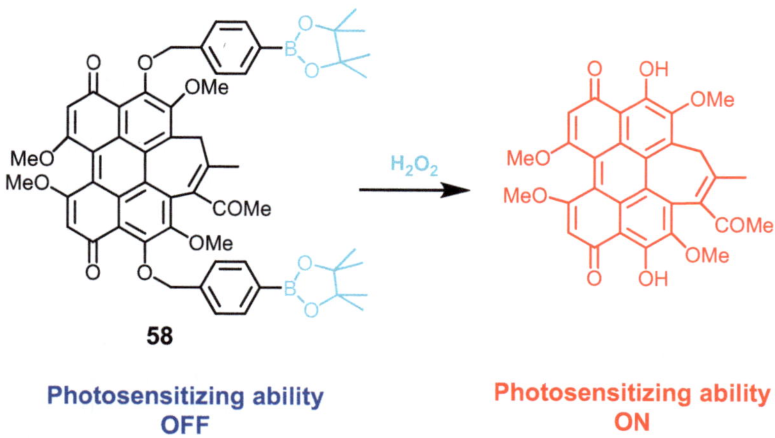

Fig. 2.25 Activation of **58** by H_2O_2

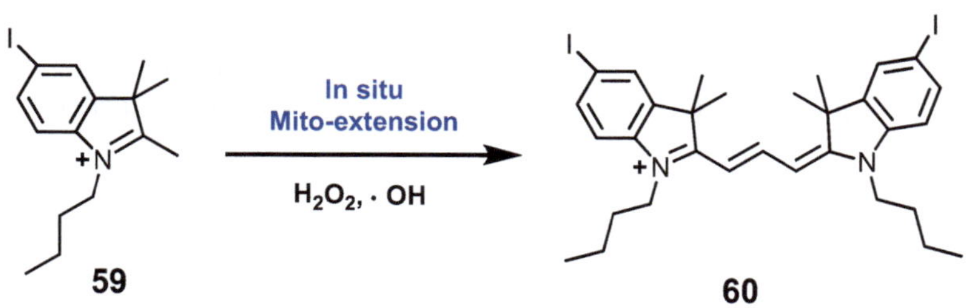

Fig. 2.26 In situ generation of **60** for PDT via a reaction with mitochondrial ROS

diminished. Upon specific cleavage of the phenylboronic esters by H_2O_2, the PS exhibits a red-shifted absorption peak within the therapeutic window (600–900 nm), and its ability to generate singlet oxygen is restored. Moreover, protein-degradation tests under light exposure revealed that hypocrellin B causes more extensive breakdown of BSA than its caged counterpart, highlighting the adjustable nature of its photosensitizing properties. Crucially, photocytotoxicity tests indicated that the caged PS demonstrates markedly higher toxicity toward H_2O_2-overproducing cancer cells (B16F10) compared to normal WI-38 cells under 660 nm light irradiation. This outcome underscores the potential of designing novel H_2O_2-activatable phototheranostics to minimize the adverse effects of cancer-specific photodynamic therapy.

Mitochondria are crucial for regulating cellular energy and metabolic processes. Novel photodynamic therapy (PDT) agents that can disrupt mitochondrial function are of significant interest in cancer treatment (Guo et al. 2021). Recently, Feng and Wang et al. discovered that iodinate indolium (**59**), a common precursor for cyanine dyes, can transform into an iodinated trimethine cyanine dye (**60**, Fig. 2.26) when it interacts with reactive oxygen species (ROS) like H_2O_2 and •OH (Heng et al. 2022). Given the elevated ROS levels in the tumor microenvironment (TME) and the ability of cationic **60** to concentrate in mitochondria, intracellular ROS can trigger the in situ formation of a cyanine-based imaging agent and photosensitizer within living cancer cells. Additionally, the presence of heavy atoms in **60** enhances its photosensitizing properties, leading to substantial production of singlet oxygen ($^1O_2$) and notable photodamage to HeLa cells under light exposure. This approach of synthesizing ROS-responsive phototheranostic agents in situ presents a promising strategy for targeted cyanine expression and PDT in tumors.

Hypochlorite and hypochlorous acids (ClO^-/HClO), generated by the intracellular enzyme myeloperoxidase, are vital in biological systems and are linked to various pathological conditions, including bacterial infections, cardiovascular and neurodegenerative disorders, arthritis, and cancer (Aratani 2018). Elevated levels of reactive oxygen species (ROS), such as ClO^-/HClO, can drive oncogenic transformation and enhance the proliferation of cancer cells (Cheung and Vousden 2022). The baseline concentrations of ClO^-/HClO in cancer cells are approximately 10 nM (Zhu et al. 2014), suggesting that targeting these cells with ClO^-/HClO-activatable phototheranostic agents is a viable approach. Wang and Zhang et al. created single-component nanodots that can localize in lysosomes, enabling the visualization and photodynamic therapy (PDT) of cervical tumors (Liu et al. 2022). These Pluronic F127-encapsulated nanodots, which include an aggregation-induced emission (AIE)-active phenothiazine derivative (**61**, Fig. 2.27), can detect increased ClO^-/

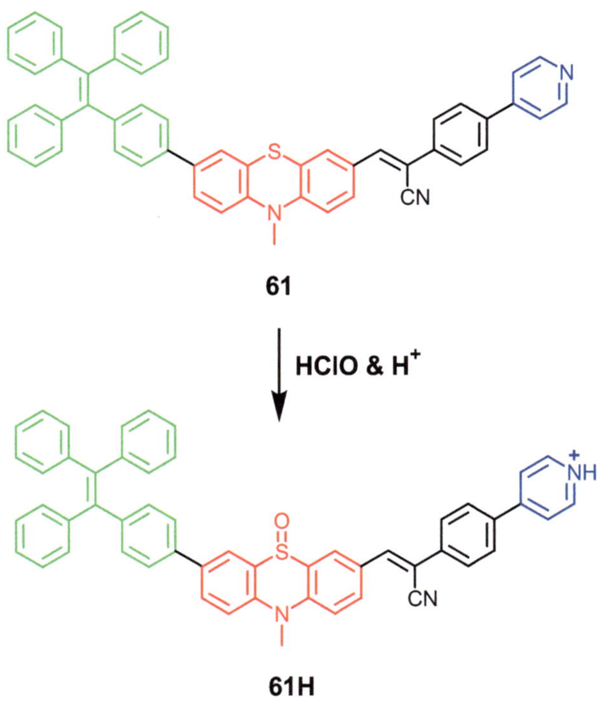

Fig. 2.27 Schematic illustration of HClO-activatable photoactivity of **61**

HClO levels in tumor tissues, leading to a nearly twofold increase in singlet oxygen ($^1O_2$) production compared to commercial Rose Bengal and significant photocytotoxicity (IC_{50} < 7.6 μM). Additionally, bioimaging studies indicate that these nanodots can track ClO^-/HClO fluctuations in human cervical tissues. This work marks a breakthrough in the development of ClO^-/HClO-responsive nanodrugs for PDT treatment of inflammation-related tumors, with minimal adverse effects.

To date, a variety of ClO^-/HClO-responsive fluorescent probes have been created and utilized in both in vivo and in vitro imaging research. However, the development of ClO^-/HClO-responsive photosensitizers (PSs) that exhibit both type I and type II photosensitizing mechanisms remains limited. Recently, Tang and Han et al. introduced a mitochondria-targeted, bioimaging-compatible photodynamic therapy (PDT) agent based on an AIE PS (**62**, Fig. 2.28), which is specifically activated by intratumoral hypochlorite (ClO^-) (Huang et al. 2023). Interestingly, before interacting with ClO^-, PS **62** exhibited type II photosensitizing characteristics, primarily producing singlet oxygen ($^1O_2$) when irradiated. After reacting with ClO^- and transforming into its oxidized form, PS **62** displayed intense fluorescence and generated type I radicals. Further investigations revealed that PS **62** could distinguish between cancerous and healthy cells, making it a promising candidate for tumor-specific phototheranostic applications.

An increased level of nitric oxide (NO) in cancerous tissues is primarily due to the enzymatic conversion of intracellular Guanidine-containing substrates by endogenous inducible nitric oxide synthase (iNOS) (Chen et al. 2023b; Li et al. 2022d) This process has been harnessed for activating anticancer phototherapy (Teng et al. 2019). NO is a key player in various inflammation-related diseases (Chen et al. 2023a; Li et al. 2022c; Greten and Grivennikov 2019). Building on these insights, Fan et al. created the first NO-activatable, two-photon-excitable fluorogenic pro-photosensitizer (**63**, Fig. 2.29) for bioimaging and photodynamic therapy (PDT). (Hu et al. 2018b). When **63** interacts with NO in cancer cells, it significantly enhances the emission quantum yield (by over 50 times), the photoinduced singlet oxygen generation quantum yield (up to 82%), and the two-photon absorption cross-section (up to 2800 GM). This discovery demonstrates the feasibility of using NO-triggered two-photon fluorescence imaging and PDT in lipopolysaccharide and interferon-γ-coactivated macrophages. Given that inflammation can

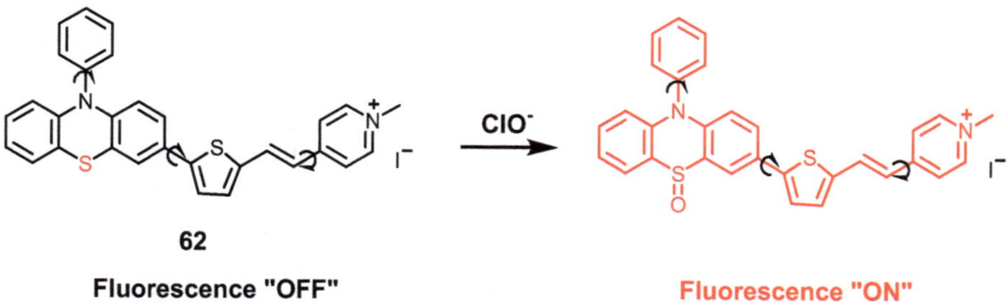

Fig. 2.28 Chemical structure of **62** and its activation by ClO^-

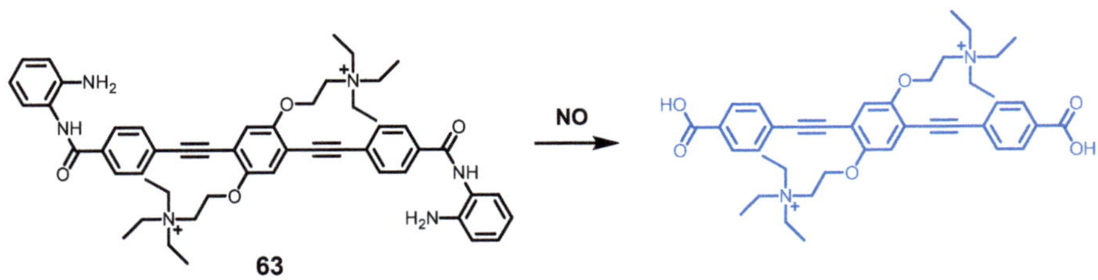

Fig. 2.29 Molecular structure and activation of **63** by NO

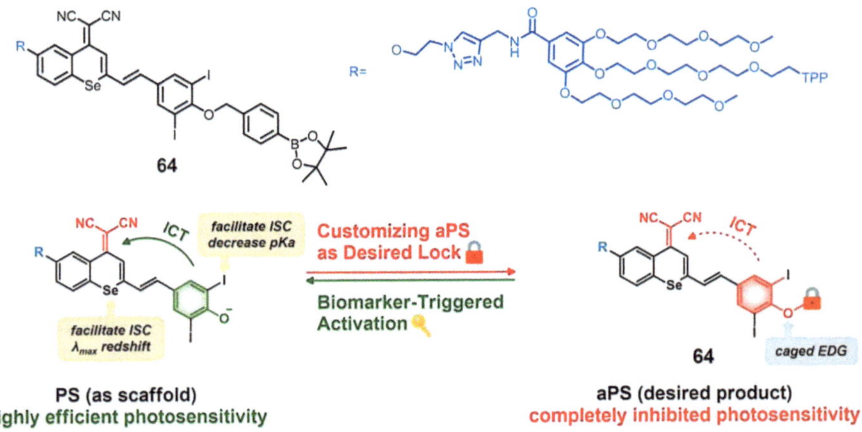

Fig. 2.30 Chemical structure of **64** and schematic illustration of the proposed PS activation strategy

predispose tissues to cancer and promote tumorigenesis, the potential of NO-activatable cancer phototheranostics remains underexplored. We expect that more sophisticated NO-activatable photosensitizers will be developed for treating inflammatory cancers in the future.

Peroxynitrite ($OONO^-$) is a highly reactive nitrogen species formed by the interaction of superoxide ($\bullet O_2^-$) and nitric oxide (NO). This molecule acts as a signaling agent with various functions in cellular activities. $OONO^-$ is also recognized as a key indicator for numerous cancer types. In a study, Li et al. proposed a novel concept that utilizes the photosensitizing inhibition mechanism to create pro-photosensitizers (pro-PS) that can regain their photosensitivity after interacting with this specific biomarker. They designed mitochondria-targeted, long-wavelength-excitable pro-PS (**64**, Fig. 2.30) capable of generating singlet oxygen ($^1O_2$) with high efficiency (Zhai et al. 2019). The phenol hydroxy group was protected using an $OONO^-$-cleavable functional group, providing a straightforward approach to developing intelligent photosensitizers. The enhancement of the photosensitizing capability of **64** was clearly shown when $OONO^-$ levels were increased in live cancer cells. To enhance the solubility of the photosensitizer, multiple polyethylene glycol

(PEG) chains with a mitochondria-targeting component were added. This resulted in excellent bioimaging performance and remarkable photodynamic therapy (PDT) efficacy in tumor-related RAW 264.7 cells.

2.3.4 Enzyme-Activatable Theranostic Fluorescent Probes in PDT

Enzymes play a pivotal role in numerous metabolic activities, and their heightened levels are thought to promote various disease mechanisms, including tumor growth, cellular infiltration, and metastasis. Scientists have identified a multitude of enzymes that are overproduced in diverse cancer types. Collaborating with biologists, researchers have pinpointed the biological substrates of these enzymes, which can be utilized to develop intelligent phototherapeutic compounds. These compounds can be activated by specific enzymes, enabling them to differentiate between normal and abnormal cells, thereby minimizing collateral damage to the skin and adjacent healthy tissues (Li et al. 2022d; Pham et al. 2021; Lu et al. 2022).

In contrast to other tumor indicators like acidic pH, redox states, and increased ROS levels, enzyme-catalyzed substrate reactions are generally rapid, gentle, and highly specific. These characteristics provide enzyme-activated photosensitizers (PSs) with exceptional sensitivity and a low rate of false positives. Consequently, the use of cancer-related enzymes to create activatable PSs has garnered significant interest. This section delves into several key enzyme-activatable "smart" PSs, such as those activated by nitroreductase (NTR), azoreductase, cathepsin B, β-galactosidase, alkaline phosphatases, tyrosinase, γ-glutamyl transpeptidase, and aminopeptidase.

Nitroreductase, an enzyme that is highly expressed in hypoxic solid tumors due to the reductive stress from low oxygen levels within cells, can efficiently convert nitroaromatics into arylamines when NADH, an electron donor, is present. This enzymatic reduction has been leveraged to develop numerous theranostic probes that are activated under hypoxic conditions, by incorporating nitroaromatic moieties into dye structures (Li et al. 2022c, d; Zhao et al. 2020a, b).

Due to its robust hydrogen-bonding and spatial fit with NTR, the p-nitrobenzoate group is considered an optimal NTR substrate, offering rapid response, high sensitivity, and excellent selectivity. By incorporating this functional group into various photosensitizer (PS) frameworks, Peng's team has engineered several NTR-activated phototheranostic agents based on PET and ICT mechanisms. For example, they developed a hypoxia-activated D-π-A PS (**65**-N) by attaching 4-nitrobenzyl bromide to the phenol hydroxyl group of an iodine-substituted hemicyanine dye (NIR PS) (Ahn et al. 2018; Xu et al. 2019). As depicted in Fig. 2.31, the addition of p-nitrobenzoate inhibits the ICT process, resulting in weak fluorescence and minimal $^1O_2$ generation. However, in hypoxic cancer cells and tumor tissues, **65**-N is converted to **65**-OH by overexpressed NTR, significantly enhancing

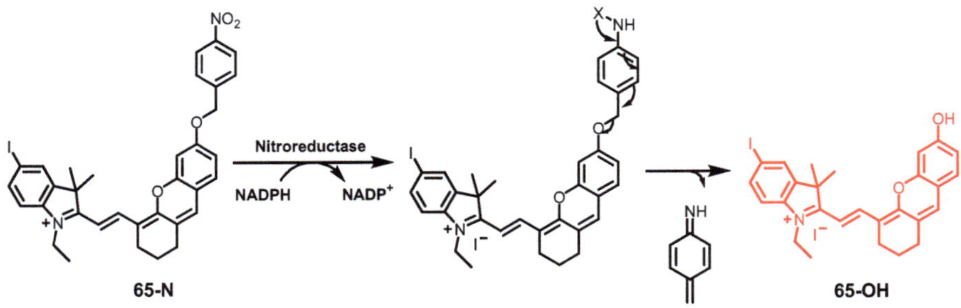

Fig. 2.31 NTR-recognition mechanism of **65**-N

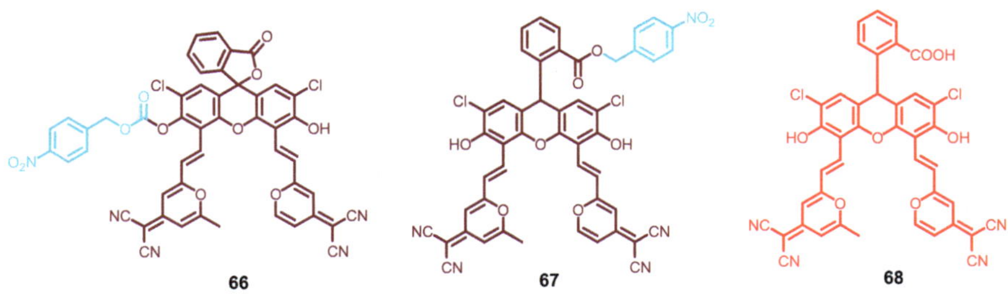

Fig. 2.32 Molecular structures of **66**–**68**

photosensitivity as ICT is restored after the caging group is removed. In normal tissues, **65**-N remains nonfluorescent and non-phototoxic. The induction of cell apoptosis and inhibition of tumor growth under 660 nm light irradiation in hypoxic conditions underscore the potential of PS for selective tumor hypoxia imaging and PDT. Additionally, two other NTR-activatable theranostic molecules (**66** and **67**), derived from TADF fluorescein derivatives, have demonstrated effective PDT in mildly hypoxic tumors (Fig. 2.32) (Liu et al. 2019a, b). Screening revealed that compound **67** outperformed compound **66** in both selectivity and NTR response. Notably, endogenous NTR in tumor cells can cleave **67** to produce **68**, achieving high PDT efficiency even at 10% oxygen levels.

Azoreductase is a well-known enzyme that responds to low oxygen conditions and is often found in higher concentrations within hypoxic cells, especially in solid tumors. This enzyme's ability to convert azo groups into amino derivatives makes it a distinctive indicator of tumor hypoxia. By utilizing the azo group as a protective element, it can shield photosensitizers from light activation, thereby creating "switch-on" photodynamic therapy (PDT) agents specifically designed to target and treat hypoxic tumors (Hao et al. 2022; Piao et al. 2017; Tang et al. 2023).

Conventional photosensitizers (PSs) suffer from limited light penetration, inadequate hypoxic sensitivity, and subpar tumor targeting, which collectively diminish the effectiveness of photodynamic therapy (PDT) (Fan et al. 2016; Li et al. 2018c). To address these challenges, Kim and colleagues engineered a PS that is responsive to hypoxia, excitable by two-photon absorption, and modified with a targeting

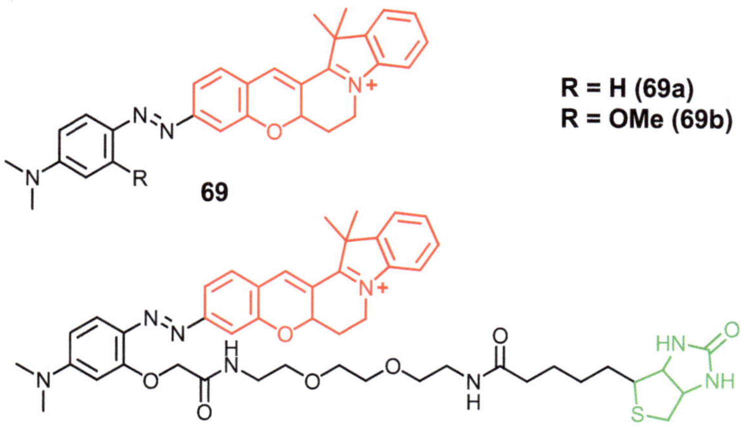

Fig. 2.33 Structures of theranostic probes **69** and **70**

moiety. Leveraging the deep tissue penetration and enhanced spatial precision of two-photon excitation, they chose a suitable type I PS. This led to the development of a series of cyclized-cyanine derivatives (**69**), as illustrated in Fig. 2.33. These compounds exhibit remarkable fluorescence (Φ_F up to 60%) and outstanding two-photon cross-sections (up to 103 GM) (Juvekar et al. 2021). Importantly, these small molecules, free of heavy atoms, generate reactive oxygen species (ROS) more effectively than 5-ALA, owing to their type I mechanism. Additionally, two-photon excitation was found to be more efficient than one-photon excitation. Azo groups were attached to the dye structure, resulting in two caged PSs (**69a** and **69b**) that can be activated under hypoxic conditions. Once internalized by cancer cells (which overexpress azoreductase), these PSs demonstrate rapid activation and potent cytotoxicity with minimal dark toxicity. In a recent advancement, the researchers coupled the PS with a biotin targeting unit, creating a highly selective two-photon PDT agent for colon cancer treatment (**70**, Fig. 2.33).

In oxygen-deprived tumor settings, the lack of adequate oxygen often hinders the effectiveness of photodynamic therapy (PDT), particularly in the core areas of solid tumors. To address this challenge, Liu and Kim et al. developed a theranostic agent, **71**-Azo, designed for hypoxia-sensitive chemotherapy and phototherapy in solid tumors (Fig. 2.34) (Xiong et al. 2022a). This compound was created by connecting a rhodol-based near-infrared (NIR) fluorophore (**71**) to nitrogen mustard via an azo bond (−N=N−). The fluorescence of **71**-Azo is suppressed by intramolecular π–π stacking and the interruption of the internal charge transfer (ICT) process. Unlike other activatable photosensitizers, **71**-Azo demonstrates robust reactive oxygen species (ROS) generation ($\Phi_\Delta = 0.14$), while its active form, **71**, does not. This unique photosensitivity, whether under hypoxic or normoxic conditions, makes **71**-Azo versatile for various anticancer treatments. On the outer layers of the tumor, **71**-Azo shows significant PDT efficacy, effectively eliminating cancer cells with low

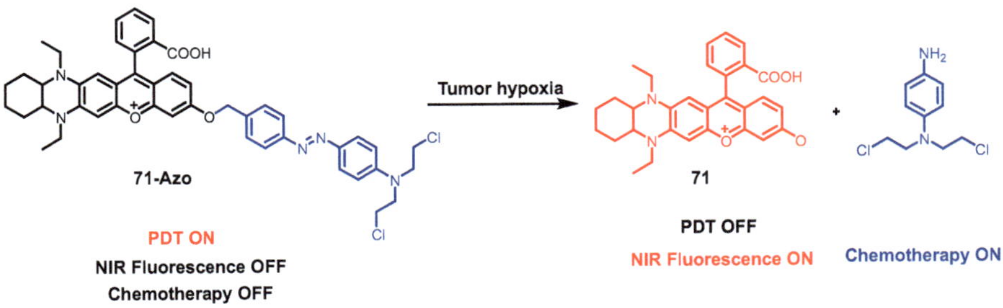

Fig. 2.34 Design and chemical structure of theranostic construct **71-Azo**

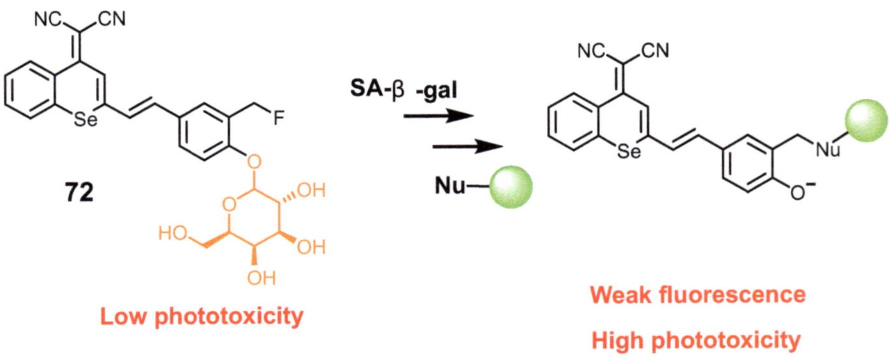

Fig. 2.35 Integrated strategy used in the design of senotherapeutic probe **72**

azoreductase levels, likely because the enzyme cannot cleave the azo bond. In deeper, hypoxic regions, the reduction of the azo bond releases **71** and active nitrogen mustard, allowing for the targeted destruction of hypoxic tumor cells through chemotherapy. This mechanism also enables real-time and on-site monitoring of drug release and kinetics. This approach effectively targets and destroys cancer cells in both well-oxygenated and hypoxic environments, optimizing therapeutic outcomes.

β-Galactosidase (β-gal) is a crucial glycoside hydrolase found within cellular lysosomes. Its primary role involves the degradation of glycosidic linkages, enabling the transformation of lactose into galactose (Xiong et al. 2022b; Zhen et al. 2018). Although it is often overexpressed in primary ovarian cancer and gliomas, β-gal is not typically recognized as a cancer biomarker. Nevertheless, recent research has highlighted its importance as a key indicator of cellular senescence, leading to extensive efforts to develop techniques for the selective removal of senescent cells.

For instance, Li and colleagues engineered a photosensitive senolytic prodrug (**72**) that can be selectively activated in the presence of SA-β-gal (Fig. 2.35) (Shi et al. 2023). To boost intersystem crossing and enhance phototherapy effectiveness, the researchers substituted a selenium atom for an oxygen atom in the dicyanomethylene-4H-pyran framework. The introduction of *E. coli* β-gal and BSA increased the Φ_Δ value of **72** from 0.07 to 0.20 and reduced its relative fluorescence quantum yields from 0.33 to 0.08. This alteration enabled **72** to demonstrate high efficacy and broad-spectrum activity against senescent cells, with minimal impact on nonirradiated regions. In both doxorubicin-induced senescence and naturally aged mouse models, **72**-mediated PDT effectively targeted and eliminated senescent cells in their tissues. This study offers a novel approach to the surveillance and selective elimination of senescent cells during the aging process.

Alkaline phosphatase (ALP) is an extracellular enzyme anchored to the plasma membrane, playing a crucial role in various cellular activities by catalyzing dephosphorylation. These activities encompass the modulation of protein phosphorylation, cell proliferation, programmed cell death, and cellular movement. Increased ALP activity has been associated with several diseases, including breast, prostate, and kidney cancers, as well as osteosarcoma. Consequently, ALP is recognized as a key biomarker for clinical diagnostics (Ji et al. 2018; Li et al. 2020e). While numerous fluorescent probes have been created to selectively visualize ALP-positive cancer cells, the development of activable photosensitizers that specifically respond to ALP remains relatively scarce.

Li and colleagues engineered a novel D-π-A photosensitizer framework (designated as **73**Se-I) by integrating selenium-substituted dicyanomethylene-4H-chromene as the electron acceptor and 2,6-diiodo phenolate as the electron donor (Fig. 2.36) (Zhai et al. 2019). The incorporation of heavy selenium atoms in **73**Se-I resulted in a notable shift toward longer wavelengths and a substantial increase in the generation of singlet oxygen, which is advantageous for effective photodynamic therapy (PDT) in deep-seated tumors. Additionally, the presence of two ortho-iodo groups not only boosted the photosensitivity but also lowered the pKa of the phenolic moiety, making it suitable for PDT at physiological pH (pH 7.4). Furthermore, by attaching a phosphate group (acting as both a protective group and an ALP recognition site) to the phenolic hydroxyl of **73**Se-I, they created an ALP-activatable photosensitizer (ALP**73**). This modification completely deactivated the photosensitivity due to the inhibition of the intramolecular charge transfer (ICT) process, as the phenolate group lost its electron-donating capability after phosphorylation. When exposed to ALP or when incubated with HeLa cells that overproduce ALP,

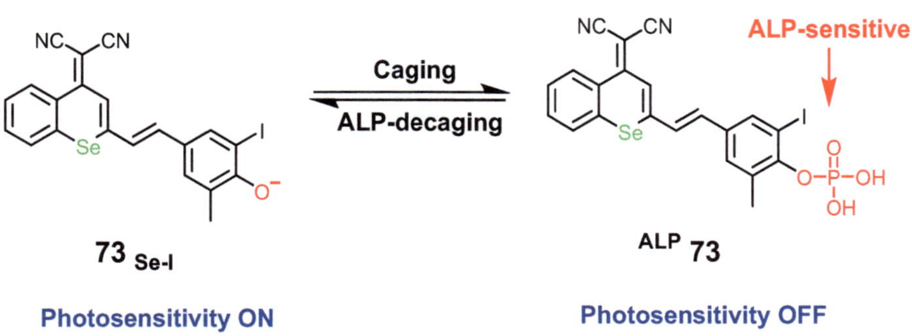

Fig. 2.36 Schematic illustration of the design of ALP-activatable PS (^{ALP}**73**)

ALP**73** rapidly transformed back into **73**Se-I, restoring both fluorescence and photosensitivity. Live/dead staining and MTT assays confirmed that ALP**73** significantly reduced the viability of HeLa cells. This study presents a robust approach for developing ALP-activated photosensitizers through the caging and uncaging of a phenol group.

Melanoma is a highly aggressive cancer characterized by its swift progression and early spread. Tyrosinase (TYR), a copper-containing enzyme, plays a crucial role in the production of melanin. This enzyme catalyzes the conversion of monophenol or catechol into o-quinone and is the key regulatory factor in the melanin synthesis pathway from tyrosine. In melanoma, TYR is significantly overexpressed, making it a vital indicator for both the detection and therapeutic targeting of this cancer (Li et al. 2018a).

For the fluorescence imaging and photodynamic therapy (PDT) of early-stage melanoma, Yoon and colleagues engineered a unique tyrosinase-activated cyclometalated Pt(II) complex (**74**-tyro). This compound integrates a 3-hydroxybenzyloxy group, which serves as the tyrosinase recognition site, with the phenyl pendant of a C^N ligand in a cyclometalated Pt(II) complex, acting as the photosensitizer core (Liu et al. 2023a, b). As illustrated in Fig. 2.37, upon interaction with tyrosinase, **74**-tyro undergoes a rapid rearrangement and elimination, transforming into **74**-OH, which triggers a significant fluorescence emission at 530 nm. Additionally, **74**-OH, a product of tyrosinase metabolism, exhibits a markedly enhanced photosensitizing capability compared to **74**-tyro, attributed to its more prolonged triplet excited state. Thus, **74**-tyro can effectively visualize endogenous tyrosinase in A375 cells (human melanoma cancer cells) and substantially suppress the proliferation of A375 cells and tumors in Balb/c mice bearing A375 xenografts.

γ-Glutamyl transpeptidase (GGT), a prominent enzyme associated with cell membranes, can specifically break the γ-glutamyl bond in GSH, transforming it into cysteinyl-glycine (Cys-Gly) to regulate the balance of GSH and cysteine within cells. Research has demonstrated that GGT levels are elevated in the cell membranes of several types of cancer, such as glioma, liver, lung, and ovarian cancers. This increased GGT activity is linked to enhanced tumor growth, invasiveness, and resistance to drugs, making GGT a crucial indicator for cancer detection and a

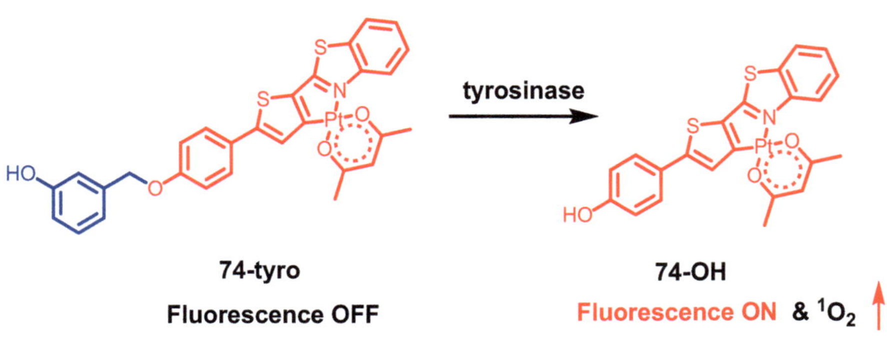

Fig. 2.37 Schematic diagram showing the mechanism of **74-tyro** in tyrosinase imaging and PDT

promising target for therapeutic strategies. L-glutamic acid, when attached via its side chain, has been employed as a GGT activator to conceal a fluorophore and detect the presence of GGT enzymes (Chen et al. 2021; Chiba et al. 2017).

To address the issue of off-target biodistribution in current photodynamic therapy (PDT) systems and to enable tumor-specific fluorescence, Han and Li et al. developed a multifunctional small-molecule probe, designated as **75**. This probe integrates a type I photosensitizer (ENBS) with a fluorescent marker (rhodamine, Rd) through a self-immolative linker, and includes an l-glutamic acid segment as a GGT-responsive trigger (Fig. 2.38). (Li et al. 2021a, b) The fluorescence and reactive oxygen species (ROS) generation of ENBS in **75** remain largely unaffected by

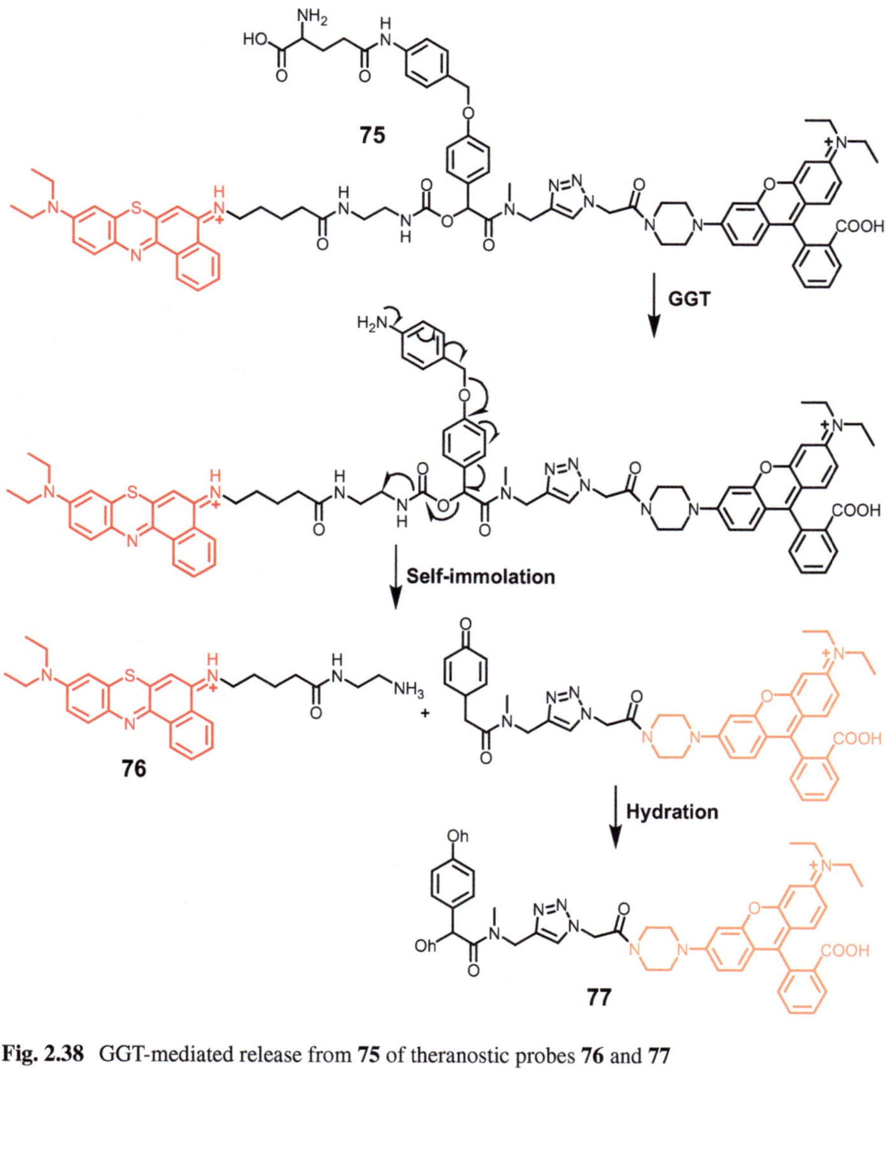

Fig. 2.38 GGT-mediated release from **75** of theranostic probes **76** and **77**

the presence of rhodamine, which is initially in a quenched, nonfluorescent state due to the interaction between ENBS and Rd. When **75** is activated by GGT, it releases both rhodamine and ENBS, thereby activating the fluorescence of rhodamine. The released components accumulate in lysosomes, key organelles for inducing cell death. Under light exposure, **75** significantly damages lysosomes and induces U87 cell (GGT+) death at a rate twice that of L02 cells (GGT−). In BALB/c nude mice with subcutaneous U87 tumors, **75** showed strong rhodamine fluorescence localized specifically within the tumor, rather than in other organs. This selective fluorescence highlights the potential of **75** to differentiate tumors from healthy tissues via a GGT-activated "Turn-On" response. Additionally, the tumor-specific fluorescence can guide precise light delivery for effective PDT. This study demonstrates that a quenched dye-photosensitizer pair can achieve tumor-activatable fluorescence independent of the photosensitizer, enhancing the precision of PDT.

Aminopeptidase N (APN/CD13) is a zinc-dependent type II metalloproteinase that typically binds to the cell membrane and can cleave the N-terminal leucine from a polypeptide. This enzyme is crucial for tumor growth, division, metastasis, and angiogenesis. Research has shown that APN is highly overexpressed on the cell membranes of various tumors, with levels approximately ten times higher than in normal tissues, particularly in colon, ovarian, breast, and liver cancers. Consequently, APN is recognized as a cancer marker and has garnered significant interest as a target for cancer-specific enzymes, facilitating the development of activatable fluorescent probes and theranostic prodrugs. Peng and colleagues pioneered the creation of an APN-activated near-infrared photosensitizer (APN-**78**) by linking an iodine-substituted hemicyanine and L-alanine via 4-aminobenzyl alcohol (Fig. 2.39) (Zhou et al. 2020). In typical cellular conditions, APN-**78** remains in an "OFF" state, with its fluorescence and phototoxicity suppressed due to an inhibited intramolecular charge transfer process. However, in tumor cells where APN is overexpressed, the enzyme specifically cleaves the amide bond between L-alanine and the 4-aminobenzyl alcohol group, followed by a spontaneous 1,6-elimination to generate **78**-OH. The NIR fluorescence and singlet oxygen production of **78**-OH are significantly enhanced compared to APN-**78**. Additionally, **78**-OH selectively accumulates in mitochondria, which are highly sensitive to reactive oxygen species. When exposed to 660 nm light, APN-**78** effectively induces apoptosis in cancer cells by damaging mitochondria and, more importantly, distinguishes between cancerous and normal cells, resulting in selective cancer cell death and a pronounced tumor suppression effect.

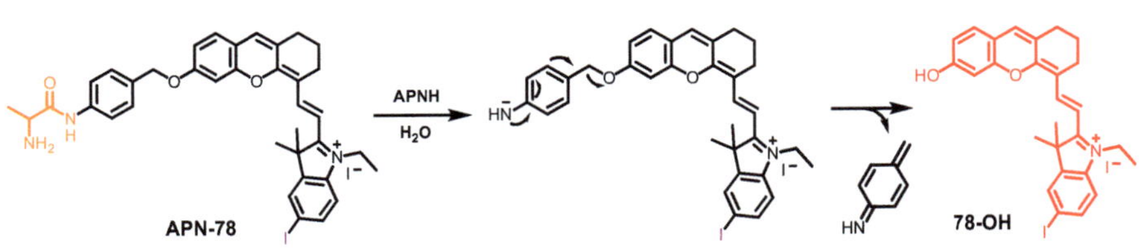

Fig. 2.39 Response mechanism of **APN-78** to APN

2.4 Theranostic Fluorescent Probes for PTT

2.4.1 Tumor Microenvironment (TME)-Activated Theranostic Probes in PTT

PTT employs a gentle yet deeply penetrating infrared light that, in the presence of a photosensitizer (PS), transforms external light into thermal energy, leading to elevated local temperatures. This photo-induced heating within a tumor can eradicate cancer cells without the use of toxic drugs. PTT has attracted considerable research attention because it is noninvasive and biologically safe, and it allows for precise control of heat generation, thereby avoiding drug resistance. The tumor microenvironment (TME) influences tumor survival, growth, and spread, making it a promising target for activatable tumor imaging and treatment. Due to the limited effectiveness of conventional photothermal agents (PTAs), there has been a significant focus on developing TME-activatable PTAs for targeted cancer therapy. These advancements leverage the unique characteristics of the TME, such as acidic pH, low oxygen levels, and specific enzymes found in cancer.

Drawing on a comparable approach, Park and colleagues engineered a pH-sensitive nanoprobe. This probe was composed of carbonized crosslinked poly(ethylene glycol-g-poly(sulfobetaine methacrylate)) and incorporated the photothermal dye IR825 (FNP-I) (Kang et al. 2018). In a related study, Cai and team utilized IR-822 as the photothermal agent, integrating a proton receptor, N1-(pyridine-4-methyl) ethane-1,2-diamine (PY), to create a fluorophore-spacer-receptor molecular probe, designated as **79**, with suppressed fluorescence (Fig. 2.40) (Meng et al. 2017). Upon protonation in the acidic tumor microenvironment, this probe effectively facilitated NIRF/PA dual-modal imaging-guided photothermal therapy.

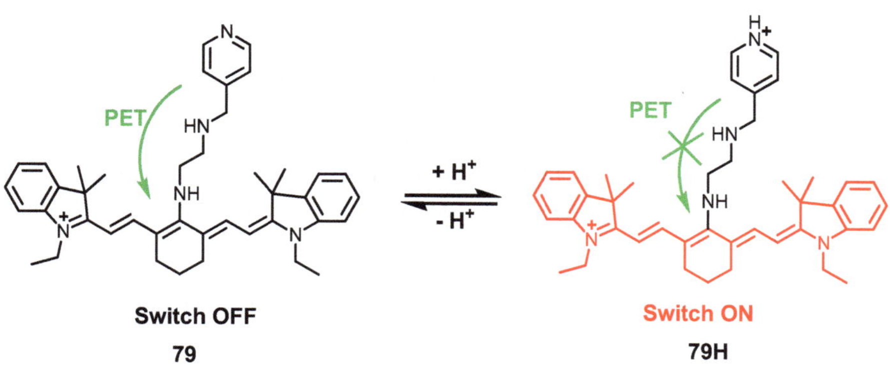

Fig. 2.40 Chemical structures and schematic illustration of **79**, a pH-responsive theranostic probe for NIRF/PA dual-modal imaging-guided PTT therapy

2.4.2 *Hypoxia-Activated Theranostic Fluorescent Probes in PTT*

Hypoxia, a condition characterized by low oxygen levels, is frequently found in the majority of solid tumors and their microenvironments. This phenomenon arises from the rapid and unregulated growth of cancer cells, along with irregular blood vessel formation and inadequate oxygen delivery. In contrast, normal tissues do not exhibit hypoxic areas, making this condition a unique target for selective cancer therapies. Several mechanisms enable tumor cells to adapt to these harsh conditions, thereby increasing their resistance to drugs and enhancing their survival rates (Chen et al. 2023a; Wei et al. 2021). Photothermal therapy (PTT) has emerged as a potential approach for treating hypoxic tumors, primarily due to its independence from oxygen and its precise spatial and temporal control (Li et al. 2020b). PTT involves the use of photothermal agents that generate localized heat when exposed to near-infrared (NIR) light, leading to the effective and irreversible destruction of cancer cells (Nomura et al. 2019; Wang et al. 2020). To address hypoxia in PTT, specific vectors can be employed to deliver photothermal agents or hypoxia-sensitive components can be incorporated (Deng et al. 2021; Li et al. 2021a). These methods are vital for activating the photothermal properties, and the creation of innovative hypoxia-activated photothermal agents is essential for regulating the activation and deactivation of PTT (Lin et al. 2022).

PTT, leveraging the principle of photothermal conversion, holds significant promise for effective cancer treatment, thanks to its high efficiency in tumor ablation and minimal impact on healthy tissues (Yang et al. 2011). Cai and colleagues engineered a novel single-molecule probe, designated as **80**, which is activated by hypoxic conditions and enzymatic reactions with nitroreductase (NTR). This probe integrates a nitroimidazole moiety, which is specifically responsive to hypoxia, and an IR-1048 dye, serving as a near-infrared II (NIR-II) and photoacoustic (PA) signal reporter. This combination facilitates high-contrast visualization of tumors via NIR-II fluorescence imaging and deep-tissue imaging through 3D PA techniques. The theranostic probe **80** demonstrates substantial photothermal effects and NIR-II fluorescence when activated in hypoxic tumor environments, enabling precise and deep tissue imaging of tumor sites and boundaries, thereby supporting effective tumor ablation and preventing recurrence. Consequently, this versatile probe **80** serves as a rapid and sensitive tool for NIR-II fluorescence and photoacoustic imaging in hypoxic conditions, making it an ideal candidate for use as an activated small molecule photosensitizer in PTT (Fig. 2.41) (Meng et al. 2018).

Activatable phototheranostics hold significant potential for the precise targeting of cancer. However, many existing probes in this field are limited to either photodynamic therapy (PDT) or photothermal therapy (PTT), which can be less effective due to factors like hypoxia and the complex tumor microenvironment (TME). To overcome these limitations, Pu et al. created a dual-locked activatable phototheranostic probe, designated as **81**. This probe is capable of emitting near-infrared fluorescence (NIRF) signals within tumors, initiating PDT when it encounters a

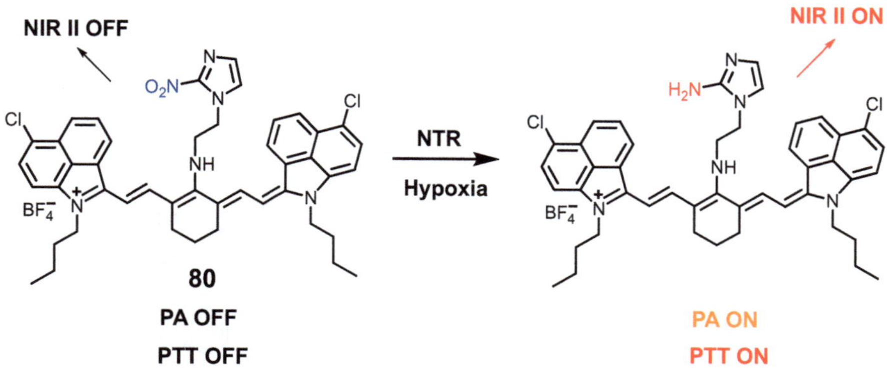

Fig. 2.41 Schematic illustration of NTR-responsive NIR-II fluorescence/PA probe **80**

biomarker at the tumor periphery, and switching to PTT upon detecting a hypoxic biomarker in the tumor's core. By using a single laser source, this self-regulating probe can produce cytotoxic singlet oxygen ($^1O_2$) around the tumor and generate heat that results in complete tumor destruction in the core. This dual-locked probe demonstrates a promising approach in the molecular design of cancer phototheranostics (Fig. 2.42) (Wei et al. 2022).

Additionally, Peng et al. utilized single-molecule Förster resonance energy transfer (smFRET) to design a smart molecule, PS BDP-CR (**82**), by integrating diiododistyrylbodipy (BDP) with croconaine (CR). This probe shows a synergistic "1 + 1 > 2" effect in combined PDT and PTT, enhancing therapeutic outcomes. It also extends the absorption range in the near-infrared (NIR) region (600–850 nm) and enhances light-harvesting capabilities. Under normal conditions, the absorbed photon energy converts oxygen into reactive oxygen species (ROS) for PDT and activates CR (via FRET) for PTT. In severely hypoxic conditions, the photoenergy primarily eradicates cancer cells through oxygen-independent PTT (Fig. 2.43) (Zou et al. 2021).

2.4.3 *Tumor-Specific Enzyme and Biomolecule-Activated Theranostic Probes in PTT*

Enzymes serve as crucial indicators in the creation of diagnostic and therapeutic methods for cancer. One such enzyme, γ-Glutamyl transpeptidase (GGT), is notably elevated in tumor cells. Liu and colleagues introduced a GGT-sensitive near-infrared nanoprobe (NRH-G-NPs) that, when interacting with GGT, transforms into NRH-NH_2-NPs via an amide bond with γ-glutamate (γ-Glu) and a cyanine dye (NRH-NH_2) (Zhou et al. 2021). This transformation not only exhibits superior photothermal therapy capabilities but also amplifies the fluorescence signal by 180 times. Additionally, this study highlights the overexpression of β-galactosidase in ovarian

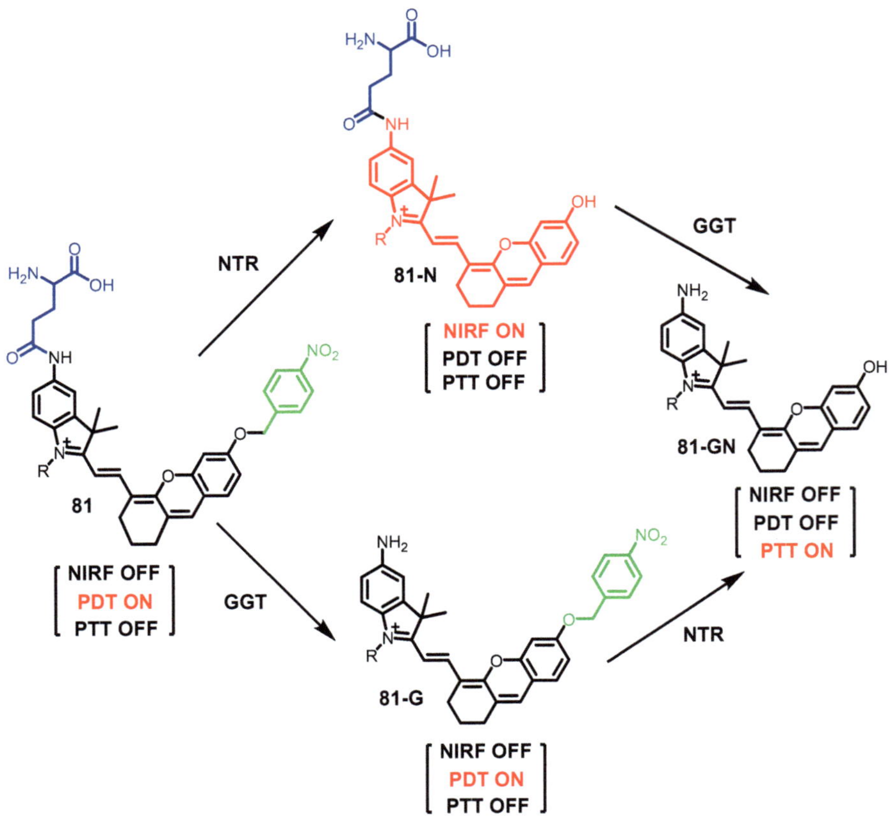

Fig. 2.42 Schematic diagram illustrating the molecular mechanism of **81** in real-time imaging of tumor and autoregulated PDT-PTT in the presence of GGT and NTR

cancers. In 2018, Pu and team created an activatable macrotheranostic probe (CyGal-P, **83**) that combines a d-galactose-capped NIR hemicyanine dye (CyOH) with a long poly(ethylene glycol) (PEG) chain (Zhen et al. 2018). Upon activation by β-galactosidase, the probe emits near-infrared fluorescence (NIRF), photoacoustic (PA), and photothermal signals, enabling imaging-guided photothermal therapy. This work paves the way for the development of β-gal-activated theranostic agents (Fig. 2.44).

Comparable research has revealed that alkaline phosphatase (ALP) is significantly upregulated in metastatic prostate cancer. Yang and colleagues engineered a mitochondria-specific probe, designated as **84**, which becomes activated by ALP to produce near-infrared fluorescence and photoacoustic signals for visualizing ALP activity. This probe undergoes in situ self-assembly, forming a supramolecular structure that enhances its photothermal therapeutic efficacy against prostate cancer (Yao et al. 2019). The combination of fluorescence and photoacoustic imaging with photothermal therapy expands the practical utility of these diagnostic tools (Fig. 2.45).

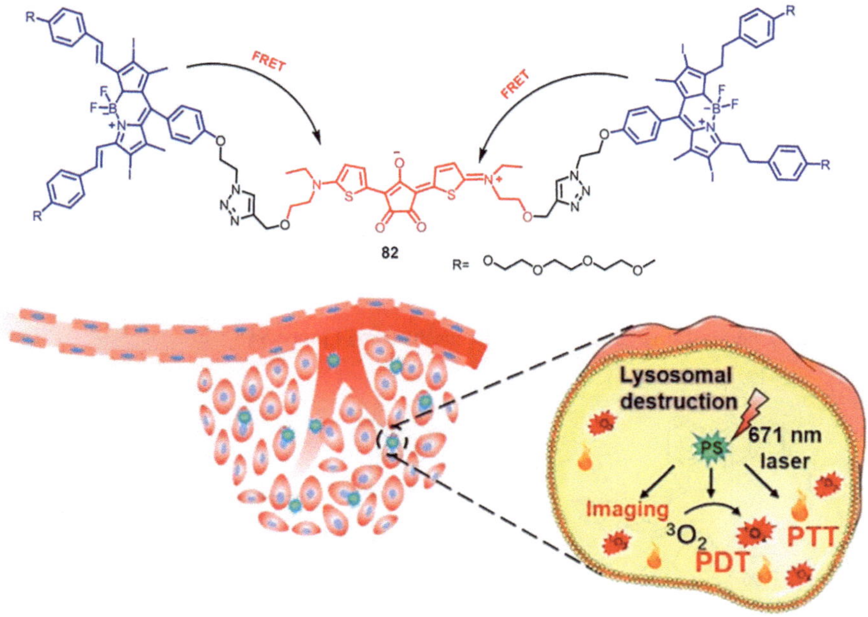

Fig. 2.43 Schematic illustration showing the smFRET-based combination phototherapy mechanism of **82** and light-triggered cancer cell death

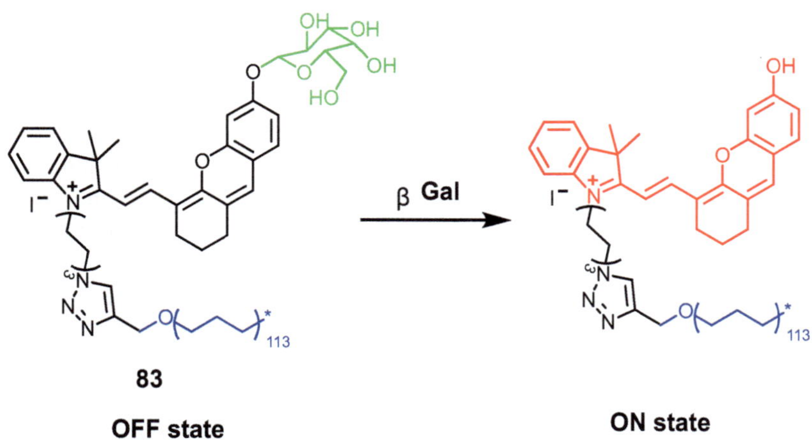

Fig. 2.44 Schematic diagram demonstrating the activation mechanism of **83**

Apart from enzymes, other biomolecules like cysteine (Cys), hydrogen sulfide (H_2S), and GSH are crucial for the advancement of diagnostic technologies. Research has demonstrated that tumor cells contain elevated levels of Cys, which is vital for maintaining redox equilibrium. Kolemen and colleagues developed a Cys-activatable chlorinated hemicyanine probe **85**, where the chlorinated hemicyanine

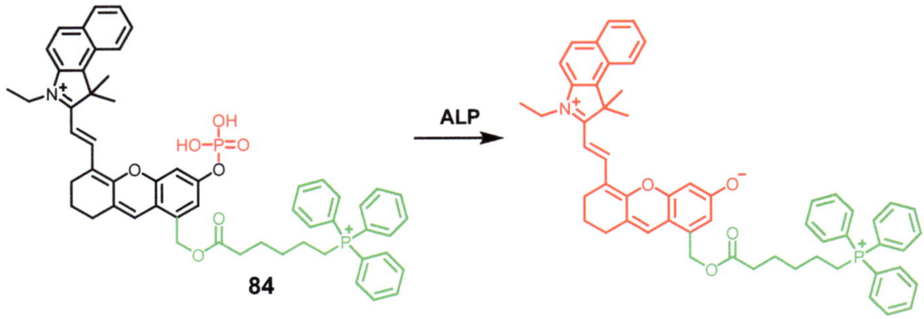

Fig. 2.45 Chemical structure and molecular mechanism of probe **84**

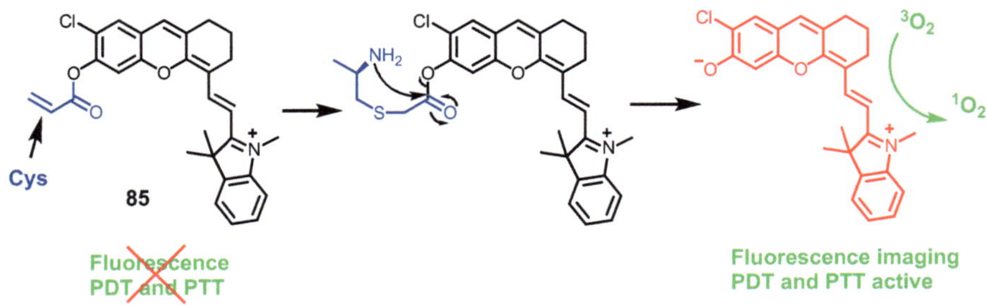

Fig. 2.46 Molecular mechanism for the interaction between **85** and Cys, which results in NIRF and "Turn-On" PPT/PDT

serves as the fluorescent core, and an acrylate unit is linked to it to recognize cysteine (Pan et al. 2023). When this probe interacts with Cys, the Cl-Cys component triggers near-infrared fluorescence, enhances its ability to generate singlet oxygen ($^1O_2$), and activates its photothermal conversion. This work marks the initial report of organic small molecules that combine photothermal and photodynamic diagnostic properties through Cys activation. It paves the way for the creation of biomolecule-activated probes for tumor diagnosis and treatment (Fig. 2.46).

Likewise, GSH is a distinctive biomarker in cancerous cells. Leveraging this, a multitude of GSH-activated "Turn-On" theranostic platforms have been devised. For instance, Jiang and colleagues engineered a GSH-sensitive self-immolative drug-dye conjugate (DDC) prodrug system (Sun et al. 2018) that utilizes 2-hydroxy-5-methyl-1,3-phenylenedimethanol as the central component, linked to both fluorescent dyes and therapeutic agents through a carbonate bridge. In the tumor milieu, the disulfide linkage is disrupted by the elevated levels of GSH, leading to the concurrent discharge of the therapeutic agent and the fluorescent dye. The liberated fluorescent dye functions as a photothermal agent, enhancing drug penetration and thereby boosting the treatment efficacy (Fig. 2.47).

To prolong the therapeutic drug's presence at the tumor site, Lan and colleagues devised an in situ self-assembly method to boost the photothermal therapy (PTT) for glioblastomas (Song et al. 2023). They created the probe ICG-PEP-c(RGD)fk by

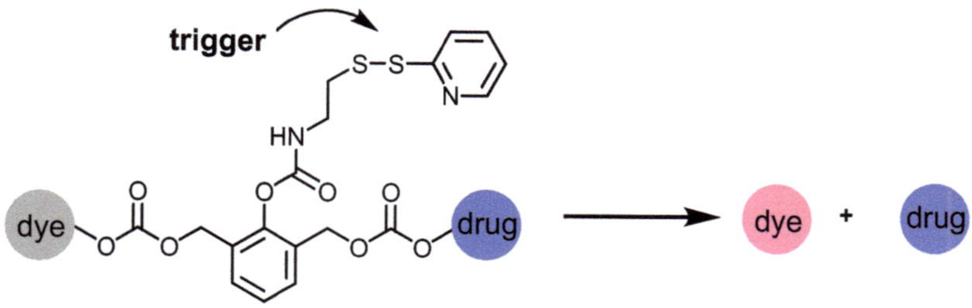

Fig. 2.47 Schematic diagram depicting the molecular mechanism of drug release, fluorescence activation, and PTT activation in response to GSH

integrating GSH-responsive self-assembling polypeptides with indocyanine green (ICG, a theranostic agent) and cyclic Arg-Gly-Asp peptides (c(RGD)fk, the targeting moiety). Within the glioblastoma microenvironment, GSH triggers the cleavage of the disulfide bond, causing the GSH-reactive self-assembling peptides to enhance the retention of the diagnostic and therapeutic system, thereby augmenting the photothermal effect.

2.5 Conclusions and Perspectives

In this chapter, we have explored the latest developments in the creation and implementation of theranostic fluorescent probes. These compact molecular tools have gained significant traction for their dual role in tumor visualization and treatment. Such probes can be triggered by the unique conditions within the tumor microenvironment, responding to a variety of stimuli. They are particularly attuned to environmental cues like acidic pH, elevated levels of specific (bio)molecules (e.g., GSH), redox imbalances (e.g., H_2S, ROS), and enzyme activities that are enhanced in hypoxic conditions (e.g., DT-diaphorase, azo-reductase, nitro-reductase). Additionally, they react to increased expression of other enzymes in cancerous tissues (e.g., esterases, proteases) and can be activated by a combination of these factors. The application of these theranostic probes holds considerable promise for minimizing drug toxicity and enabling real-time, spatial monitoring of drug delivery. Many of these probes are engineered to be responsive to external triggers, such as light, heat, or sound, allowing for precise regulation of therapeutic interventions.

Despite the remarkable progress in the development of diagnostic and therapeutic fluorescent probes, several obstacles still impede their clinical application. One major challenge is the limited ability of fluorescence to penetrate deep tissues, which restricts its use for internal body areas. Additionally, the long-term safety of these integrated probes for in vivo use needs thorough assessment. The efficacy of theranostic probes is another critical factor that hinders their clinical adoption. Thus, creating probes with enhanced accuracy and superior therapeutic outcomes remains a key objective in this field. To facilitate clinical translation, it is crucial to develop a substantial number of high-performance probes. Through ongoing interdisciplinary collaboration and the relentless dedication of researchers, we anticipate that fluorescent theranostic probes will eventually become a standard tool for tumor diagnosis and treatment in clinical practice.

References

Ahn B, Seong-Gyu L, Ryeon YH et al (2018) Four-fold channel-nicked human ferritin nanocages for active drug loading and pH-responsive drug release. Angew Chem Int Ed Engl 57(11):2909–2913

Alfadda A, Sallam Reem M (2012) Reactive oxygen species in health and disease. Biomed Res Int 2012(1):936486

Allison RR (2016) Fluorescence guided resection (FGR): a primer for oncology. Photodiagn Photodyn Ther 13:73–80

Almeida-Marrero V, De Winckel V, Eveline A-PE et al (2018) Porphyrinoid biohybrid materials as an emerging toolbox for biomedical light management. Chem Soc Rev 47(19):7369–7400

An R, Xiaoyang C, Shixuan W et al (2020) Smart magnetic and fluorogenic photosensitizer nanoassemblies enable redox-driven disassembly for photodynamic therapy. Angew Chem Int Ed Engl 59(46):20636–20644

Aranda J, Cerqueira NMFSA, Fernandes PA et al (2014) The catalytic mechanism of carboxylesterases: a computational study. Biochemistry 53(36):5820–5829

Aratani Y (2018) Myeloperoxidase: its role for host defense, inflammation, and neutrophil function. Arch Biochem Biophys 640:47–52

Asanuma D, Masayo S, Mako K et al (2015) Sensitive β-galactosidase-targeting fluorescence probe for visualizing small peritoneal metastatic tumours in vivo. Nat Commun 6(1):6463

Bandyopadhyay U, Dipak D, Banerjee Ranajit K (1999) Reactive oxygen species: oxidative damage and pathogenesis. Curr Sci 77:658–666

Banerjee S, Bond JS (2008) Prointerleukin-18 is activated by meprin β in vitro and in vivo in intestinal inflammation. J Biol Chem 283(46):31371–31377

Bardhan R, Surbhi L, Amit J et al (2011) Theranostic nanoshells: from probe design to imaging and treatment of cancer. Acc Chem Res 44(10):936–946

Barrios B, Benjamin M, Doskey PV et al (2021) Mechanistic insight into the reactivities of aqueous-phase singlet oxygen with organic compounds. Environ Sci Technol 55(12):8054–8067

Bayr H (2005) Reactive oxygen species. Crit Care Med 33(12):S498–S501

Benjamin DI, Alyssa C, Xiaodan J et al (2013) Ether lipid generating enzyme AGPS alters the balance of structural and signaling lipids to fuel cancer pathogenicity. Proc Natl Acad Sci 110(37):14912–14917

Bhuniya S, Sukhendu M, Eun-Joong K et al (2014) An activatable theranostic for targeted cancer therapy and imaging. Angew Chem Int Ed Engl 53(17):4469–4474

Bian H, Dandan M, Xinfu Z et al (2021) Tailored engineering of novel xanthonium polymethine dyes for synergetic PDT and PTT triggered by 1064 nm laser toward deep-seated tumors. Small 17(21):2100398

Bian H, Dandan M, Fei P et al (2022) Cardiolipin-targeted NIR-II fluorophore causes "avalanche effects" for re-engaging cancer apoptosis and inhibiting metastasis. J Am Chem Soc 144(49):22562–22573

Blesa J, Ines T-D, Anna Q-V et al (2015) Oxidative stress and Parkinson's disease. Front Neuroanat 9:91

Bobba KN, Giridharan S, Sujai PT et al (2019) Endogenous H_2S-assisted cancer-cell-specific activation of theranostics with emission readout. ACS Appl Bio Mater 2(3):1322–1330

Bonnet D, Dick JE (1997) Human acute myeloid leukemia is organized as a hierarchy that originates from a primitive hematopoietic cell. Nat Med 3(7):730–737
Bray F, Ferlay J, Soerjomataram I et al (2020) Erratum: Global cancer statistics 2018: GLOBOCAN estimates of incidence and mortality worldwide for 36 cancers in 185 countries. CA Cancer J Clin 70(4):313
Bukowski K, Mateusz K, Renata K (2020) Mechanisms of multidrug resistance in cancer chemotherapy. Int J Mol Sci 21(9):3233
Cao Y, Jerry Y (2014) Development of a folate receptor (FR)-targeted indenoisoquinoline using a pH-sensitive N-ethoxybenzylimidazole (NEBI) bifunctional cross-linker. Bioconjug Chem 25(5):873–878
Cao D, Zhiqiang L, Peter V et al (2019) Coumarin-based small-molecule fluorescent chemosensors. Chem Rev 119(18):10403–10519
Chatterjee SK, Bhattacharya M, Barlow JJ (1979) Glycosyltransferase and glycosidase activities in ovarian cancer patients. Cancer Res 39(6 Pt 1):1943–1951
Chen N-T, Chia-Yan W, Chao-Yu C et al (2012) Probing the dynamics of doxorubicin-DNA intercalation during the initial activation of apoptosis by fluorescence lifetime imaging microscopy (FLIM). PLoS One 7(9):e44947
Chen S, Hui D, Shiming Y et al (2017) Cathepsins in digestive cancers. Oncotarget 8:25
Chen Y, Yuping J, Weiguo S et al (2018) Therapeutic potential of nitrogen mustard based hybrid molecules. Front Pharmacol 9:1453
Chen Y, Xueze Z, Tao X et al (2021) NIR photosensitizers activated by γ-glutamyl transpeptidase for precise tumor fluorescence imaging and photodynamic therapy. Sci China Chem 64(5):808–816
Chen H, Ting L, Zhiyong L et al (2023a) A nitric-oxide driven chemotactic nanomotor for enhanced immunotherapy of glioblastoma. Nat Commun 14(1):941
Chen Z, Fangfang H, Yan D et al (2023b) Hypoxic microenvironment in cancer: molecular mechanisms and therapeutic interventions. Signal Transduct Target Ther 8(1):70
Cheng R, Fang F, Fenghua M et al (2011) Glutathione-responsive nano-vehicles as a promising platform for targeted intracellular drug and gene delivery. J Control Release 152(1):2–12
Cheung EC, Vousden KH (2022) The role of ROS in tumour development and progression. Nat Rev Cancer 22(5):280–297
Chiba M, Yuki I, Mako K et al (2017) An activatable photosensitizer targeted to γ-glutamyltranspeptidase. Angew Chem Int Ed Engl 56(35):10418–10422
Cho H, Kyu SM, Suah Y et al (2022) Cathepsin B-overexpressed tumor cell activatable albumin-binding doxorubicin prodrug for cancer-targeted therapy. Pharmaceutics 14(1):83
Choi HY, Ai-Ming Y (2014) ABC transporters in multidrug resistance and pharmacokinetics, and strategies for drug development. Curr Pharm Des 20(5):793–807
Choi PJ, Thomas P, Elizabeth C et al (2020) Heptamethine cyanine dye mediated drug delivery: hype or hope. Bioconjug Chem 31(7):1724–1739
Chung SW, Choi Jeong UK, Seok CY et al (2018) Self-triggered apoptosis enzyme prodrug therapy (STAEPT): enhancing targeted therapies via recurrent bystander killing effect by exploiting caspase-cleavable linker. Adv Sci 5(7):1800368
Circu ML, Yee AT (2010) Reactive oxygen species, cellular redox systems, and apoptosis. Free Rad Bio Med 48(6):749–762
Cleland NRW, Al-Juboori SI, Evgenia D et al (2021) Altered substrate metabolism in neurodegenerative disease: new insights from metabolic imaging. J Neuroinflammation 18(1):248
Cortés-Funes H, Cyntia C (2007) Role of anthracyclines in the era of targeted therapy. Cardiovasc Toxicol 7(2):56–60
Cui Y, Haiqing D, Xiaojun C et al (2012) Mesoporous silica nanoparticles capped with disulfide-linked PEG gatekeepers for glutathione-mediated controlled release. ACS Appl Mater Interfaces 4(6):3177–3183
Cui X, Lianhua L, Guanghai Y et al (2015) High expression of NQO1 is associated with poor prognosis in serous ovarian carcinoma. BMC Cancer 15(1):244

Dallavalle S, Vladimir D, Loretta L et al (2020) Improvement of conventional anti-cancer drugs as new tools against multidrug resistant tumors. Drug Resist Updat 50:100682
Dasari M, Acharya AP, Dongin K et al (2013) H-Gemcitabine: a new gemcitabine prodrug for treating cancer. Bioconjug Chem 24(1):4–8
De Boer E, Harlaar NJ, Taruttis A et al (2015) Optical innovations in surgery. J Br Surg 102(2):e56–e72
Deepagan VG, Seunglee K, Gil YD et al (2016) In situ diselenide-crosslinked polymeric micelles for ROS-mediated anticancer drug delivery. Biomaterials 103:56–66
Deng X, Zengwu S, Yanli Z (2021) Solutions to the drawbacks of photothermal and photodynamic cancer therapy. Adv Sci 8(3):2002504
Denkova AG, De Kruijff RM, Pablo S-C (2018) Combined photo-therapies: nanocarrier-mediated photochemotherapy and photoradiotherapy. Adv Healthc Mater 7(8):1870033
Di MS, Reed Tanea T, Paola V et al (2016) Role of ROS and RNS sources in physiological and pathological conditions. Oxidative Med Cell Longev 2016(1):1245049
Dieci MV, Arnedos M, Andre F et al (2013) Fibroblast growth factor receptor inhibitors as a cancer treatment: from a biologic rationale to medical perspectives. Cancer Discov 3(3):264–279
Dinand E, Zloh M, Brocchini S (2002) Competitive reactions during amine addition to cis-aconityl anhydride. Aust J Chem 55(7):467–474
Dongbang S, Mi JH, Hee LM et al (2014) Camptothecin delivery into hepatoma cell line by galactose-appended fluorescent drug delivery system. RSC Adv 4(36):18744–18748
Du J, Ning X, Jiangli F et al (2019) Carbon dots for in vivo bioimaging and theranostics. Small 15(32):1805087
Duclos C, Christine L, Jean-Bernard D (2017) Caspases rule the intracellular trafficking cartel. FEBS J 284(10):1394–1420
Dutta D, Alex SM, Naidu BK et al (2016) New insight into a cancer theranostic probe: efficient cell-specific delivery of SN-38 guided by biotinylated poly(vinyl alcohol). ACS Appl Mater Interfaces 8(49):33430–33438
Engin K, Leeper DB, Cater JR et al (1995) Extracellular pH distribution in human tumours. Int J Hyperth 11(2):211–216
Escobedo JO, Oleksandr R, Soojin L et al (2010) NIR dyes for bioimaging applications. Curr Opin Chem Biol 14(1):64–70
Estrela JM, Angel O, Elena O (2006) Glutathione in cancer biology and therapy. Crit Rev Clin Lab Sci 43(2):143–181
Euhus DM, Charles H, Laregina MC et al (1986) Tumor measurement in the nude mouse. J Surg Oncol 31(4):229–234
Fan W, Peng H, Xiaoyuan C (2016) Overcoming the achilles' heel of photodynamic therapy. Chem Soc Rev 45(23):6488–6519
Fang FC (2004) Antimicrobial reactive oxygen and nitrogen species: concepts and controversies. Nat Rev Microbiol 2(10):820–832
Fernandez A, Matthieu V, Duncan H et al (2017) Chemical modulation of in vivo macrophage function with subpopulation-specific fluorescent prodrug conjugates. ACS Central Sci 3(9):995–1005
Finkel T (2011) Signal transduction by reactive oxygen species. J Cell Biol 194(1):7–15
Frangioni JV (2008) New technologies for human cancer imaging. J Clin Oncol 26(24):4012–4021
Fujita E, Egashira J, Urase K et al (2001) Caspase-9 processing by caspase-3 via a feedback amplification loop in vivo. Cell Death Differ 8(4):335–344
Fukami T, Tsuyoshi Y (2012) The emerging role of human esterases. Drug Metab Pharmacokinet 27(5):466–477
Gamcsik MP, Kasibhatla MS, Teeter SD et al (2012) Glutathione levels in human tumors. Biomarkers 17(8):671–691
Ge C, Xiaojiao D, Siqi H et al (2021a) Hydrogen sulfide triggered molecular agent for imaging and cancer therapy. Chem Commun 57(15):1931–1934

Ge C, Ji L, Liu L et al (2021b) A self-immolated fluorogenic agent triggered by H_2S exhibiting potential anti-glioblastoma activity. Analyst 146(11):3510–3515

Giang I, Boland EL, Poon GM, K. (2014) Prodrug applications for targeted cancer therapy. AAPS J 16(5):899–913

Gocheva V, Joyce JA (2007) Cysteine cathepsins and the cutting edge of cancer invasion. Cell Cycle 6(1):60–64

Greten FR, Grivennikov SI (2019) Inflammation and cancer: triggers, mechanisms, and consequences. Immunity 51(1):27–41

Gu H, Weijian L, Haidong L et al (2022) 2,1,3-benzothiadiazole derivative AIEgens for smart phototheranostics. Coord Chem Rev 473:214803

Gui L, Zhenwei Y, Habtamu K et al (2018) A tumor-targeting probe based on a mitophagy process for live imaging. Chem Commun 54(69):9675–9678

Guo X, Lin W, Kayla D et al (2018) Dimeric drug polymeric micelles with acid-active tumor targeting and FRET-traceable drug release. Adv Mater 30(3):1705436

Guo X, Naidi Y, Wenhui J et al (2021). Mito-bomb: targeting mitochondria for cancer therapy) Adv Mater 33(43):2170340

Gutteridge JMC (1984) Reactivity of hydroxyl and hydroxyl-like radicals discriminated by release of thiobarbituric acid-reactive material from deoxy sugars, nucleosides and benzoate. Biochem J 224(3):761–767

Haefeli RH, Michael E, Gemperli AC et al (2011) NQO1-dependent redox cycling of idebenone: effects on cellular redox potential and energy levels. PLoS One 6(3):e17963

Hagen H, Paul M, Elmar J et al (2012) Aminoferrocene-based prodrugs activated by reactive oxygen species. J Med Chem 55(2):924–934

Haigis MC, Yankner BA (2010) The aging stress response. Mol Cell 40(2):333–344

Halliwell B (2006) Oxidative stress and neurodegeneration: where are we now? J Neurochem 97(6):1634–1658

Hameed S, Hong C, Muhammad I et al (2019) Fluorescence guided sentinel lymph node mapping: from current molecular probes to future multimodal nanoprobes. Bioconjug Chem 30(1):13–28

Han X, Jinling C, Mengjuan J et al (2016) Paclitaxel–paclitaxel prodrug nanoassembly as a versatile nanoplatform for combinational cancer therapy. ACS Appl Mater Interfaces 8(49):33506–33513

Han H-H, Han-Min W, Paramesh J et al (2023) The design of small-molecule prodrugs and activatable phototherapeutics for cancer therapy. Chem Soc Rev 52(3):879–920

Hao B, Jiaxin W, Chao W et al (2022) Bridging D–A type photosensitizers with the azo group to boost intersystem crossing for efficient photodynamic therapy. Chem Sci 13(14):4139–4149

Heng H, Gang S, Xuetong C et al (2022) Intrinsic mitochondrial reactive oxygen species (ROS) activate the in situ synthesis of trimethine cyanines in cancer cells. Angew Chem Int Ed Engl 61(38):e202203444

Hoffmann MH, Griffiths HR (2018) The dual role of Reactive Oxygen Species in autoimmune and inflammatory diseases: evidence from preclinical models. Free Rad Bio Med 125:62–71

Hou M, Peng X, Gao Y-E et al (2017) Gemcitabine–camptothecin conjugates: a hybrid prodrug for controlled drug release and synergistic therapeutics. Biomater Sci 5(9):1889–1897

Hu Y, Fang Z (2017) A theranostic prodrug based on FRET for real-time drug release monitoring in response to biothiols. Mater Sci Eng C 72:77–85

Hu M, Chao Y, Yi L et al (2018a) A hypoxia-specific and mitochondria-targeted anticancer theranostic agent with high selectivity for cancer cells. J Mater Chem B 6(16):2413–2416

Hu W, Meng X, Hui Z et al (2018b) Nitric oxide activatable photosensitizer accompanying extremely elevated two-photon absorption for efficient fluorescence imaging and photodynamic therapy. Chem Sci 9(4):999–1005

Hua Y, Sreejayan N (2015) Proteases in cardiometabolic diseases: pathophysiology, molecular mechanisms and clinical applications. Biochim Biophys Acta 1852(2):195–208

Huang B, Ankur D, Shengzhuang T et al (2010) The synthesis of a c(RGDyK) targeted SN38 prodrug with an indolequinone structure for bioreductive drug release. Org Lett 12(7):1384–1387

Huang Q, Fang L, Xinjian L et al (2011) Caspase 3–mediated stimulation of tumor cell repopulation during cancer radiotherapy. Nat Med 17(7):860–866

Huang K, Song X, Weijun W et al (2022) A hydrogen sulfide-responsive prodrug for monitoring real-time release and improving therapeutic effects of anticancer drug SN-38. Sens Actuators B Chem 373:132750

Huang T, Heng J, Shirong Y et al (2023) A hypochlorite-activated strategy for realizing fluorescence turn-on, type I and type II ROS-combined photodynamic tumor ablation. Biomaterials 297:122108

Hwang B, Tae-Il K, Hyunjin K et al (2021) Ubiquinone-BODIPY nanoparticles for tumor redox-responsive fluorescence imaging and photodynamic activity. J Mater Chem B 9(3):824–831

Ishikawa K, Keizo T, Miho A et al (2008) ROS-generating mitochondrial DNA mutations can regulate tumor cell metastasis. Science 320(5876):661–664

Ji S, Heqi G, Wancen M et al (2018) Enzyme-instructed self-assembly leads to the activation of optical properties for selective fluorescence detection and photodynamic ablation of cancer cells. J Mater Chem B 6(17):2566–2573

Jiang F, Biao L, Jun L et al (2015) Progress and challenges in developing aptamer-functionalized targeted drug delivery systems. Int J Mol Sci 16(10):23784–23822

Jin Y, Lian S, Yue S et al (2011) Oxime linkage: a robust tool for the design of pH-sensitive polymeric drug carriers. Biomacromolecules 12(10):3460–3468

Jin S, Jiaxun W, Lizheng M et al (2015a) Biodegradation and toxicity of protease/redox/pH stimuli-responsive PEGlated PMAA nanohydrogels for targeting drug delivery. ACS Appl Mater Interfaces 7(35):19843–19852

Jin Y, Yanyan H, Hua Y et al (2015b) A peptide-based pH-sensitive drug delivery system for targeted ablation of cancer cells. Chem Commun 51(77):14454–14457

Jo S-M, Seok KH, Miae W et al (2022) Dual-targeted nanoreactors and prodrugs: hydrogen peroxide triggers oxidative damage and prodrug activation for synergistic elimination of cancer cells. Adv Funct Mater 32(26):2200791

Jones LR, Goun EA, Rajesh S et al (2006) Releasable luciferin–transporter conjugates: tools for the real-time analysis of cellular uptake and release. J Am Chem Soc 128(20):6526–6527

Juers DH, Matthews BW, Huber RE (2012) LacZ β-galactosidase: structure and function of an enzyme of historical and molecular biological importance. Protein Sci 21(12):1792–1807

Jung HS, Xiaoqiang C, Seung KJ et al (2013) Recent progress in luminescent and colorimetric chemosensors for detection of thiols. Chem Soc Rev 42(14):6019–6031

Juvekar V, Lim Chang S, Joon LD et al (2021) An azo dye for photodynamic therapy that is activated selectively by two-photon excitation. Chem Sci 12(1):427–434

Kalia J, Raines RT (2008) Hydrolytic stability of hydrazones and oximes. Angew Chem Int Ed Engl 47(39):7523–7526

Kanamala M, Wilson WR, Mimi Y et al (2016) Mechanisms and biomaterials in pH-responsive tumour targeted drug delivery: a review. Biomaterials 85:152–167

Kang EB, Eun LJ, Islamy MZA et al (2018) pH-Responsible fluorescent carbon nanoparticles for tumor selective theranostics via pH-turn on/off fluorescence and photothermal effect in vivo and in vitro. Nanoscale 10(5):2512–2523

Kilic E, Zubeyir E, Toghrul A et al (2022) Activity-based photosensitizers with optimized triplet state characteristics toward cancer cell selective and image guided photodynamic therapy. ACS Appl Bio Mater 5(6):2754–2767

Kim JH, Peter V, Miae W et al (2021) A small molecule strategy for targeting cancer stem cells in hypoxic microenvironments and preventing tumorigenesis. J Am Chem Soc 143(35):14115–14124

Kitamura T, Hirotaka N, Daisuke T et al (2022) Hypocrellin B-based activatable photosensitizers for specific photodynamic effects against high H_2O_2-expressing cancer cells. Chem Commun 58(2):242–245

Kohnz RA, Mulvihill MM, Won CJ et al (2015) Activity-based protein profiling of oncogene-driven changes in metabolism reveals broad dysregulation of PAFAH1B2 and 1B3 in cancer. ACS Chem Biol 10(7):1624–1630

Kong F, Ziye L, Dongrui L et al (2016) A glutathione (GSH)-responsive near-infrared (NIR) theranostic prodrug for cancer therapy and imaging. Anal Chem 88(12):6450–6456

Kong Y, Jacqueline S, Kongwen L et al (2017) Development of a novel near-infrared fluorescent theranostic combretastain A-4 analogue, YK-5-252, to target triple negative breast cancer. Bioorg Med Chem 25(7):2226–2233

Kratz F, Müller IA, Claudia R et al (2008) Prodrug strategies in anticancer chemotherapy. ChemMedChem 3(1):20–53

Krishna DR, Bernhard S, Peter F et al (1999) Does pH 6 β-galactosidase activity indicate cell senescence? Mech Ageing Dev 109(2):113–123

Kubo H, Yasutoshi M, Soichiro O et al (2021) β-Galactosidase is a target enzyme for detecting peritoneal metastasis of gastric cancer. Sci Rep 11(1):10664

Kumar R, Jiyou H, Hee-Joung L et al (2014) Mitochondrial induced and self-monitored intrinsic apoptosis by antitumor theranostic prodrug: in vivo imaging and precise cancer treatment. J Am Chem Soc 136(51):17836–17843

Kumari R, Dhanya S, Ningthoujam RS (2020) Hypoxia-responsive nanoparticle based drug delivery systems in cancer therapy: an up-to-date review. J Control Release 319:135–156

Kunimoto T, Nitta K, Tanaka T et al (1987) Antitumor activity of 7-ethyl-10-[4-(1-piperidino)-1-piperidino]carbonyloxy-camptothec in, a novel water-soluble derivative of camptothecin, against murine tumors. Cancer Res 47(22):5944–5947

Lankelma J, Dekker H, Luque FR et al (1999) Doxorubicin gradients in human breast cancer. Clin Cancer Res 5(7):1703–1707

Lassègue B, Alejandra SM, Griendling KK (2012) Biochemistry, physiology, and pathophysiology of NADPH oxidases in the cardiovascular system. Circ Res 110(10):1364–1390

Lee MH, Young KJ, Hye HJ et al (2012) Direct fluorescence monitoring of the delivery and cellular uptake of a cancer-targeted RGD peptide-appended naphthalimide theragnostic prodrug. J Am Chem Soc 134(30):12668–12674

Lee MH, Eun-Joong K, Hyunseung L et al (2016) Acid-triggered release of doxorubicin from a hydrazone-linked Gd^{3+}-texaphyrin conjugate. Chem Commun 52(69):10551–10554

Lee S, Jun L, Xin Z et al (2018) Recent progress on the development of glutathione (GSH) selective fluorescent and colorimetric probes. Coord Chem Rev 366:29–68

Lewis AM, Matthew O, Hinkhouse MM et al (2005) Targeting NAD(P)H:quinone oxidoreductase (NQO1) in pancreatic cancer. Mol Carcinog 43(4):215–224

Li J, Sausville EA, Klein PJ et al (2009) Clinical pharmacokinetics and exposure-toxicity relationship of a folate-vinca alkaloid conjugate EC145 in cancer patients. J Clin Pharmacol 49(12):1467–1476

Li S-Y, Li-Han L, Hui-Zhen J et al (2014) A pH-responsive prodrug for real-time drug release monitoring and targeted cancer therapy. Chem Commun 50(80):11852–11855

Li S-Y, Li-Han L, Lei R et al (2015a) A dual-FRET-based versatile prodrug for real-time drug release monitoring and in situ therapeutic efficacy evaluation. Adv Funct Mater 25(47):7317–7326

Li Z, Yue Z, Tiefeng J et al (2015b) NQO1 protein expression predicts poor prognosis of non-small cell lung cancers. BMC Cancer 15(1):207

Li B, Lisheng L, Huiyun L et al (2016) Photosensitized singlet oxygen generation and detection: recent advances and future perspectives in cancer photodynamic therapy. J Biophotonics 9(11-12):1314–1325

Li S, Xueyan J, Rongrong Z et al (2018a) An azobenzene-based heteromeric prodrug for hypoxia-activated chemotherapy by regulating subcellular localization. Chem Commun 54(57):7983–7986

Li X, Nahyun K, Tian G et al (2018b) Innovative strategies for hypoxic-tumor photodynamic therapy. Angew Chem Int Ed Engl 57(36):11522–11531

Li Z, Yi-Feng W, Chengchu Z et al (2018c) Ultrasensitive tyrosinase-activated turn-on near-infrared fluorescent probe with a rationally designed urea bond for selective imaging and photodamage to melanoma cells. Anal Chem 90(6):3666–3669

Li X, Mingyang J, Songjun Z et al (2019) Polydopamine coated multifunctional lanthanide theranostic agent for vascular malformation and tumor vessel imaging beyond 1500 nm and imaging-guided photothermal therapy. Theranostics 9(13):3866

Li H, Qichao Y, Wen S et al (2020a) Aminopeptidase N activatable fluorescent probe for tracking metastatic cancer and image-guided surgery via in situ spraying. J Am Chem Soc 142(13):6381–6389

Li T, Shuojiong P, Shiqian G et al (2020b) Diselenide–pemetrexed assemblies for combined cancer immuno-, radio-, and chemotherapies. Angew Chem Int Ed Engl 59(7):2700–2704

Li X, Yutong P, Huan C et al (2020c) Specific near-infrared probe for ultrafast imaging of lysosomal β-galactosidase in ovarian cancer cells. Anal Chem 92(8):5772–5779

Li Y, Song H, Xue C et al (2020d) A self-immobilizing near-infrared fluorogenic probe for sensitive imaging of extracellular enzyme activity in vivo. Chem Sci 11(23):5889–5894

Li X, Lovell JF, Juyoung Y et al (2020e) Clinical development and potential of photothermal and photodynamic therapies for cancer. Nat Rev Clin Oncol 17(11):657–674

Li J, Tingting W, Feng J et al (2021a) A fluorescence-activatable tumor-reporting probe for precise photodynamic therapy. J Mater Chem B 9(29):5829–5836

Li J, Wei Z, Wenhui J et al (2021b) Near infrared photothermal conversion materials: mechanism, preparation, and photothermal cancer therapy applications. J Mater Chem B 9(38):7909–7926

Li H, Yujun K, Hyoje J et al (2022a) Near-infrared (NIR) fluorescence-emitting small organic molecules for cancer imaging and therapy. Chem Soc Rev 51(21):8957–9008

Li T, Zhiyong L, Jinglei H et al (2022b) A universal chemotactic targeted delivery strategy for inflammatory diseases. Adv Mater 34(47):2206654

Li M, Hiluf GK, Dandan M et al (2022c) Conditionally activatable photoredox catalysis in living systems. J Am Chem Soc 144(1):163–173

Li X, Mingwang Y, Jianfang C et al (2022d) H-aggregates of prodrug-hemicyanine conjugate for enhanced photothermal therapy and sequential hypoxia-activated chemotherapy. ACS Mater Lett 4(4):724–732

Liberti MV, Locasale JW (2016) The Warburg effect: how does it benefit cancer cells? Trends Biochem Sci 41(3):211–218

Lim J, Abdellatif C, Su-Tang L et al (2009) Design, synthesis, characterization, and biological evaluation of triazine dendrimers bearing paclitaxel using ester and ester/disulfide linkages. Bioconjug Chem 20(11):2154–2161

Lin H-H, Xiao-Zhen Y, Ming-Yue Y et al (2022) A switchable phototherapeutic strategy based on a hypoxia-responsive photosensitizer for heterogeneous tumor ablation. Dyes Pigments 201:110217

Liu H-W, Tadhg B (2020) Comprehensive natural products III. Elsevier, Amsterdam

Liu P, Jiangsheng X, Donghang Y et al (2015) A DT-diaphorase responsive theranostic prodrug for diagnosis, drug release monitoring and therapy. Chem Commun 51(46):9567–9570

Liu Y, Qing P, Li C et al (2016a) Reduction-responsive fluorescence off–on BODIPY–camptothecin conjugates for self-reporting drug release. J Mater Chem B 4(13):2332–2337

Liu L-H, Wen-Xiu Q, Bin L et al (2016b) A red light activatable multifunctional prodrug for image-guided photodynamic therapy and cascaded chemotherapy. Adv Funct Mater 26(34):6257–6269

Liu Y, Shaojia Z, Kaizhi G et al (2017) GSH-activated NIR fluorescent prodrug for podophyllotoxin delivery. ACS Appl Mater Interfaces 9(35):29496–29504

Liu R, Jie T, Yunxue X et al (2019a) Bioluminescence imaging of inflammation in vivo based on bioluminescence and fluorescence resonance energy transfer using nanobubble ultrasound contrast agent. ACS Nano 13(5):5124–5132

Liu Z, Fengling S, Wenbo S et al (2019b) Nitroreductase-activatable theranostic molecules with high PDT efficiency under mild hypoxia based on a TADF fluorescein derivative. ACS Appl Mater Interfaces 11(17):15426–15435

Liu H, Ling-Hong X, Kin KRT et al (2020a) AIE bioconjugates for biomedical applications. Adv Opt Mater 8(14):2000162

Liu Y, Chengyan X, Lili T et al (2020b) pH stimulus-disaggregated BODIPY: an activated photodynamic/photothermal sensitizer applicable to tumor ablation. Chem Commun 56(13):1956–1959
Liu G, Zhen J, Lovell JF et al (2021a) Design of a thiol-responsive, traceless prodrug with rapid self-immolation for cancer chemotherapy. ACS Appl Bio Mater 4(6):4982–4989
Liu S, Guangxue F, Zhong TB et al (2021b) Recent advances of AIE light-up probes for photodynamic therapy. Chem Sci 12(19):6488–6506
Liu Y, Yibing L, Le Y et al (2022) Monocomponent nanodots with dichromatic output regulated by synergistic dual-stimuli for cervical cancer tissue imaging and photodynamic tumor therapy. Anal Chem 94(2):811–819
Liu Y, Xueqian C, Xiaoting L et al (2023a) Aggregation-induced emission-active micelles: synthesis, characterization, and applications. Chem Soc Rev 52(4):1456–1490
Liu Y, Han Z, Li L et al (2023b) A tyrosinase-activated Pt(II) complex for melanoma photodynamic therapy and fluorescence imaging. Sensors Actuators B Chem 374:132836
Lock FE, Mcdonald PC, Lou Y et al (2013) Targeting carbonic anhydrase IX depletes breast cancer stem cells within the hypoxic niche. Oncogene 32(44):5210–5219
López-Otín C, Bond JS (2008) Proteases: multifunctional enzymes in life and disease. J Biol Chem 283(45):30433–30437
Lu S-B, Luyao W, Wenjing H et al (2022) Recent advances in the enzyme-activatable organic fluorescent probes for tumor imaging and therapy. ChemistryOpen 11(10):e202200137
Lu Y, Wen S, Jianjun D et al (2023) Immuno-photodynamic therapy (IPDT): organic photosensitizers and their application in cancer ablation. JACS Au 3(3):682–699
Lucero MY, Jefferson C (2021) Photoacoustic imaging of elevated glutathione in models of lung cancer for companion diagnostic applications. Nat Chem 13(12):1248–1256
Luo S, Erlong Z, Yongping S et al (2011) A review of NIR dyes in cancer targeting and imaging. Biomaterials 32(29):7127–7138
Luo D, Carter KA, Dyego M et al (2017) Chemophototherapy: an emerging treatment option for solid tumors. Adv Sci 4(1):1600106
Luqmani YA (2008) Mechanisms of drug resistance in cancer chemotherapy. Med Princ Pract 14(Suppl. 1):35–48
Lyngsie G, Lelde K, Anders T et al (2018) Generation of hydroxyl radicals from reactions between a dimethoxyhydroquinone and iron oxide nanoparticles. Sci Rep 8(1):10834
Ma N, Ying L, Huaping X et al (2010) Dual redox responsive assemblies formed from diselenide block copolymers. J Am Chem Soc 132(2):442–443
Ma X, Wen S, Mengjie Y et al (2023) Endogenous/exogenous stimulies inspired polyprodrug nanoinducer switches pyroptosis path for promoting antitumor immunity. Nano Today 48:101727
Mahato R, Wanyi T, Kun C (2011) Prodrugs for improving tumor targetability and efficiency. Adv Drug Deliv Rev 63(8):659–670
Maiti S, Nayoung P, Hye HJ et al (2013) Gemcitabine–coumarin–biotin conjugates: a target specific theranostic anticancer prodrug. J Am Chem Soc 135(11):4567–4572
Mariaule V, Aicha K, Souha S et al (2021) Digestive inflammation: role of proteolytic dysregulation. Int J Mol Sci 22(6):2817
Maroney MJ, Hondal RJ (2018) Selenium versus sulfur: Reversibility of chemical reactions and resistance to permanent oxidation in proteins and nucleic acids. Free Rad Bio Med 127:228–237
Matters GL, Andrea M, Bond JS (2005) Inhibitors of polyamine biosynthesis decrease the expression of the metalloproteases meprin α and MMP-7 in hormone-independent human breast cancer cells. Clin Exp Metastasis 22(4):331–339
Meng F, Hennink WE, Zhiyuan Z (2009) Reduction-sensitive polymers and bioconjugates for biomedical applications. Biomaterials 30(12):2180–2198
Meng X, Wenjun L, Zhihong S et al (2017) Tumor-targeted small molecule for dual-modal imaging-guided phototherapy upon near-infrared excitation. J Mater Chem B 5(47):9405–9411
Meng X, Jiali Z, Zhihong S et al (2018) Hypoxia-triggered single molecule probe for high-contrast NIR II/PA tumor imaging and robust photothermal therapy. Theranostics 8(21):6025–6034

Montero D, Christine T, Jakob RW et al (2013) Intracellular glutathione pools are heterogeneously concentrated. Redox Biol 1(1):508–513
Moris D, Michael S, Eleftherios S et al (2017) The role of reactive oxygen species in the pathophysiology of cardiovascular diseases and the clinical significance of myocardial redox. Ann Transl Med 5(16):326
Motlagh NSH, Parviz P, Fatemah G et al (2016) Fluorescence properties of several chemotherapy drugs: doxorubicin, paclitaxel and bleomycin. Biomed Opt Express 7(6):2400–2406
Nagaya T, Nakamura YA, Choyke Peter L et al (2017) Fluorescence-guided surgery. Front Oncol 7:314
Nayak MK, Nirav D, Prakash D et al (2018) Dichloroacetate, an inhibitor of pyruvate dehydrogenase kinases, inhibits platelet aggregation and arterial thrombosis. Blood Adv 2(15):2029–2038
Needham DM, Cohen JA, Barrett AM (1947) The mechanism of damage to the bone marrow in systemic poisoning with mustard gas. Biochem J 41(4):631–639
Newton AD, Predina Jarrod D, Shuming N et al (2018) Intraoperative fluorescence imaging in thoracic surgery. J Surg Oncol 118(2):344–355
Nguyen V-N, Zheng Z, Zhong TB et al (2022) Organic photosensitizers for antimicrobial phototherapy. Chem Soc Rev 51(9):3324–3340
Niu R, Hua J, Zhao C et al (2012) Differentiating malignant colorectal tumor patients from benign colorectal tumor patients by assaying morning urinary arylsulfatase activity. Asia Pac J Clin Oncol 8(4):362–367
Niu L-Y, Yu-Zhe C, Hai-Rong Z et al (2015) Design strategies of fluorescent probes for selective detection among biothiols. Chem Soc Rev 44(17):6143–6160
Nomura DK, Long JZ, Niessen S et al (2010) Monoacylglycerol lipase regulates a fatty acid network that promotes cancer pathogenesis. Cell 140(1):49–61
Nomura S, Yuji M, Hironori T et al (2019) Highly reliable, targeted photothermal cancer therapy combined with thermal dosimetry using indocyanine green lactosome. bioRxiv: 659334
Ouyang J, Lihe S, Fang Z et al (2022) Biomarker-activatable probes based on smart AIEgens for fluorescence and optoacoustic imaging. Coord Chem Rev 458:214438
Pacher P, Beckman JS, Lucas L (2007) Nitric oxide and peroxynitrite in health and disease. Physiol Rev 87(1):315–424
Pan Z, Yakun W, Niping C et al (2023) Aggregation-induced emission photosensitizer with lysosomal response for photodynamic therapy against cancer. Bioorg Chem 132:106349
Park S, Eunjin K, Young KW et al (2015) Biotin-guided anticancer drug delivery with acidity-triggered drug release. Chem Commun 51(45):9343–9345
Parodi WP (2007) A role for milk proteins and their peptides in cancer prevention. Curr Pharm Des 13(8):813–828
Paul BD, Snyder SH (2015) H_2S: a novel gasotransmitter that signals by sulfhydration. Trends Biochem Sci 40(11):687–700
Pei P, Chunyang S, Wei T et al (2019) ROS-sensitive thioketal-linked polyphosphoester-doxorubicin conjugate for precise phototriggered locoregional chemotherapy. Biomaterials 188:74–82
Peng J, Ke D, Jian S et al (2023) Photocatalytic generation of hydrogen radical (H·) with GSH for photodynamic therapy. Angew Chem Int Ed Engl 62(9):e202214991
Petrova V, Margherita A-P, Gerry M et al (2018) The hypoxic tumour microenvironment. Oncogene 7(1):10
Pham TC, Van-Nghia N, Yeonghwan C et al (2021) Recent strategies to develop innovative photosensitizers for enhanced photodynamic therapy. Chem Rev 121(21):13454–13619
Piao W, Kenjiro H, Tomotsumi F et al (2017) Development of an Azo-based photosensitizer activated under mild hypoxia for photodynamic therapy. J Am Chem Soc 139(39):13713–13719
Pollock JA, Parker Hannah K (2022) Advances in the development of prodrugs as selective modulators of estrogen receptors. J Endocr Soc 6:12
Porret E, Xavier LG, Jean-Luc C (2020) Gold nanoclusters for biomedical applications: toward in vivo studies. J Mater Chem B 8(11):2216–2232

Pourova J, Kottova M, Voprsalova M et al (2010) Reactive oxygen and nitrogen species in normal physiological processes. Acta Physiol 198(1):15–35

Qin H, Zhi-Hua R (2014) The role of monoacylglycerol lipase (MAGL) in the cancer progress. Cell Biochem Biophys 70(1):33–36

Quinones MM, Gallegos AM, Vankee LF et al (2020) Dysregulation of inflammation, neurobiology, and cognitive function in PTSD: an integrative review. Cogn Affect Behav Neurosci 20(3):455–480

Redy-Keisar O, Shiran F, Ronit S-F et al (2015) NIR fluorogenic dye as a modular platform for prodrug assembly: real-time in vivo monitoring of drug release. ChemMedChem 10(6):999–1007

Rehman T, Asim SM, Muhammad I-U-R et al (2020) Cysteine and homocysteine as biomarker of various diseases. Food Sci Nutr 8(9):4696–4707

Reiffenstein RJ, Hulbert WC, Roth SH (1992) Toxicology of hydrogen sulfide. Annu Rev Pharmacol Toxicol 32:109–134

Reiser J, Brian A, Thomas R (2010) Specialized roles for cysteine cathepsins in health and disease. J Clin Invest 120(10):3421–3431

Robinson PK (2015) Enzymes: principles and biotechnological applications. Essays Biochem 59:1–41

Ross D, David S (2021) The diverse functionality of NQO1 and its roles in redox control. Redox Biol 41:101950

Sabharanjak S, Satyajit M (2004) Folate receptor endocytosis and trafficking. Adv Drug Deliv Rev 56(8):1099–1109

Saito G, Swanson JA, Kyung-Dall L (2003) Drug delivery strategy utilizing conjugation via reversible disulfide linkages: role and site of cellular reducing activities. Adv Drug Deliv Rev 55(2):199–215

Sakabe M, Daisuke A, Mako K et al (2013) Rational design of highly sensitive fluorescence probes for protease and glycosidase based on precisely controlled spirocyclization. J Am Chem Soc 135(1):409–414

Santra S, Charalambos K, Santiesteban OJ et al (2011) Cell-specific, activatable, and theranostic prodrug for dual-targeted cancer imaging and therapy. J Am Chem Soc 133(41):16680–16688

Savani S, Kubra O, Hande G et al (2022) Development of a cysteine responsive chlorinated hemicyanine for image-guided dual phototherapy. Bioorg Chem 122:105725

Schafer FQ, Buettner GR (2001) Redox environment of the cell as viewed through the redox state of the glutathione disulfide/glutathione couple. Free Rad Bio Med 30(11):1191–1212

Sen N (2017) Functional and molecular insights of hydrogen sulfide signaling and protein sulfhydration. J Mol Biol 429(4):543–561

Seo J, Mikyoung P (2020) Molecular crosstalk between cancer and neurodegenerative diseases. Cell Mol Life Sci 77(14):2659–2680

Seung LJ, Jiwon K, Yeong-Sinn Y et al (2022) Materials and device design for advanced phototherapy systems. Adv Drug Deliv Rev 186:114339

Sharma A, Min-Goo L, Shi H et al (2018a) Overcoming drug resistance by targeting cancer bioenergetics with an activatable prodrug. Chem 4(10):2370–2383

Sharma A, Eun-Joong K, Shi H et al (2018b) Development of a theranostic prodrug for colon cancer therapy by combining ligand-targeted delivery and enzyme-stimulated activation. Biomaterials 155:145–151

Sharma A, Arambula JF, Seyoung K et al (2019) Hypoxia-targeted drug delivery. Chem Soc Rev 48(3):771–813

Shetty Y, Priyanka P, Bala P (2019) Emerging vistas in theranostic medicine. Int J Pharm 558:29–42

Shi D, Wenwen L, Ying G et al (2023) Photoactivatable senolysis with single-cell resolution delays aging. Nat Aging 3(3):297–312

Shin WS, Jiyou H, Peter V et al (2016) Cancer targeted enzymatic theranostic prodrug: precise diagnosis and chemotherapy. Bioconjug Chem 27(5):1419–1426

Shindy AH (2012) Basics, mechanisms and properties in the chemistry of cyanine dyes: a review paper. Mini-Rev Org Chem 9(4):352–360

Siegel D, Kepa JK, David R (2012) NAD(P)H: quinone oxidoreductase 1 (NQO1) localizes to the mitotic spindle in human cells. PLoS One 7(9):e44861

Sies H (1999) Glutathione and its role in cellular functions. Free Rad Bio Med 27(9):916–921

Siriwibool S, Nantawat K, Kantapat C et al (2020) Near-infrared fluorescent pH responsive probe for targeted photodynamic cancer therapy. Sci Rep 10(1):1283

Smith CV, Jones DP, Guenthner TM et al (1996) Compartmentation of glutathione: implications for the study of toxicity and disease. Toxicol Appl Pharmacol 140(1):1–12

Sonawane SJ, Kalhapure RS, Thirumala G (2017) Hydrazone linkages in pH responsive drug delivery systems. Eur J Pharm Sci 99:45–65

Song W, Xiao Z, Yangmeihui S et al (2023) Enhancing photothermal therapy efficacy by in situ self-assembly in glioma. ACS Appl Mater Interfaces 15(1):57–66

Su D, Shuai M, Lin S et al (2018) Ubiquitin-specific protease 7 sustains DNA damage response and promotes cervical carcinogenesis. J Clin Invest 128(10):4280–4296

Sufian A, Debojit B, Pallavi B et al (2022) Stimuli-responsive prodrug of non-steroidal anti-inflammatory drug diclofenac: self-immolative drug release with turn-on near-infrared fluorescence. Chem Commun 58(56):7833–7836

Sugimoto W, Hideki I, Katsunori Y et al (2005) Proton and electron conductivity in hydrous ruthenium oxides evaluated by electrochemical impedance spectroscopy: the origin of large capacitance. J Phys Chem B 109(15):7330–7338

Sun T, Guangping Z, Qingbing W et al (2018) A targeting theranostics nanomedicine as an alternative approach for hyperthermia perfusion. Biomaterials 183:268–279

Szabo C (2016) Gasotransmitters in cancer: from pathophysiology to experimental therapy. Nat Rev Drug Discov 15(3):185–203

Takahashi S (2018) Molecular functions of SIRPα and its role in cancer (review). Biomed Rep 9(1):3–7

Tan GJ, Peng ZK, Lu JP et al (2013) Cathepsins mediate tumor metastasis. World J Biol Chem 4(4):91–101

Tanabe K, Hiroshi H, Michiko N et al (2009) Monitoring of biological one-electron reduction by 19F NMR using hypoxia selective activation of an 19F-labeled indolequinone derivative. J Am Chem Soc 131(44):15982–15983

Tang Y, Xing W, Guanqun Z et al (2023) Hypoxia-responsive photosensitizer targeting dual organelles for photodynamic therapy of tumors. Small 19(1):2205440

Teng L, Guosheng S, Yongchao L et al (2019) Nitric oxide-activated "dual-key–one-lock" nanoprobe for in vivo molecular imaging and high-specificity cancer therapy. J Am Chem Soc 141(34):13572–13581

Thambi T, Hyung PJ, Sung LD (2016) Hypoxia-responsive nanocarriers for cancer imaging and therapy: recent approaches and future perspectives. Chem Commun 52(55):8492–8500

Thorn CF, Connie O, Sharon M et al (2011) Doxorubicin pathways: pharmacodynamics and adverse effects. Pharmacogenet Genomics 21(7):440–446

Tian X, Zhao L, Ning D et al (2020) Near-infrared ratiometric self-assembled theranostic nanoprobe: imaging and tracking cancer chemotherapy. Chem Commun 56(25):3629–3632

Trachootham D, Jerome A, Peng H (2009) Targeting cancer cells by ROS-mediated mechanisms: a radical therapeutic approach? Nat Rev Drug Discov 8(7):579–591

Tsolekile N, Simphiwe N, Samuel OO (2019) Porphyrin as diagnostic and therapeutic agent. Molecules 24(14):2669

Ueno M, Kakinuma Y, Yuhki K et al (2006) Doxorubicin induces apoptosis by activation of caspase-3 in cultured cardiomyocytes in vitro and rat cardiac ventricles in vivo. J Pharmacol Sci 101(2):151–158

Upchurch E, Shelly G, Gavin-Rhys L et al (2017) Developments in optical imaging for gastrointestinal surgery. Future Oncol 13(26):2363–2382

Van Gisbergen MW, Marcus C, Jie Z et al (2016) Chemical reactivity window determines prodrug efficiency toward glutathione transferase overexpressing cancer cells. Mol Pharm 13(6):2010–2025
Vander HMG, Cantley LC, Thompson CB (2009) Understanding the Warburg effect: the metabolic requirements of cell proliferation. Science 324(5930):1029–1033
Vaupel P, Arnulf M (2007) Hypoxia in cancer: significance and impact on clinical outcome. Cancer Metastasis Rev 26(2):225–239
Verwilst P, Jiyou H, Jiyeong L et al (2017) Reconsidering azobenzene as a component of small-molecule hypoxia-mediated cancer drugs: a theranostic case study. Biomaterials 115:104–114
Wagner J, Nathan D, Bing Y et al (2015) Overexpression of the novel senescence marker β-galactosidase (GLB1) in prostate cancer predicts reduced PSA recurrence. PLoS One 10(4):e0124366
Wallace JL, Rui W (2015) Hydrogen sulfide-based therapeutics: exploiting a unique but ubiquitous gasotransmitter. Nat Rev Drug Discov 14(5):329–345
Wang B, Nicolaou MG, Siming L et al (1996) Structural analysis of a facile lactonization system facilitated by a "trimethyl lock". Bioorg Chem 24(1):39–49
Wang C, Huan X, Chao L et al (2013) Iron oxide @ polypyrrole nanoparticles as a multifunctional drug carrier for remotely controlled cancer therapy with synergistic antitumor effect. ACS Nano 7(8):6782–6795
Wang M, Jingzhou Z, Lishen Z et al (2017a) Role of tumor microenvironment in tumorigenesis. J Cancer 8(5):761–773
Wang J, Nicole S, Huirong S et al (2017b) Novel strategies to prevent the development of multidrug resistance (MDR) in cancer. Oncotarget 8(48):84559
Wang S-B, Cheng Z, Jing-Jie Y et al (2020) Near-infrared light responsive nanoreactor for simultaneous tumor photothermal therapy and carbon monoxide-mediated anti-inflammation. ACS Central Sci 6(4):555–565
Wang S, Yingcui B, Xin L et al (2022a) Molecular engineering to construct specific cancer cell lysosome targeting photosensitizer by adjusting the proton binding ability. Sensors Actuators B Chem 371:132546
Wang R, Xiang X, Yanjun Y et al (2022b) A glutathione activatable photosensitizer for combined photodynamic and gas therapy under red light irradiation. Adv Healthc Mater 11(4):2102017
Wei D, Yingjie Y, Xingcai Z et al (2020) Breaking the intracellular redox balance with diselenium nanoparticles for maximizing chemotherapy efficacy on patient-derived xenograft models. ACS Nano 14(12):16984–16996
Wei X, Yunhua C, Xianjie J et al (2021) Mechanisms of vasculogenic mimicry in hypoxic tumor microenvironments. Mol Cancer 20(1):7
Wei X, Chi Z, Shasha H et al (2022) A dual-locked activatable phototheranostic probe for biomarker-regulated photodynamic and photothermal cancer therapy. Angew Chem Int Ed Engl 61(26):e202202966
Weiss RB (1992) The anthracyclines: will we ever find a better doxorubicin? Semin Oncol 19(6):670–686
Wells A, Rubin GJ (2003) Phospholipase C-γ1 in tumor progression. Clin Exp Metastasis 20(4):285–290
Whang C-H, Eunsoo Y, Kwon HS et al (2018) A highly GSH-sensitive SN-38 prodrug with an "OFF-to-ON" fluorescence switch as a bifunctional anticancer agent. Chem Commun 54(65):9031–9034
Wilkinson SR, Meyer DJ, Taylor MC et al (2002) The trypanosoma cruzi enzyme TcGPXI is a glycosomal peroxidase and can be linked to trypanothione reduction by glutathione or tryparedoxin. J Biol Chem 277(19):17062–17071
Willems PGM, Rodrigue R, Dieteren Cindy EJ et al (2015) Redox homeostasis and mitochondrial dynamics. Cell Metab 22(2):207–218
Wu X, Xuanrong S, Zhiqian G et al (2014a) In vivo and in situ tracking cancer chemotherapy by highly photostable NIR fluorescent theranostic prodrug. J Am Chem Soc 136(9):3579–3588

Wu Q, Zhiping Y, Yongzhan N et al (2014b) Multi-drug resistance in cancer chemotherapeutics: Mechanisms and lab approaches. Cancer Lett 347(2):159–166
Xiao Y, Tian Z, Xianbin M et al (2021) Microenvironment-responsive prodrug-induced pyroptosis boosts cancer immunotherapy. Adv Sci 8(24):2101840
Xie Z, Taojian F, Jusung A et al (2020) Emerging combination strategies with phototherapy in cancer nanomedicine. Chem Soc Rev 49(22):8065–8087
Xiong H, Kejin Z, Yunfeng Y et al (2018) Tumor-activated water-soluble photosensitizers for near-infrared photodynamic cancer therapy. ACS Appl Mater Interfaces 10(19):16335–16343
Xiong J, Pengzhan W, Subin S et al (2022a) Engineering a theranostic platform for synergistic hypoxia-responsive photodynamic therapy and chemotherapy. Matter 5(5):1502–1519
Xiong J, Chu Jacky CH, Wing-Ping F et al (2022b) Specific activation of photosensitizer with extrinsic enzyme for precisive photodynamic therapy. J Am Chem Soc 144(23):10647–10658
Xu W, Jianxun D, Chunsheng X et al (2015) Versatile preparation of intracellular-acidity-sensitive oxime-linked polysaccharide-doxorubicin conjugate for malignancy therapeutic. Biomaterials 54:72–86
Xu F, Haidong L, Qichao Y et al (2019) Hypoxia-activated NIR photosensitizer anchoring in the mitochondria for photodynamic therapy. Chem Sci 10(45):10586–10594
Yan C, Limin S, Zhiqian G et al (2019) Molecularly near-infrared fluorescent theranostics for in vivo tracking tumor-specific chemotherapy. Chin Chem Lett 30(10):1849–1855
Yang J, Jihye C, Doyeon B et al (2011) Convertible organic nanoparticles for near-infrared photothermal ablation of cancer cells. Angew Chem Int Ed Engl 50(2):441–444
Yang Z, Hong LJ, Mi JH et al (2013) Folate-based near-infrared fluorescent theranostic gemcitabine delivery. J Am Chem Soc 135(31):11657–11662
Yang Y, Yan Z, Qunying W et al (2014) Clinical implications of high NQO1 expression in breast cancers. J Exp Clin Cancer Res 33(1):14
Yang N, Wanyue X, Xuejiao S et al (2020) Recent advances in tumor microenvironment hydrogen peroxide-responsive materials for cancer photodynamic therapy. Nano-Micro Lett 12(1):15
Yang Y, Weiwei Z, Ping H et al (2021) Smart materials for drug delivery and cancer therapy. VIEW 2(2):20200042
Yang Y, Yifeng Z, Ran W et al (2022) A glutathione activatable pro-drug-photosensitizer for combined chemotherapy and photodynamic therapy. Chin Chem Lett 33(10):4583–4586
Yao D, Shuyan Y, Yanshu W et al (2019) An ALP-activatable and mitochondria-targeted probe for prostate cancer-specific bimodal imaging and aggregation-enhanced photothermal therapy. Nanoscale 11(13):6307–6314
Ye M, Xiaohang W, Jianbin T et al (2016) Dual-channel NIR activatable theranostic prodrug for in vivo spatiotemporal tracking thiol-triggered chemotherapy. Chem Sci 7(8):4958–4965
Zahreddine H, Katherine B (2013) Mechanisms and insights into drug resistance in cancer. Front Pharmacol 4:28
Zhai W, Yongkang Z, Ming L et al (2019) Universal scaffold for an activatable photosensitizer with completely inhibited photosensitivity. Angew Chem Int Ed Engl 58(46):16601–16609
Zhang H, Zhijie F (2018) A novel glutathione-triggered theranostic prodrug for anticancer and imaging in living cells. RSC Adv 8(21):11419–11423
Zhang S, Choon-Nam O, Han-Ming S (2004) Critical roles of intracellular thiols and calcium in parthenolide-induced apoptosis in human colorectal cancer cells. Cancer Lett 208(2):143–153
Zhang Z, Kazuhito T, Hiroshi H et al (2005) Bioreduction activated prodrugs of camptothecin: molecular design, synthesis, activation mechanism and hypoxia selective cytotoxicity. Org Biomol Chem 3(10):1905–1910
Zhang A, Lan Y, Ming A (2017a) Reversing the undesirable pH-profile of doxorubicin via activation of a di-substituted maleamic acid prodrug at tumor acidity. Chem Commun 53(95):12826–12829
Zhang F, Guizhi Z, Orit J et al (2017b) Transformative nanomedicine of an amphiphilic camptothecin prodrug for long circulation and high tumor uptake in cancer therapy. ACS Nano 11(9):8838–8848

Zhang D, Junbao D, Chaoshu T et al (2017c) H_2S-induced sulfhydration: biological function and detection methodology. Front Pharmacol 8:608
Zhang Y-H, Xue L, Li H et al (2020) AIE based GSH activatable photosensitizer for imaging-guided photodynamic therapy. Chem Commun 56(71):10317–10320
Zhao X-B, Wei H, Kun G et al (2020a) Precisely traceable drug delivery of azoreductase-responsive prodrug for colon targeting via multimodal imaging. Anal Chem 92(13):9039–9047
Zhao X, Saran L, Mingle L et al (2020b) Oxygen-dependent regulation of excited-state deactivation process of rational photosensitizer for smart phototherapy. J Am Chem Soc 142(3):1510–1517
Zhao J, Zihua W, Miao Z et al (2021a) Integration of a diselenide unit generates fluorogenic camptothecin prodrugs with improved cytotoxicity to cancer cells. J Med Chem 64(24):17979–17991
Zhao H, Lei W, Guifang Y et al (2021b) Inflammation and tumor progression: signaling pathways and targeted intervention. Signal Transduct Target Ther 6(1):263
Zhen X, Jianjian Z, Jiaguo H et al (2018) Macrotheranostic probe with disease-activated near-infrared fluorescence, photoacoustic, and photothermal signals for imaging-guided therapy. Angew Chem Int Ed Engl 57(26):7804–7808
Zheng Z-B, Guozhang Z, Heekyung T et al (2005) N-(2-hydroxypropyl)methacrylamide copolymers of a glutathione (GSH)-activated glyoxalase I inhibitor and DNA alkylating agent: synthesis, reaction kinetics with GSH, and in vitro antitumor activities. Bioconjug Chem 16(3):598–607
Zheng Y, Yang S, Xiaoting M et al (2019) Stabilizing p-dithiobenzyl urethane linkers without rate-limiting self-immolation for traceless drug release. ChemMedChem 14(12):1196–1203
Zhong H, Huan Y, Junjue C et al (2020) Hydrogen sulfide and endoplasmic reticulum stress: a potential therapeutic target for central nervous system degeneration diseases. Front Pharmacol 11:702
Zhou X, Haidong L, Chao S et al (2020) An APN-activated NIR photosensitizer for cancer photodynamic therapy and fluorescence imaging. Biomaterials 253:120089
Zhou F, Shikui Y, Chao Z et al (2021) γ-Glutamyl transpeptidase-activatable near-infrared nanoassembly for tumor fluorescence imaging-guided photothermal therapy. Theranostics 11(14):7045–7056
Zhu H, Jiangli F, Jingyun W et al (2014) An "enhanced PET"-based fluorescent probe with ultra-sensitivity for imaging basal and elesclomol-induced HClO in cancer cells. J Am Chem Soc 136(37):12820–12823
Zou J, Lei X, Nan Y et al (2019) A glutathione responsive pyrrolopyrrolidone nanotheranostic agent for turn-on fluorescence imaging guided photothermal/photodynamic cancer therapy. Mater Chem Front 3(10):2143–2150
Zou Y, Saran L, Tao X et al (2021) Single-molecule förster resonance energy transfer-based photosensitizer for synergistic photodynamic/photothermal therapy. ACS Central Sci 7(2):327–334

Chapter 3
Application of Metal-Organic Frameworks Materials in Tumor Diagnosis and Treatment

Chuanchen Wu and Pengfei Shi

Abstract Metal-organic frameworks (MOFs) consist of metal ions and organic ligands. Due to the customizable nature of ligands and metal nodes, MOFs exhibit a high degree of plasticity in terms of their structure and pore size, allowing them to satisfy various application requirements. Furthermore, the MOFs exhibit a notable combination of elevated specific surface area and porosity. Since the 1990s, extensive research has been conducted and implemented in various cutting-edge areas, such as the identification of cancer-related biomarkers, medical imaging, and the precise delivery of drugs.

Keywords Metal-organic frameworks · Tumor · Optical imaging · Multimodal imaging · Photodynamic therapy · Photothermal therapy · Drug delivery

3.1 Introduction

Metal-organic frameworks (MOFs) play a significant role in the diagnosis and treatment of tumors because their organic ligand moieties frequently have intrinsic fluorescence characteristics that offer MOFs promise for fluorescence detection and imaging technologies. By employing a methodical approach to designing and choosing organic ligands, the fluorescence capabilities of MOFs can be boosted, leading to higher detection sensitivity and imaging resolution. Simultaneously, the metallic nodes present in MOFs can serve as contrast agents in medical imaging techniques like computed tomography (CT) and magnetic resonance imaging (MRI). In order to improve the imaging signal and, thus, the imaging quality, heavy metal components are frequently utilized. Nanoscale MOFs possess notable

C. Wu · P. Shi (✉)
Shandong Provincial Key Laboratory of Tumor Imaging Equipment Development and Tumor Diagnosis and Treatment Integration Technology, School of Medicine, Linyi University, Linyi, Shandong, People's Republic of China
e-mail: wuchuanchen@lyu.edu.cn

© The Author(s), under exclusive license to Springer Nature Singapore Pte Ltd. 2026

P. Shi et al. (eds.), *Functional Organic Materials for Tumor Diagnostics and Therapy*, https://doi.org/10.1007/978-981-96-9478-5_3

benefits compared to conventional drug carriers. They can notably enhance the bioavailability of pharmaceuticals because of their enhanced cellular absorption and prolonged circulation inside the body. Not only may this enhance the effectiveness of medications, but it can also effectively mitigate the adverse effects resulting from high drug concentrations, thereby enhancing patient safety and comfort. Furthermore, the drug-loading capacity and slow-release performance of MOFs can be enhanced through functionalized alteration. Functionalization modification often entails the incorporation of particular functional groups or ligands onto the surface of MOFs, which can establish robust connections with drug molecules, hence enhancing drug loading and stability. Simultaneously, by modulating the intensity of these interactions, precise manipulation of the drug release rate can be attained, thereby accomplishing the sustained release effect of the drug. This ensures that the drug is discharged at the suitable rate and concentration within the body, consequently enhancing the therapeutic outcome. In this chapter, we will systematically discuss the synthesis methods of MOFs materials and their applications in the imaging and treatment of tumors.

3.2 Preparation of the MOFs

In the last 20 years, researchers have employed numerous synthetic techniques to create new MOFs with distinct crystal structures, large surface areas, specific particle and pore sizes, and varied morphologies (Bavykina et al. 2020; Daglar et al. 2021). These advancements have given researchers an exceptional opportunity to modify and enhance the performance of MOFs. Metal-organic frameworks (MOFs) are constructed by combining metal centers with organic ligands (Espallargas and Eugenio 2018; Thorarinsdottir and David 2020; Tsai et al. 2021). The specific characteristics of the produced nanoparticles are influenced by several factors such as different metal ions and ligands, reaction solvents, and reaction durations and temperatures. The interaction between the metal ions and the ligands is a crucial factor in determining the nucleation, production, and final shape of the nanostructures (Xie et al. 2020; Xu et al. 2021; Yaghi et al. 2003). The temperature and solubility of the reactants in the solution have a significant impact on the formation of the reaction site during nucleation. Enhanced solubility of the reactants often promotes the formation of several smaller nanoparticles by increasing the availability of reaction sites and facilitating the development of nanoscale protobodies (Zhao et al. 2021b). Additionally, emulsion or template approaches can restrict the size of self-assembled molecules to the nanoscale (Zhong et al. 2020).

3.2.1 Solvent Thermal Method

The solvent thermal method, which is a variation of the hydrothermal method, is widely used for constructing MOFs materials (Kuppler et al. 2009). The chemical reaction between organic ligands and inorganic components of metal-organic frameworks (MOFs) transpires in sealed containers at elevated pressures and temperatures. The resultant product must subsequently undergo many washing to eliminate surface contaminants (Novaković et al. 2015). The nucleation and growth process comprises four stages: first, the concentration of the reaction monomer in the solution escalates rapidly; second, the concentration exceeds the critical nucleation threshold; third, the swift decline in monomer concentration induces further nucleation events; finally, the resultant crystals elongate to yield MOFs with high crystallinity and uniform morphology. The solvent heating approach facilitates the creation of various MOFs using air-sensitive precursors while simultaneously mitigating the volatilization of harmful chemicals due to the enclosed reaction environment. Furthermore, it can provide a product with reasonably consistent particle size and excellent dispersion. The dimensions and morphology of the reaction product can be modified by varying the concentration and ratio of reactants, duration of the reaction, and temperature (Qian et al. 2012).

Given the elevated temperature and pressure circumstances, it is advisable to use polar solvents with a high boiling point, such as acetonitrile, methanol, dimethylformamide (DMF), dimethyl sulfoxide, and certain mixed solutions (Railey et al. 2017; Yusuf et al. 2022). Behrens and colleagues assessed the impact of these factors on the Zr-MOFs. By incorporating varying amounts of benzoic acid into the solution, they demonstrated the ability to control the size of Zr-MOFs. As the amount of benzoic acid increases, scanning electron microscopy (SEM) imaging reveals a steady accumulation of very small crystals into a larger, fully formed nanocrystal. This nanocrystal has an improved crystal structure and contains around 30 times the amount of $ZrCl_4$ compared to particles of around 200 nm.

3.2.2 Microwave Synthesis Method

Microwave (MW) radiation is a highly promising alternative to the conventional approach and has gained extensive use in synthesizing MOF due to its several advantages (Bag et al. 2015; Beldon et al. 2010). One of the primary benefits of this technology is its rapid heating time, energy efficiency, high productivity, and high level of purity (Choi et al. 2008). These advantages surpass those of previous methods in terms of energy and time savings. By utilizing electromagnetic waves of the suitable frequency, the molecules within the solution can undergo collisions, thereby augmenting the kinetic energy of the molecules and elevating the temperature of the entire system (Głowniak et al. 2021; Hafizovic et al. 2007). The direct interaction of radiation with the solution enhances the efficiency of the heating process (Klinowski et al. 2011). Consequently, microwave synthesis can enhance a quick formation of nuclei and accelerate the process of crystallization (Krusenbaum et al. 2022; Kwon et al. 2020). Unlike traditional solvent thermal methods that necessitate extended annealing, microwave synthesis enables the rapid production of MOF in a matter of minutes (Klinowski et al. 2011; Klimakow et al. 2010). Choi et al. examined the structure and surface characteristics of MOF-5 through the application of microwave heating and solvent heating (Choi et al. 2008). By varying the power levels, radiation time, temperature, solvent concentration, and matrix composition, uniform cubic MOF-5 crystals were achieve in about 30 min. These crystals had an average size of 20-25 μm. Conventional solvent thermal procedures, on the other hand, necessitated a 24-h synthesis period at a temperature of 105 °C to produce MOF-5 crystals ranging in size from 400 to 500 μm. Aside from its quick crystallization capabilities, microwave-assisted synthesis has several other benefits like phase selectivity, decreased particle size, and control over shape. However, it does not facilitate the creation of massive crystals necessary for obtaining accurate structural data.

3.2.3 Mechanochemical Synthesis Method

The mechanochemical synthesis method for producing MOF has significant potential because of its ecologically favorable and economically advantageous characteristics (Głowniak et al. 2021). During this procedure, metal salts and organic ligands are subjected to mechanical grinding in a ball mill, resulting in the disruption of intermolecular connections. Subsequently, new bonds are formed through chemical modification. A single-phase MOF with a high degree of crystallinity, including guest molecules within its empty spaces, have been successfully synthesized (Krusenbaum et al. 2022). This approach offers advantages such as solvent-free operation, generation of safe by-products, rapid reaction rates, and the ability to achieve high product yields at ambient temperature. Nevertheless, the inclusion of a small amount of solvent would enhance the mechanochemical reactions by augmenting the mobility of the reactants at the molecular scale (Klimakow et al. 2010). The drawbacks of this process include the challenge of precisely regulating the synthesis, which necessitates the use of a specialized grinder or mill. Additionally, this method consumes a significant amount of energy, exhibits low working efficiency, and may result in the production of contaminants in the grinding reactor.

3.2.4 Post-Synthesis Modification (PSM) Synthesis

The technique of post-synthetic modification (PSM) was extensively utilized to enhance the range and adaptability of the functional groups integrated into the original MOF structure (Wang et al. 2017a). Cohen explains the idea of "post-synthetic modification" (PSM) and applies the PSM technique to chemically modify the MOF using a single or several reactions, resulting in a MOF that has the same structure but different functions. In order to create MOFs with various functional groups, one can employ multiple PSM pathways involving metal exchange, ligand exchange, guest displacement, and metallization of open coordination sites in ligands (Kalaj and Cohen Seth 2020; Mandal et al. 2021). These routes can be carried out individually or in combination. PSM can only be achieved if the MOF is stable throughout the reaction and does not have any negative impact on its structure, crystallinity, and porosity (Kou et al. 2018). Additionally, it enables and initially discerns places to regulate the chiral selectivity in the MOF. Bennet's team discovered that the amine group in ZIF MOF not only decreased the melting temperature, but also modified the surface characteristics from being water-attracting to water-repelling in the presence of octyl by PSM isocyanate (Bumstead et al. 2022). This strategy is a highly efficient method for synthesizing new MOF compounds under gentle reaction conditions. It offers advantages such as easy handling, versatility, precise alterations, and the ability to generate heterogeneous membrane structures without any noticeable interference. The approach has drawbacks such as the creation of additional flaws within or between crystals, as well as a loss in crystallographic structure.

3.2.5 Other Methods

Synthesis at Room Temperature MOF is synthesized or crystallized without the necessity for heating, as it belongs to a unique category of solvent thermal methods that involve the reaction of an organic ligand with isolated metal ions (Shi et al. 2019). The solvent within the reaction mixture is kept at room temperature to promote the crystallization of the MOF under mild conditions. Triethylamine and other bases are employed in the procedure to deprotonate the organic ligand, leading to the precipitation of the MOF (Yuan et al. 2015). This method is generally preferred for the preparation of MOFs due to its simplicity, high yield, and lack of additional energy requirements. It is a time-consuming process that can range from several hours to several days, and certain reactions necessitate the use of hazardous substances such as DMF (a solvent) and hydrofluoric acid (an acid regulator).

Electrochemical Biosynthesis Electrochemical techniques enable rapid MOF synthesis under mild conditions, preventing the inclusion of unwanted anions (Martinez et al. 2012). Metal ions in solution replace metal salts as the metal source, and are introduced into the reaction mixture via anodic lysis (Van et al. 2012). Proton solvents are used to avoid metal buildup on the cathode. This method allows for faster synthesis at lower temperatures without anionic residue, but reaction time and solvent amount affect MOF structure and morphology. Long durations can lead to crystal coatings that damage the support grid (Sachdeva et al. 2016). The main drawback is low efficiency and incompatibility with large-scale synthesis due to the need for complex equipment and large metal electrodes.

Acoustic Chemical Synthesis High-energy ultrasound induces acoustic cavitation in liquids, creating bubbles that generate high temperatures and pressures, leading to rapid heating and cooling (Głowniak et al. 2023; Qiu et al. 2008; Li et al. 2009). This method consumes less power and time than microwave processes and produces smaller nanocrystals (Son et al. 2008; Khan and Hwa 2015; Yao et al. 2021). For large-scale production, using solvent 1-methyl-2-pyrrolidone (NMP) is more cost-effective than diethylformamide (DEF). However, limitations in penetration depth and product output hinder the method's use in large-scale MOF production.

Microfluidic-Assisted Synthesis The microfluidic technique, which manipulates fluid flow at microscopic levels using a chip, enables efficient and precise synthesis of MOFs (Koryakina et al. 2023; Echaide-Górriz et al. 2018). It enhances the efficiency, adaptability, and expandability of MOF use, as demonstrated by Mao and colleagues in their encapsulation and extraction of DNA for data storage. This process can be completed in minutes, significantly faster than traditional solvent thermal methods. Microfluidic-assisted synthesis provides precise control, high reproducibility, and the ability to produce high-quality MOF-based hollow fiber membranes while conserving reactants (Mao et al. 2023). However, challenges include the need for large space, high power consumption, expensive equipment, and the production of environmentally harmful waste.

3.3 The MOFs Materials in Diagnosis of Tumors

The importance of tumor diagnosis is self-evident, as it directly affects the survival rate and quality of life of patients. The World Health Organization(WHO) reports that cancer is the second foremost cause of mortality globally, resulting in around 9.6 million fatalities annually. Early detection and diagnosis of tumors are crucial for improving treatment outcomes, as the 5-year survival rate for early-stage cancer can be as high as 90%, while it significantly drops to about 10% for late-stage cancer. However, tumor diagnosis faces many challenges, including tumor heterogeneity, the complexity of the tumor microenvironment, and the low abundance of early tumor markers. Traditional tumor diagnostic methods, such as X-rays, CT scans, and MRIs, have limitations in resolution and sensitivity, while emerging biomarker detection technologies, although highly specific, often lack sufficient sensitivity. Therefore, the development of new tumor diagnostic tools and methods is particularly urgent. MOFs as a new type of porous material, show great potential in the field of tumor diagnosis due to their unique structure and functional properties (Zhao et al. 2022a; Zhang et al. 2021; Wang et al. 2021a). The high specific surface area and tunable pore structure of MOFs allow them to carry a large amount of diagnostic agents, such as contrast agents and drugs, thereby improving the efficiency of diagnosis and treatment (Sun et al. 2022). In addition, the surface of MOFs can be functionalized to enhance their targeting ability toward tumor cells, which is significant for improving the sensitivity and specificity of tumor marker detection

(Hou et al. 2024; Mohan et al. 2022; Vodyashkin Andrey et al. 2023). Therefore, research on the application of MOFs in tumor diagnosis can not only promote the advancement of diagnostic technology but may also bring earlier, more accurate diagnoses and more personalized treatment plans for cancer patients.

3.3.1 Fluorescence-Based Tumor Diagnosis

Fluorescence imaging is key for early cancer detection and treatment monitoring. Advancements in medical imaging have highlighted its high sensitivity and resolution. The IARC reported over 19 million new cancer cases and nearly ten million deaths in 2020, emphasizing the need for improved tumor detection methods. MOFs, with their unique structures and tunable properties, are promising for tumor fluorescence imaging. These materials have high specific surface areas and can be tailored for drug delivery and bioimaging. Some MOFs exhibit excellent fluorescence in the near-infrared region, which is beneficial for imaging due to its deeper tissue penetration and less background interference. The versatility in synthesizing MOFs allows for customization of their fluorescence properties to suit various imaging needs. Research into MOF applications for tumor imaging could enhance early diagnosis rates and contribute to personalized medicine and precise treatments.

MOFs serve as effective fluorescent probes for tumor imaging due to their porous structure and adjustable luminescence. They are synthesized from metal ions or clusters and organic ligands, resulting in high surface areas and customizable porosity for loading fluorescent agents. Specific MOF materials can encapsulate dyes like rhodamine and fluorescein, enhancing imaging contrast. Adjusting the metal ions and ligands can optimize luminescence, as demonstrated by Zhang et al. with near-infrared-emitting MOFs for deep tissue imaging. The utility of MOFs as probes also depends on their stability, biocompatibility, and targeting in vivo. Their application in tumor imaging offers new avenues for early diagnosis and treatment monitoring, merging materials science with biomedical research.

A H_2S-activatable nano-photosensitizer (MOF) was engineered to facilitate accurate and highly effective photodynamic treatment (PDT) for colorectal cancer (CRC) (Fig. 3.1) (Li et al. 2022a). The fluorescence and photosensitivity of the MOF decreased during storage and blood circulation as a result of aggregation-caused quenching (ACQ). Upon accumulation of the MOF in the tumor areas of CRC, fluorescence and photosensitivity were triggered, coinciding with the degradation and release of the free photosensitizer (TCPP). Initially, it functioned as a specialist fluorescence imaging probe for colorectal cancer, pinpointing tumor sites, which was crucial for precise irradiation area planning and minimizing phototoxicity to healthy tissues. Secondly, as an H_2S-activatable nano-photosensitizer, it was particularly suitable for photodynamic therapy (PDT) of cancers, including colorectal cancer (CRC) and breast cancer, and shown promise for the treatment of specific malignancies. As a new photosensitizer distinguished by easy synthesis, simple

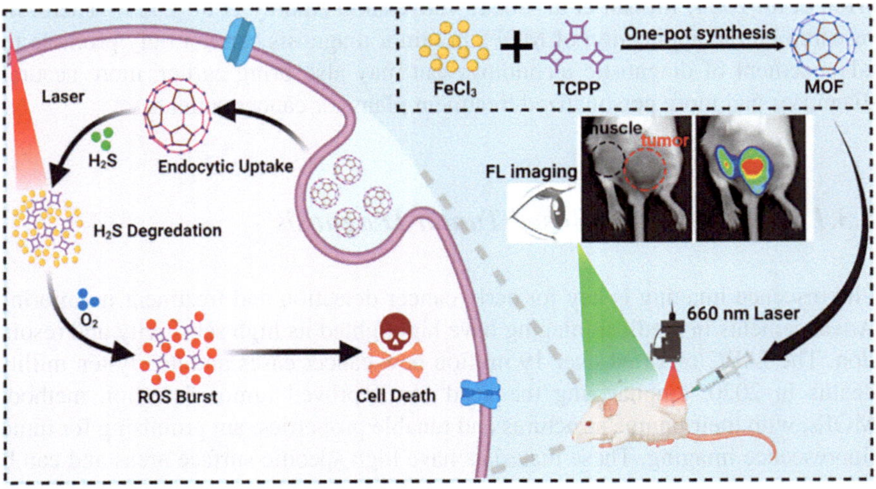

Fig. 3.1 The illustration of MOF synthesis and schematic mechanism of MOF-mediated fluorescence imaging-guided PDT (Li et al. 2022a). Reprinted with permission. Copyright © 2022 Li, Huang, Wei, He, Ma, Lu, Jin, Wang and Wen

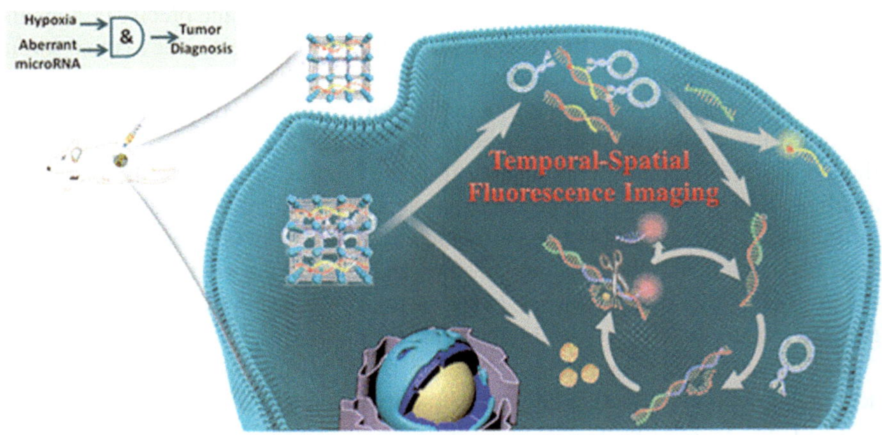

Fig. 3.2 Schematic illustration of hypoxia-responsive DNA@Cu-MOF nanoprobes for aberrant miRNA and hypoxic tumor imaging using Cu2+ self-powered DNAzyme-assisted amplification. Reprinted with permission from Meng et al. (2020). Copyright © 2020, American Chemical Society

formulation, and enhanced biological safety, it has considerable promise for therapeutic use.

Meng et al. presented a tumor microenvironment (TME) hypoxia-responsive and self-powered DNA@Cu-MOF (metal-organic framework) nanosystem for in vivo temporal-spatial fluorescence imaging of aberrant microRNAs (miRNAs) and hypoxic tumors (Fig. 3.2) (Meng et al. 2020). Cu-MOF nanosheets are infused with numerous DNA nanoprobes, such as signal strand-sealed Cu^{2+}-specific DNAzyme precursors and their substrate strands, to form the DNA@Cu-MOF nanosystem. The TME hypoxia-responsive properties of the Cu-MOF and the viability of the DNAzyme-mediated signal amplification method for the sensitive and selective detection of synthetic miRNA in vitro were confirmed, with Cu2+ specifically facilitating the process. The superior biocompatibility of Cu-MOF was confirmed. A successful in vitro distinction was made between normal and cancer cells that exhibited a variety of tumor miRNA expression levels. The in vivo detection of hypoxic tumors was effectively achieved through the use of fluorescence that was dependent on aberrant miRNA expression. The proposed nanosystem offers a novel diagnostic method for hypoxic lesions that are linked to aberrant miRNA and can be readily modified to regulate and modulate biological activities.

Liang and colleagues developed a fluorescent probe, MIL-101(Fe)/PEG-FA/Cy, to initially assess the specific cellular absorption and in vivo targeting efficacy of the nanocarriers MIL-101(Fe)/PEG-FA (Liang et al. 2023). Subsequently, Liang and colleagues employed the radiopharmaceutical 177Lu-MIL-101(Fe)/PEG-FA to assess the in vivo targeting efficacy demonstrated by fluorescence imaging (FI) and to establish an appropriate dosing regimen for radiation therapy in glioma. The in vitro and in vivo anticancer efficacy of 177Lu-MIL-101(Fe)/PEG-FA on glioblastoma was assessed by multimodal imaging data. Research indicates that synthetic radiopharmaceuticals can effectively suppress cancer cell proliferation and tumor growth, stimulate cancer cells to release cytokines, and activate immune responses, resulting in augmented inhibitory effects. Multimodal imaging has demonstrated efficiency and feasibility in the creation and preliminary assessment of innovative metal-organic framework (MOF)-based radiopharmaceuticals. Subsequent research will concentrate on refining treatment protocols, encompassing injectable dosages, fractional dosing, and evaluations of long-term toxicity.

Zhang and group present a flexible platform for the selective imaging of cancer cells and photodynamic therapy (PDT) through miRNA-guided fluorescence (Fig. 3.3) (Zhang et al. 2022a). The platform utilizes UiO-66-NH2 NMOF, which is loaded with Zn(II)-PPIX and gated by two hairpin structures, Hi and Hj. These hairpins are attached to vacant Zr4 ion sites on the NMOF via phosphate units linked to the DNA backbone. The hairpin structure Hi/Hj is engineered to include miRNA recognition sequences within the stem region of Hi and G-quadruplex units within its partially locked stem domain. Upon entering the cell, phosphate ions displace the hairpin unit, liberating Zn(II)-PPIX. The intracellular miRNA-triggered opening of the hairpin structure Hi and the subsequent hybridization chain reaction (HCR) between the hairpin structures produce a highly fluorescent Zn(II)-PPIX-loaded G-quadruplex strand. This mechanism facilitates selective miRNA-guided imaging of cancer cells and selective PDT by generating reactive oxygen species (ROS). The precise imaging and photodynamic therapy of MDA-MB-231 breast cancer cells and OVCAR-3 ovarian cancer cells were illustrated utilizing their respective miRNA-21 and miRNA-221 as biomarkers. This idea can theoretically be utilized for imaging and PDT treatment of any sick cells marked by the overexpression of particular miRNAs. Although the current imaging and PDT treatment process relies on hairpin structures tailored to the corresponding miRNA, the

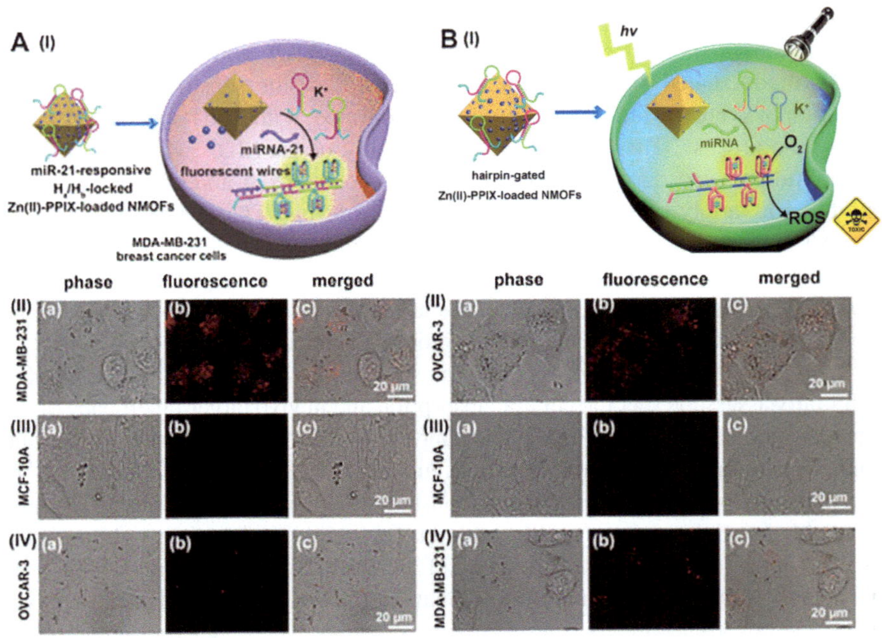

Fig. 3.3 Schematic imaging of MDA-MB-231 cells that consist of overexpressed miRNA-21 using miRNA-21-responsive Ha/Hb-gated Zn(II)-PPIX-loaded NMOFs. Confocal microscopy images of: (II) MDA-MB-231 cells, (III) MCF-10A cells, and (IV) OVCAR-3 cells. (**a**) miR-21-responsive Zn(II)-PPIX-loaded NMOFs in MDA-MB-231 breast cancer cells. (**b**) miR-221-responsive Zn(II)-PPIX-loaded NMOFs in OVCAR-3 ovarian cancer cells. Reprinted with permission from Zhang et al. (2022a). Copyright © 2022, American Chemical Society

potential exists to expand this approach by incorporating other recognition sequences, such as biomarker-specific aptamers, within the hairpin domain.

By efficiently encapsulating a therapeutic DNAzyme agent within ZIF-8 nanoparticles, Wang and colleagues have created a versatile nanoplatform, Ce6-DNAzyme@ZIF-8, intended for imaging-guided combined cancer gene-photodynamic therapy (Fig. 3.4) (Wang et al. 2019a). This nanosystem exhibits effective intracellular transport and pH-responsive drug release, while also ensuring a self-sustaining supply of cofactors. The concurrent administration of functional DNA and the photosensitizer Ce6 to the tumor locale initiates DNAzyme-mediated suppression of EGR-1, alongside cell death prompted by photo-generated reactive oxygen species, culminating in efficient imaging-guided combined gene-photodynamic therapy. This technology is expected to be modified to encapsulate and distribute various biocatalysts that require cofactors, such as apoenzymes, while concurrently supplying additional cofactors (metal ions or organic coenzymes) from MOF scaffolds. This autonomous drug delivery system provides an intelligent and adaptable approach to initiate targeted intracellular biocatalytic processes of therapeutic importance and is anticipated to be crucial in future cancer diagnosis and prognosis.

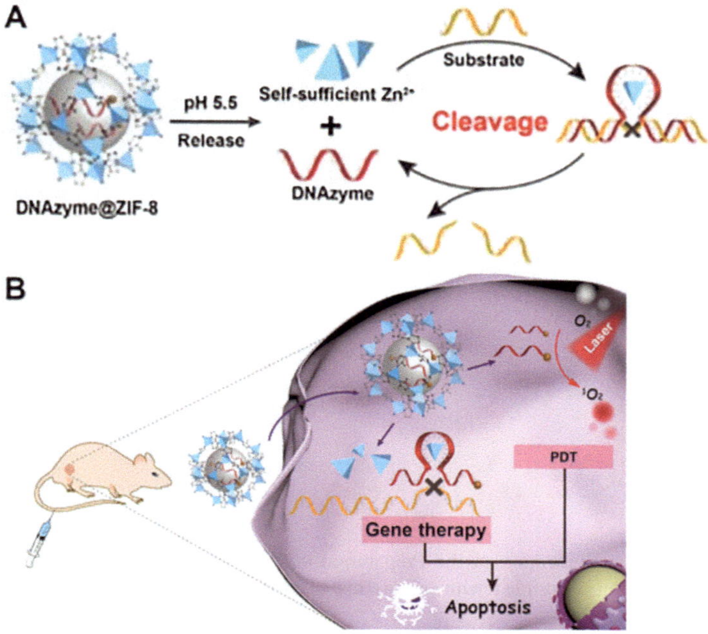

Fig. 3.4 (**a**) The pH-triggered release of DNAzyme and the corresponding Zn^{2+}-ion cofactors for efficient gene silencing therapy. (**b**) An all-in-one DNAzyme@ZIF-8 nanosystem for fluorescence imaging-guided combined gene-photodynamic therapy. Reprinted with permission from (Wang et al. 2019a). Copyright © 2019 Wiley-VCH Verlag GmbH & Co. KGaA, Weinheim

Gao and colleagues have created two novel carbon nanoprobes derived from covalent organic frameworks (COFs) for the purpose of imaging intracellular messenger RNA. These probes are named C-COF-survivin and C-COF-TK1, respectively (Fig. 3.5) (Gao et al. 2021a). The graphene-inspired mesoporous carbon structure of C-COFs provides exceptional UV-Vis absorption, photothermal conversion efficiency, and improved water stability. Compared to traditional COF nanoparticles, these porous C-COF nanoparticles exhibit a greater nucleic acid-loading capacity and an improved fluorescence quenching effect. The resultant nanoprobes exhibit negligible background fluorescence. Upon interaction with the target molecule, the dye-labeled recognition sequence is promptly liberated, therefore "activating" the fluorescent signal. Due to their remarkable selectivity, strong stability, and superior biocompatibility, these nanoprobes have become essential instruments for imaging intracellular TK1 and survivin messenger RNA. The remarkable bioimaging capabilities of these nanoprobes, combined with the porous structure and effective photothermal conversion properties of C-COFs, indicate that the carbonization method not only mitigates the potential biological toxicity linked to the aromatic rigid building blocks of COFs but also enhances their functionality and applicability. The engineered carbon nanoprobes also exhibit potential for further uses, including medication delivery, photothermal therapy, and photoacoustic imaging. This

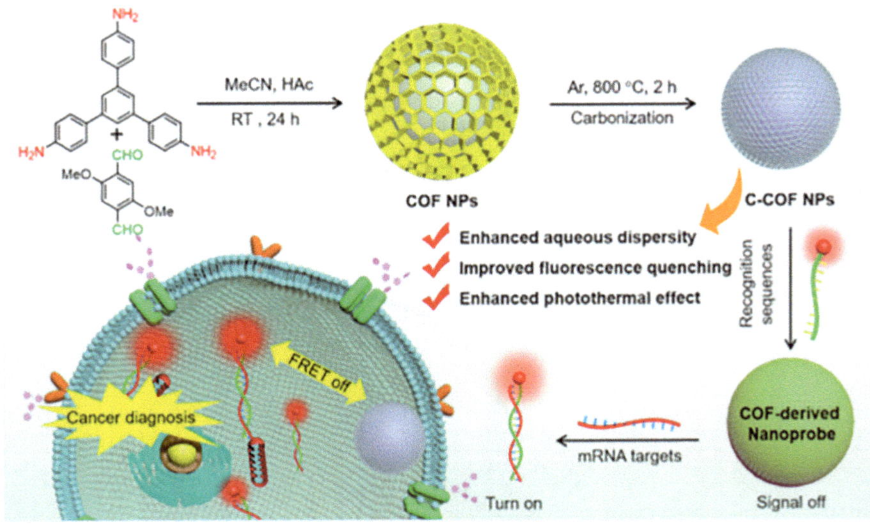

Fig. 3.5 Schematic Illustration of the Preparation of Carbonized COF-Based Nanoprobes for Cancer Cell Imaging. Reprinted with permission from (Gao et al. 2021a). Copyright © 2021, American Chemical Society

research will stimulate the development of more adaptable instruments for the biomedical field.

A biosensor was constructed for the detection of miRNA-122 using Zn^{2+}-triggered AIE and synergistic chemo-photothermal treatment (Fig. 3.6) (Tang et al. 2021). The biosensor is a rationally designed multifunctional PDA-coated MOFs biosensor. In this case, the signal amplification caused by AIE guaranteed extremely high detection sensitivity and was able to successfully differentiate cells that had various amounts of miRNA-122 expression. More crucially, the insertion of a PDA coating resulted in outstanding photothermal conversion efficiency, allowing the created multifunctional PDA-coated MOFs biosensor to achieve synergistic cancer chemo-photothermal therapy. As a result, this technique is anticipated to demonstrate its promise in the implementation of precise diagnosis and synergistic treatment of cancer.

Dong and research team have devised an effective strategy for constructing a MOF hybrid AIEgen anticancer theranostic platform, known as A-NUiO@DCDA@ZIF-Cu (Fig. 3.7). This platform involves the encapsulation of A within a hydrophobic core-shell MOF, a method that allows for substantial tumor accumulation in response to stimuli in the TME, alongside fluorescence imaging functionalities, and a dual therapeutic effect combining photodynamic therapy (PDT) and chemodynamic therapy (CDT) (Dong et al. 2022). Not only does the incorporation of hydrophobic DCDA molecules increase the fluorescence of A by protecting it from aqueous interference, but it also facilitates the self-assembly of A-NUiO@DCDA centers into larger aggregates within tumor cells, thereby increasing its accumulation and retention in tumors. Furthermore, the incorporation of a ZIF-Cu shell with

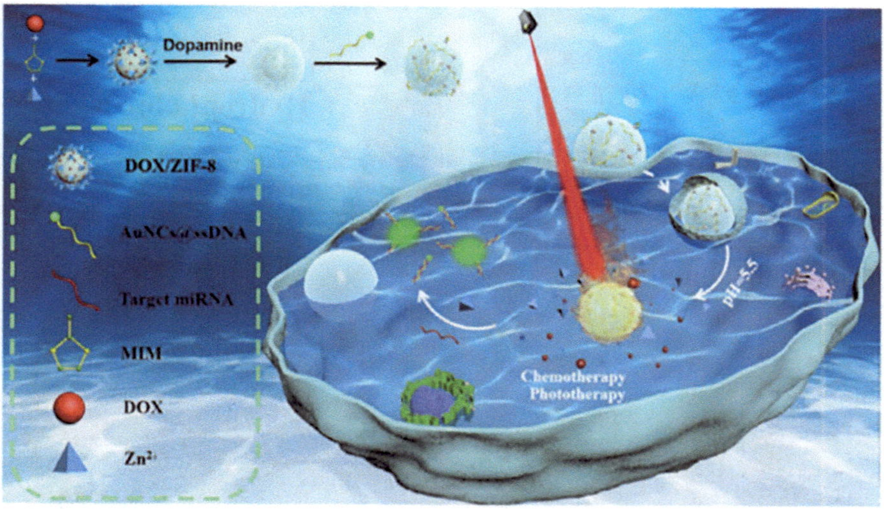

Fig. 3.6 Multifunctional polydopamine (PDA)-coated metal-organic frameworks (MOFs) biosensors for detection of miRNA-122 with Zn2þ-triggered aggregation-induced enhancement (AIE) and synergistic chemo-photothermal therapy in vitro. Reprinted with permission from Tang et al. (2021). Copyright © 2021 Elsevier BV

stimuli-responsive degradation properties facilitates the depletion of glutathione (GSH) and CDT within tumors, while simultaneously decreasing the efficacy of PDT and fluorescence emission of A in non-tumor cells. Consequently, the MOF hybrid AIEgens are capable of precisely targeting and eliminating tumors with minimal adverse effects.

PCN CuS-FA-ICG is a theranostic agent that has been developed for the treatment of breast cancer, which has multiple functions (Fig. 3.8) (Hu et al. 2020). PCN is pivotal in PDT, generating a substantial amount of singlet oxygen ($^1O_2$) upon activation by a 650 nm laser. The CuS nanoparticles, integrated with PCN, exhibit a potent PTT effect. Indocyanine green (ICG) significantly boosts the photothermal conversion efficiency of CuS by enhancing the absorption of 808 nm laser radiation. Furthermore, once PCN-CuS-FA-ICG accumulates in tumor sites, facilitated by the folate receptor, it enables dual-modal imaging through fluorescence and thermal imaging, signifying its substantial potential in tumor diagnosis. Cellular tests have confirmed the effectiveness of monotherapy (either photodynamic therapy or photothermal therapy) and the synergistic effect of their combination, as demonstrated by MTT and live/dead cell staining assays. The concurrent administration of PDT and PTT led to total suppression of tumor development with little adverse effects in a tumor xenograft model. Immunohistochemical staining for caspase-3 expression demonstrated that the combination phototherapy promoted apoptosis in breast cancer cells, resulting in a synergistic anticancer effect. Given these findings, PCN-CuS-FA-ICG emerges as a simple yet effective phototherapeutic agent, offering significant potential for the clinical diagnosis and treatment of breast cancer.

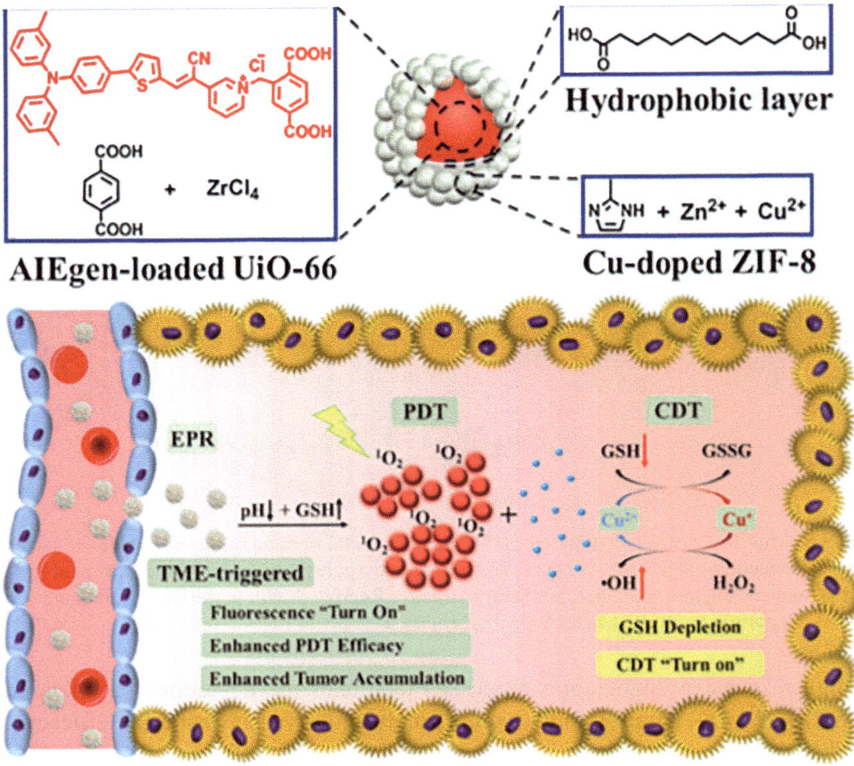

Fig. 3.7 A metal-organic framework (MOF) hybrid AIEgen theranostic platform is designed, termed A-NUiO@DCDA@ZIF-Cu, composed of an AIEgen-loaded hydrophobic UiO-66 (A-NUiO@DCDA) core and a Cu-doped hydrophilic ZIF-8 (ZIFCu) shell. Reprinted with permission from (Dong et al. 2022). Copyright © 2022, American Chemical Society

Shang and et al. have synthesized NMOFs-RGD, which exhibit intense fluorescence, coupled with minimal biotoxicity and outstanding biocompatibility (Fig. 3.9) (Shang et al. 2022). The innovative nanocomposites, HCPT@NMOFs-RGD, not only enhance the hydrophobicity of HCPT but also enable tumor targeting by specifically binding to cells expressing integrin $\alpha v\beta 3$. Additionally, HCPT@NMOFs-RGD combine the therapeutic effects of chemotherapy and photodynamic therapy, demonstrating improved efficacy in treating hepatocellular carcinoma (HCC) with minimal side effects. HCPT@NMOFs-RGD presents an imaging-guided drug delivery and theranostic platform, potentially offering a promising strategy for HCC treatment.

Zou and et al. offered an innovative approach to create a turn-on fluorescence probe for visualizing the hypoxic conditions within solid tumors (Fig. 3.10) (Zou et al. 2020). Initially, an organic, enzyme-responsive small molecule probe, referred to as Probe 1, which exhibited rapid and selective reactivity toward nitroreductase, was successfully synthesized. Subsequently, Probe 2 (Probe 1@ZIF-90@PEG) was

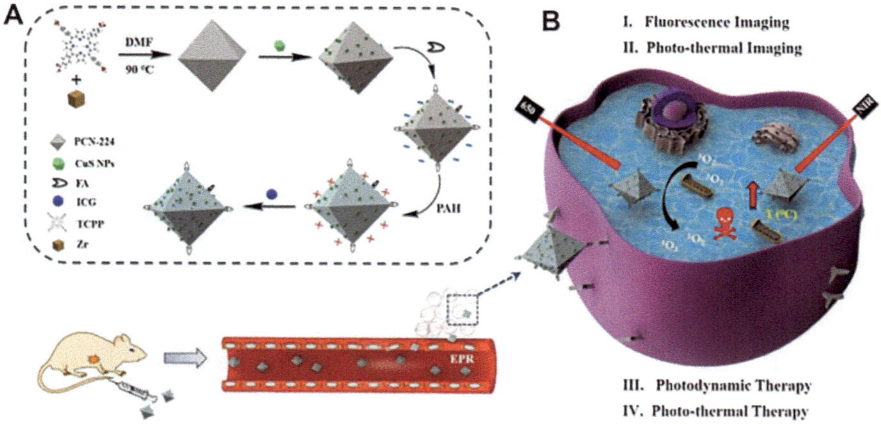

Fig. 3.8 A simple but effective method is reported to fabricate a novel PCN-CuS-FA-ICG-based nanoplatform for dual-modal imaging-guided synergistic photothermal/photodynamic therapy. (**a**) Chemical synthesis of PCN CuS-FA-ICG. (**b**) Conceptual models of therapy and image of PCN-CuS-FA-ICG-based nanoplatform. Reproduced from Hu et al. (2020) with permission from the Royal Society of Chemistry

engineered by embedding the aforementioned Probe 1 into a newly synthesized ZIF-90@PEG matrix. This composite retained the enzyme-responsive properties of nitroreductase and also maintained pH and adenosine triphosphate (ATP) sensitivity. Both synthesized probes were effectively utilized for hypoxia-responsive fluorescence imaging in both in vitro and in vivo settings. This marked the first instance where an enzyme-responsive organic small molecule probe had been integrated into metal-organic frameworks (MOFs) without compromising the probe's enzyme recognition capabilities. The ZIF-90@PEG utilized there may serve as a versatile nanoplatform for the loading of small molecular probes in turn-on fluorescence imaging. Their findings were poised to significantly influence the development of new enzymatic response probes for imaging applications both in vitro and in vivo.

A novel theranostic platform has been introduced, featuring a hairpin probe and photosensitizing metal-organic frameworks (HPMOF). This system consists of photosensitizing MOFs (PMOFs) and hairpin probes that are labeled with a fluorophore and quencher. In this configuration, the PMOF functions as both the photosensitizer and the nanocarrier for the hairpin probe (Fig. 3.11) (Yang et al. 2023a). A layer of ZIF-8 was applied to the HPMOF to create the dual-layered nanotheranostics (HPMOF@ZIF-8). The HPMOF@ZIF-8 demonstrated significant DNA-loading capacity and effective intracellular delivery for imaging tumor-related miRNA. Furthermore, HPMOF@ZIF-8 demonstrated a high efficiency in generating reactive oxygen species, which triggered cell apoptosis, resulting in effective photodynamic therapy. The distinct expression of miRNA in normal versus cancer cells allows HPMOF@ZIF-8 to identify cancer cells via miRNA imaging, facilitating more precise cancer treatment and presenting a promising theranostic nanoplatform.

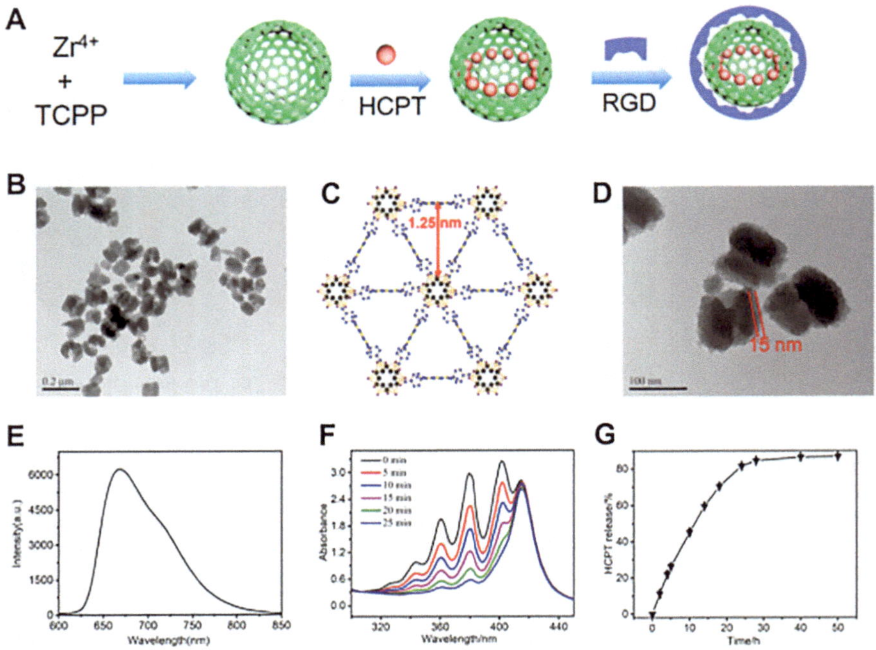

Fig. 3.9 Synthesis and characterization of HCPT@NMOFs-RGD. (**a**) Schematic illustration of the synthetic route for HCPT@NMOFs-RGD nanocomposites. (**b**) Transmission electron microscopy (TEM) image of NMOFs. (**c**) Morphological features of NMOFs showing uniform triangular one-dimensional channels in the structure. (**d**) TEM image of NMOFs-RGD, revealing an RGD layer uniformly coating the NMOFs surface with a thickness of 15 nm. (**e**) Fluorescence spectra of the NMOFs-RGD aqueous solution (excitation at 530 nm). (**f**) Absorbance changes of ABDA in the NMOFs-RGD system over different time periods under 655 nm laser irradiation. (**g**) HCPT release profile of HCPT@NMOFs-RGD nanocomposites in PBS buffer. Reproduced from Shang et al. (2022). Copyright © 2022 Shang et al.

A multifunctional theranostic system, designated as MOF-derived theranostic system (DAPM), was meticulously engineered by harnessing the properties of photosensitive metal-organic frameworks (PMOF, PM) and a DNA AND logic gate (DA) (Fig. 3.12) (Yang et al. 2023b). DAPM exhibited exceptional biostability and demonstrated its ability to detect miR-21 and miR-155 with a low limit of detection (LOD) of 89.10 pM for miR-21 and 54.02 pM for miR-155. The DAPM probe's superior capability for tumor cell recognition was underscored by the emission of a fluorescence signal that was specifically targeted to tumor cells where miR-21 and miR-155 were concurrently present. Furthermore, DAPM was capable of producing efficient reactive oxygen species (ROS) generation and exhibited cytotoxicity in a concentration-dependent manner upon light irradiation, offering potent photodynamic therapy for combating tumors.

Sun et al. reported a novel multifunctional ZIF-8, embedded within a sodalite (sod) framework, was engineered using a straightforward one-pot synthesis technique for the specific targeting of tumors, as well as for bioimaging and drug

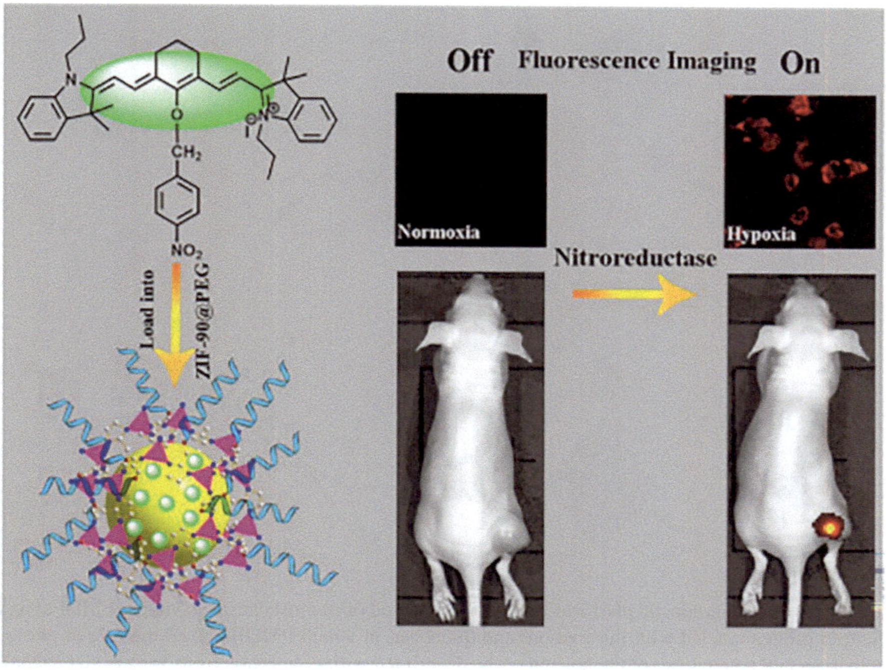

Fig. 3.10 A new organic small molecular probe and a novel fluorescence imaging system developed through loading the organic probe into ZIF-90@PEG were successfully synthesized for off-on fluorescence imaging via enzyme response to nitroreductase in vitro and in vivo. Reproduced from (Zou et al. 2020) with permission from the Royal Society of Chemistry

delivery purposes (Fig. 3.13) (Sun et al. 2023). The synthesized DOX/Fe_3O_4@ZIF-8 showcases a pronounced hexagonal morphology with particle dimensions ranging from 50 to 100 nm. The inclusion of Fe_3O_4 endows DOX/Fe_3O_4@ZIF-8 with superior magnetic tumor-targeting capabilities and the ability to perform magnetic resonance imaging in vivo. Furthermore, the presence of DOX within the structure imparts exceptional fluorescence imaging properties to the material. As anticipated, DOX is capable of being released from DOX/Fe_3O_4@ZIF-8, demonstrating pronounced antitumor effects in vivo. Additionally, the apoptotic process of DOX/Fe_3O_4@ZIF-8 on HepG2 cells has been thoroughly investigated.

The NMOF Cu-Mi-UiO-66, a critical component of the dual-responsive fluorescence nanoprobe Cu-Mi-UiO-66/aptamer, was synthesized by Liu and his coworkers with ease through a straightforward one-step solvothermal method (Fig. 3.14) (Liu et al. 2024a). This process utilized $ZrCl_4$, $Cu(NO_3)_2$, and the maleimide-functionalized ligand H2L1 as co-precursors. The integration of H2L1 within the Cu-Mi-UiO-66 framework and the assembly of Cy5-labeled ATP aptamer onto its surface equip the nanoprobe Cu-Mi-UiO-66/aptamer with the capability to quantify both GSH and ATP. This ingeniously designed nanoprobe exhibits high selectivity and sensitivity, coupled with excellent biocompatibility, and operates on an

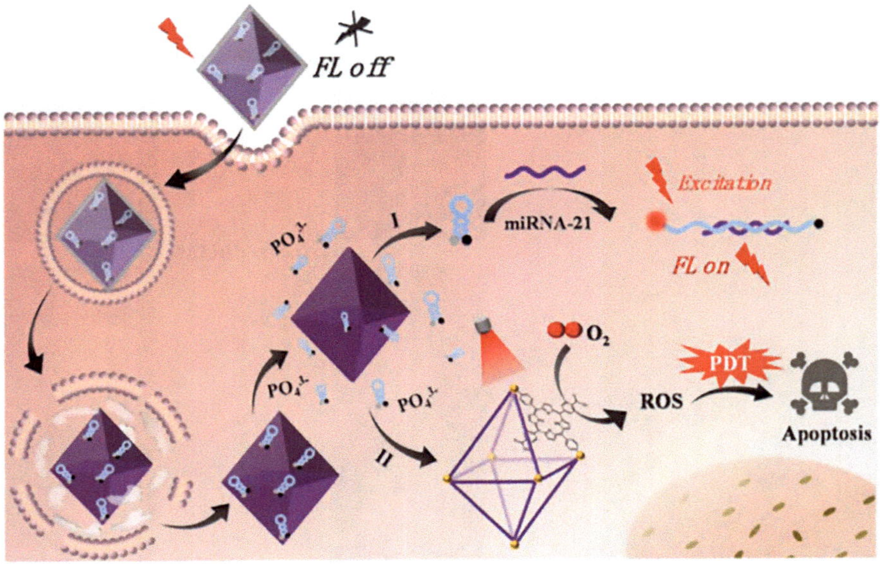

Fig. 3.11 New theranostic platform HPMOF, composed of photosensitizing MOFs (PMOFs) and hairpin probes labeled with fluorophore and quencher, in which PMOF played the role of photosensitizer and nanocarrier of the hairpin probe. Reprinted with permission from Yang et al. (2023a). Copyright © 2022 Elsevier B.V

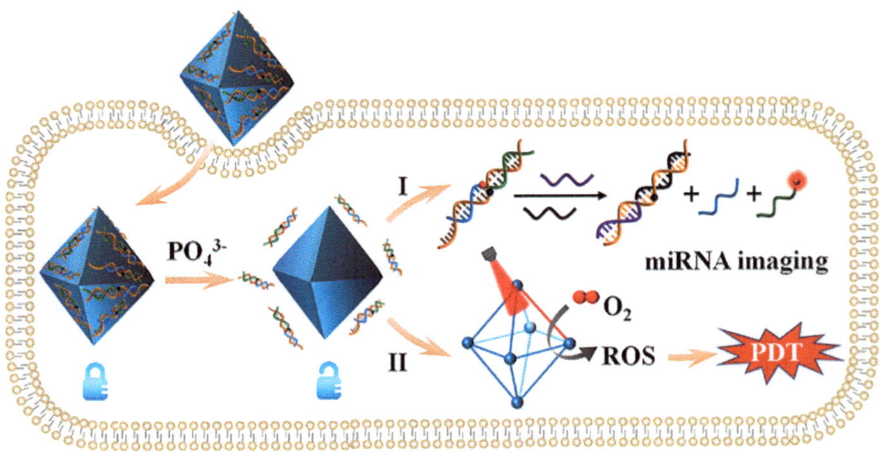

Fig. 3.12 Preparation process of the designed DAPM theranostic nanosystem and its application in miRNA imaging and photodynamic therapy. Reprinted with permission from Yang et al. (2023b). Copyright © 2023 Elsevier B.V

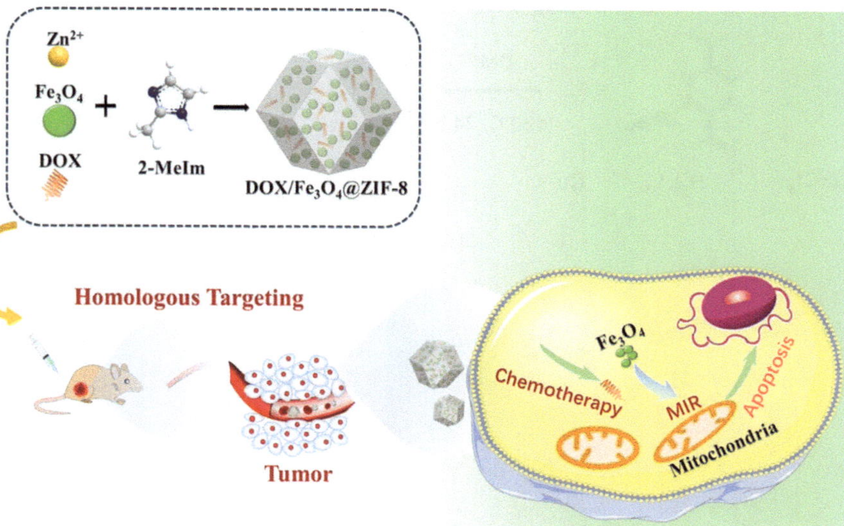

Fig. 3.13 Schematic of synthesis of $DOX/Fe_3O_4@ZIF-8$. Reproduced from Sun et al. (2023) with permission from the Royal Society of Chemistry

"off-to-on" mechanism in response to GSH and ATP, both in aqueous solutions and within living cells, with minimal cross-talk between the two analytes. Confocal imaging results have shown that the nanoprobe can be readily internalized by HepG2 cells and can detect intracellular GSH and ATP, allowing for the tracking of their concentration changes under various stimulations.

Qin and colleagues have successfully developed GSH/pH dual-response biodegradable nanoparticles (FZIF-8/DOX-MIPs), which demonstrate targeted accumulation at tumor sites and effective penetration of intracellular barriers for drug release (Fig. 3.15) (Qin et al. 2020). Importantly, the epitope-imprinted polymer provides FZIF-8/DOX-MIPs with two key characteristics. Initially, these nanoparticles have the capability to selectively target and gather in tumors that exhibit high levels of CD59 glycoproteins, utilizing the binding of the N-terminal epitope of these glycoproteins with the epitope-imprinted cavities. The red fluorescence of the embedded CDs facilitates targeted fluorescence imaging. Furthermore, the incorporation of disulfide bonds and DMAEMA into the backbone of the epitope-imprinted polymer guarantees that the MIP coating on FZIF-8/DOX undergoes degradation within the tumor microenvironment. The encapsulated FZIF-8/DOX undergoes further degradation when exposed to a mildly acidic environment, resulting in the release of DOX for targeted and controllable treatment of tumors that overexpress CD59 glycoproteins. The nanoparticles exhibited remarkable targeted fluorescence imaging and tumor treatment effectiveness in both in vitro and in vivo experiments, while significantly reducing side effects on normal tissues. The multifunctional FZIF-8/DOX-MIPs are expected to be applied in clinical oncology for upcoming GSH/pH dual-response drug release. Furthermore, as shown in Table S1, the

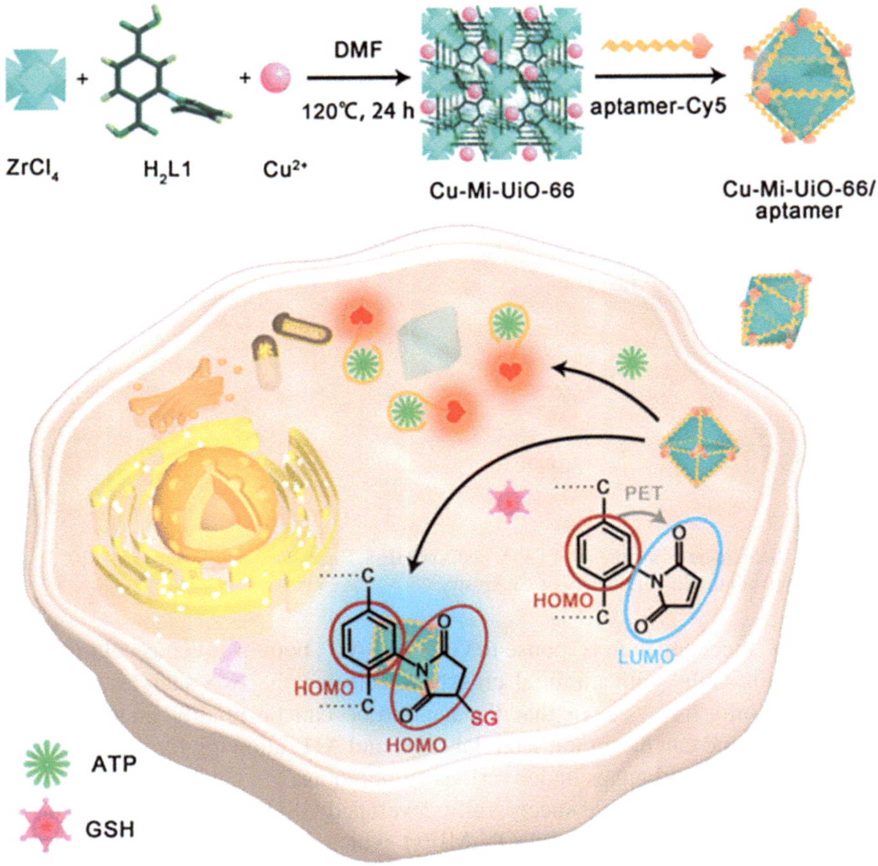

Fig. 3.14 Schematic of synthesis of MOF Cu-Mi-UiO-66. Reproduced from Liu et al. (2024a) with permission from the Royal Society of Chemistry

majority of current MIPs for drug delivery are nondegradable; therefore, the MIP nanoparticles developed in this study also open up a new path for the progress of degradable molecular imprinting polymers.

Wan and coworkers have engineered a biomimetic and oxygen self-supplying nanotheranostic system, designated as MMCC@EM, using a straightforward one-pot method for optical molecular imaging-guided photodynamic therapy (PDT) (Fig. 3.16) (Wan et al. 2024). The coating of EM significantly enhances tumor accumulation, thereby extending the blood circulation duration. Zeolitic imidazolate-based metal-organic framework (MOF) with high drug-loading capacity and pH-responsive characteristics was selected as the nanocarrier. This MOF is designed to efficiently release the photosensitizer Chlorin e6 (Ce6) and catalase (CAT) within

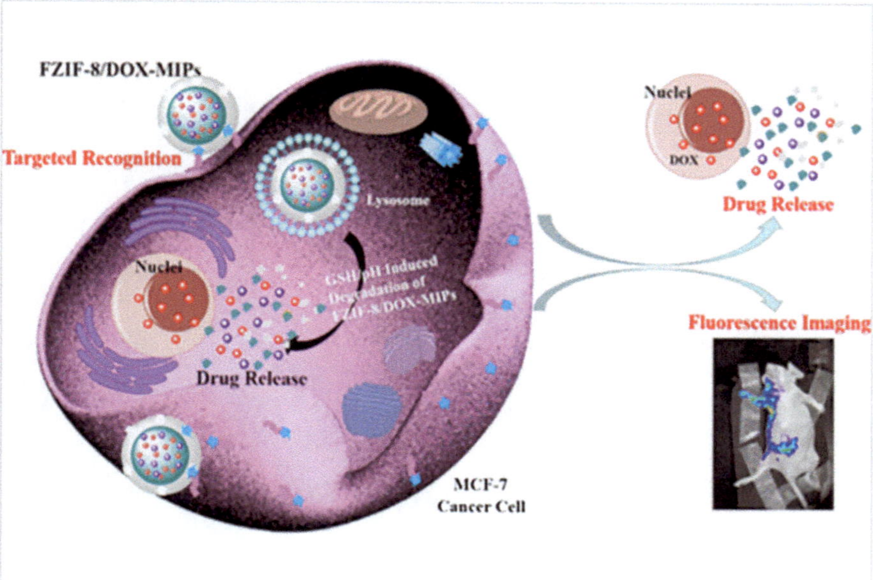

Fig. 3.15 Schematic illustration of targeted imaging and GSH/pH-responsive drug delivery of FZIF-8/DOX-MIPs. Reprinted with permission from (Qin et al. 2020). Copyright © 2020, American Chemical Society

the acidic tumor microenvironment, a process that can be monitored by fluorescence (FL) imaging. Once accumulated at the tumor site, the catalytic activity of CAT decomposes endogenous hydrogen peroxide into water and oxygen. This alleviates the hypoxic conditions, a change that can be quantified by photoacoustic (PA) imaging through the measurement of oxygen saturation. The guidance provided by FL/PA dual-mode real-time imaging allows for precise optimization of light irradiation time. Upon laser exposure, the copious singlet oxygen ($^1O_2$) generated by Ce6 substantially decreases the intracellular mitochondrial membrane potential ($\Delta\Psi m$), leading to caspase-3-mediated cellular apoptosis. Additionally, this apoptotic process triggers a pronounced immunogenic cell death (ICD) effect, characterized by the exposure of calreticulin (CRT), the release of high mobility group box 1 (HMGB1), and the secretion of adenosine triphosphate (ATP). The potent ICD effect induces dendritic cell (DC) maturation and cytotoxic T lymphocyte (CTL) infiltration, thereby enhancing therapeutic efficacy. Notably, the MMCC@EM plus laser treatment effectively suppresses tumor growth in vivo with minimal side effects. Consequently, the developed MMCC@EM nanotheranostics offer a translational paradigm for spatially and temporally precise optical molecular imaging-guided, self-oxygenating PDT.

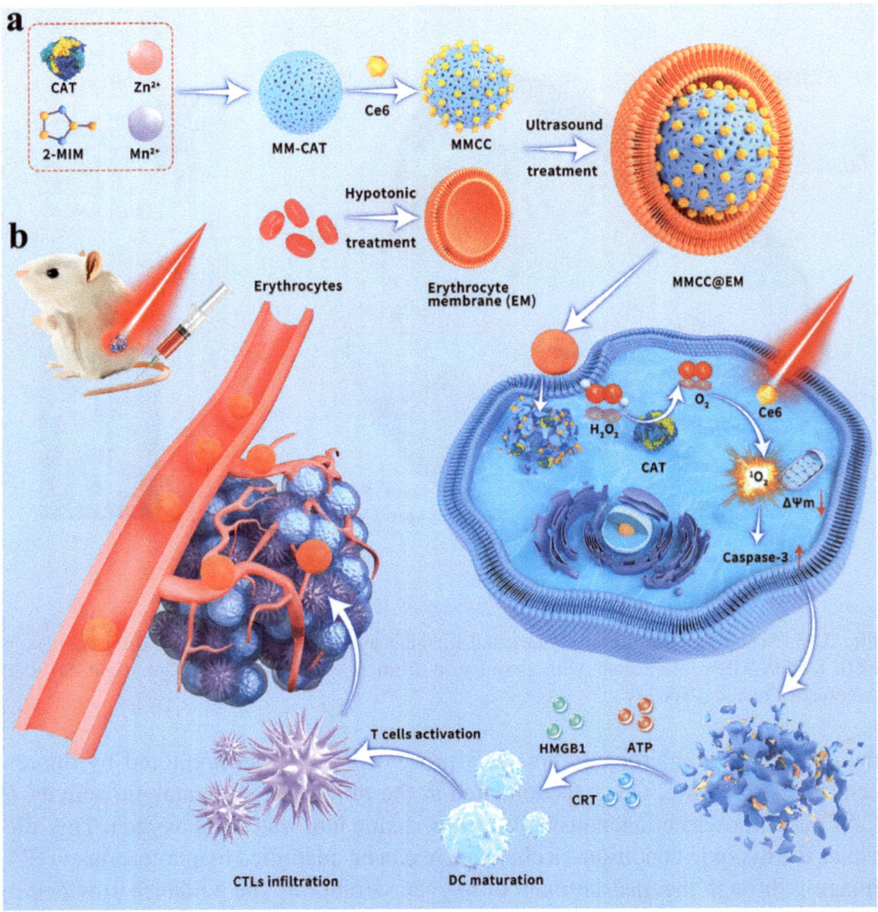

Fig. 3.16 Schematic illustration of MMCC@EM nanotheranostics for FL/PA duplex imaging-escorted self-oxygenation PDT. (**a**) The synthetic process of MMCC@EM. (**b**) The application in a mouse model, where erythrocytes undergo hypotonic treatment to form erythrocyte membranes. Reproduced from Wan et al. (2024) with permission. Copyright © 2024 Wiley-VCH GmbH

3.3.2 Near-Infrared Fluorescence-Based Tumor Diagnosis

Chen and colleagues have created an optimal nanoplatform, AR-ZS/ID-PNPs, that is targeted to tumors and responsive to pH changes. This innovative platform integrates photothermal therapy (PTT), photodynamic therapy (PDT), and chemotherapy to facilitate a synergistic approach to cancer treatment (Fig. 3.17) (Chen et al. 2023). Through the modification of the silk fibroin bio-template concentration, Chen and colleagues successfully produced AR-ZS/ID-PNPs that exhibit an appropriate size and outstanding dispersity, along with a high encapsulation efficiency for indocyanine green (ICG) and doxorubicin (DOX). The pH-sensitive characteristics

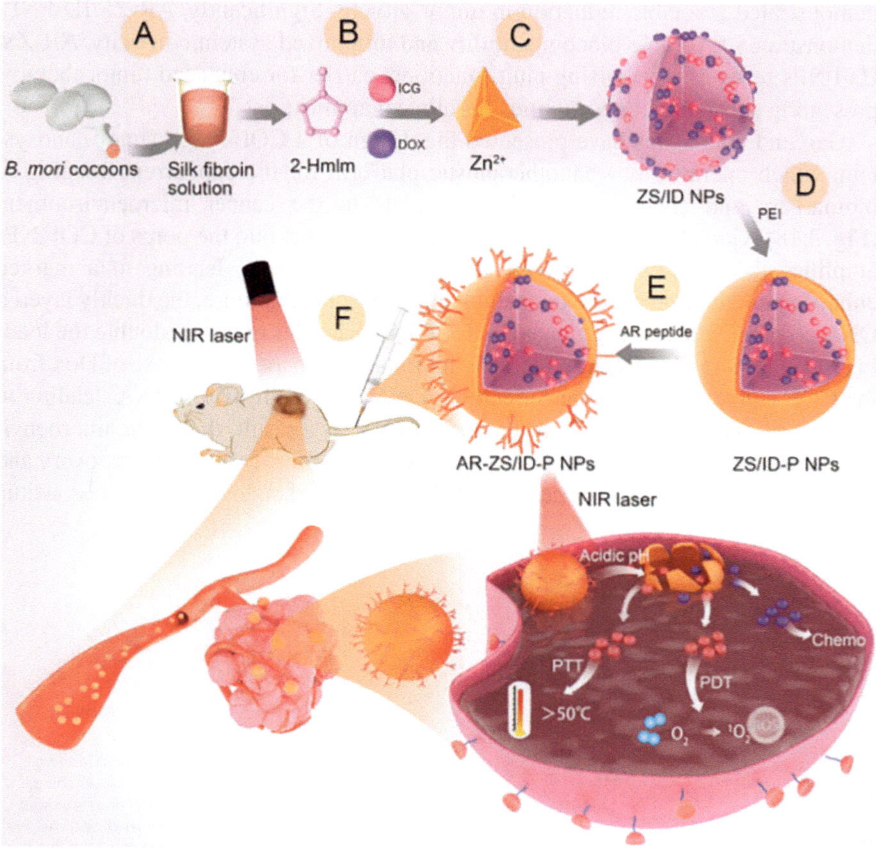

Fig. 3.17 Schematic illustration for the preparation of AR-ZS/ID-P NPs and target peptide-guided breast cancer therapy. (**a–c**) SF was dissolved in 2-HmIm aqueous solution as a bio-template(a), then mixed with DOX and ICG(b). $Zn(NO_3)_2{\cdot}6H_2O$ was added to form ZS/ID NPs (drug-loaded ZIF-8) via one-pot synthesis (c). (**d**) PEI-coated ZS/ID NPs (ZS/ID-P NPs) enhanced stability and peptide conjugation. (**e**) The MCF-7-targeting peptide AR was linked to ZS/ID-P NPs using EDC/NHS. (**f**) Intravenously injected AR-ZS/ID-P NPs accumulated in tumors via AR-mediated targeting. In the acidic tumor microenvironment, the NPs degraded, releasing DOX and ICG. Subsequent 808 nm laser irradiation induced cancer cell death. Reproduced from Chen et al. (2023) with permission. Copyright © 2023 Chen et al. Advanced Science published by Wiley-VCH GmbH

of ZIF-8 were retained in the AR-ZS/ID-PNPs, facilitating controlled drug release in the acidic tumor microenvironment while ensuring stability under neutral conditions. The modification of the AR peptide enabled the selective targeting and recruitment of AR-ZS/ID-PNPs to tumor cells. Following laser irradiation, ICG generated harmful heat and reactive oxygen species (ROS) to enhance photothermal therapy (PTT) and photodynamic therapy (PDT), while also allowing for thermal and fluorescence imaging to effectively direct DOX-triggered chemotherapy. The in vivo experiments validated the increased tumor accumulation of AR-ZS/ID-PNPs and demonstrated a notable reduction in tumor growth. Significantly, AR-ZS/ID-PNPs demonstrated favorable biocompatibility and minimized systemic toxicity. AR-ZS/ID-PNPs serve as a promising multifunctional carrier for enhanced tumor therapy, presenting substantial opportunities for clinical applications.

Gao and colleagues have presented the design of a COF-based smart nanosystem, which operates as a nanotheranostic platform for the concurrent imaging of biomarkers and chemotherapy that responds to the cancer microenvironment (Fig. 3.18) (Gao et al. 2021b). The incorporation of Dox into the pores of COF NPs amplifies the interaction between ssDNA and COF NPs, leading to a marked enhancement in fluorescence quenching efficiency. Furthermore, the thickly layered DNA coating, which is augmented by COF nanoparticles to nearly double the loading of single-stranded DNA, effectively inhibits the nonspecific release of Dox from the COF nanoparticles. In cancer cells, ssDNA pairs with TK1 mRNA, leading to the restoration of the fluorescence signal of Cy5; as a result, the acidic microenvironment triggers the release of Dox. Conversely, in typical cells, signal recovery and Dox release are minimal because of reduced biomarker levels and a less acidic

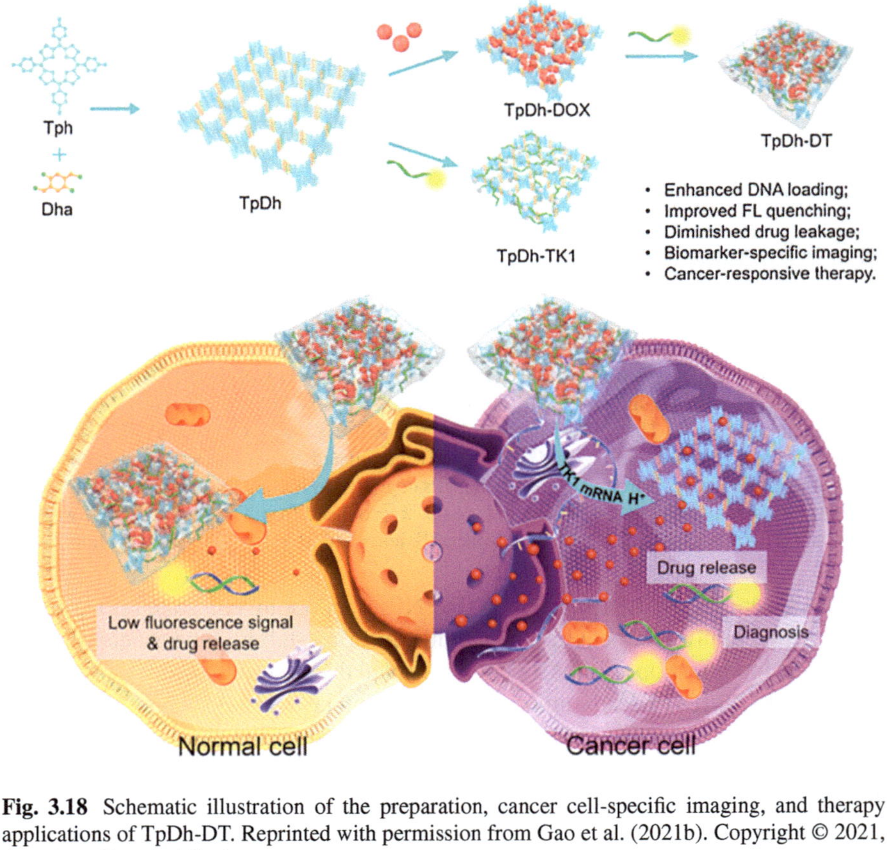

Fig. 3.18 Schematic illustration of the preparation, cancer cell-specific imaging, and therapy applications of TpDh-DT. Reprinted with permission from Gao et al. (2021b). Copyright © 2021, American Chemical Society

microenvironment. This advanced nanosystem shows potential as an effective platform for cancer theranostics, and this study offers new perspectives on the thoughtful design of COF-nucleic acid nanosystems for use in analytical and healthcare sectors.

Shu and coworkers have engineered PLNPs@MIL-100(Fe) employing a conventional layer-by-layer approach, designed for dual-mode imaging through near-infrared (NIR) luminescence and magnetic resonance (MR) imaging, as well as for drug delivery (Fig. 3.19) (Shu et al. 2023). The capability of sustained renewal is demonstrated by the NIR luminescence of PLNPs@MIL-100(Fe) at 700 nm, which enables protracted in vivo imaging. These nanoparticles have demonstrated a transverse relaxivity of 3.95 mg mL^{-1} s^{-1}, which is highly advantageous for in vivo magnetic resonance imaging. Consequently, the essential imaging data required for tumor treatment is provided by the persistent NIR luminescence and MR imaging capabilities of tumor-bearing rodents. Additionally, PLNPs@MIL-100(Fe) exhibit a significant drug-loading capacity and enable pH-responsive drug release exclusively at the tumor site, thereby enhancing the efficacy of drug delivery. The PLNPs@MIL-100(Fe) technology ensures efficient drug delivery and enables autofluorescence-free dual-modal tumor imaging. This advancement enables the

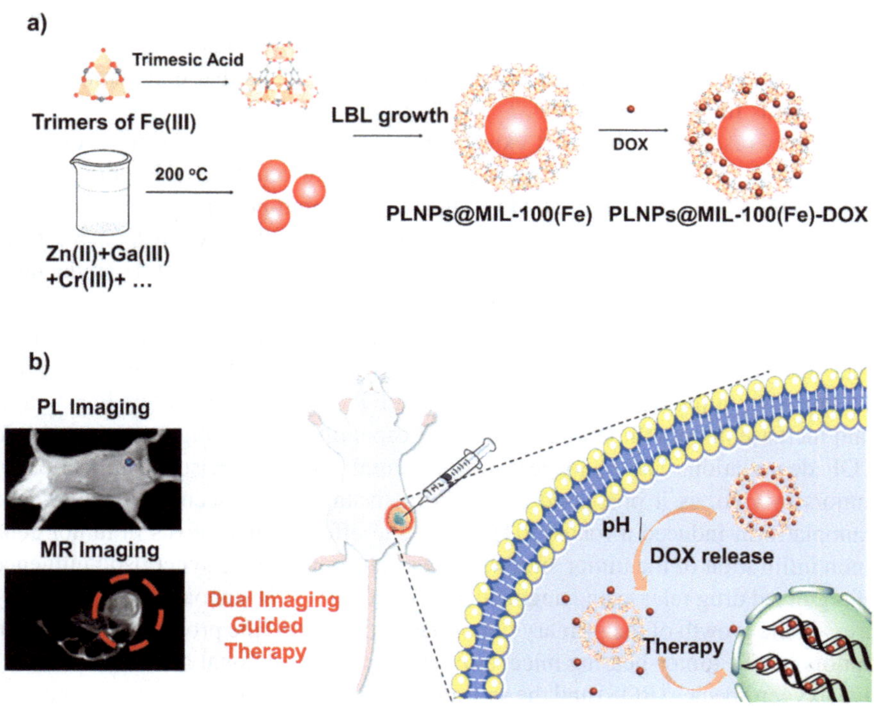

Fig. 3.19 Schematic illustration of the preparation of PLNPs@MIL-100(Fe). (**a**) Schematic illustration for the preparation of PLNPs@MIL-100(Fe) nanoparticles. (**b**) Luminescent (PL)/magnetic resonance (MR) dual-modal imaging and drug delivery. Reproduced from Shu et al. (2023) with permission from the Royal Society of Chemistry

precise, multimodal imaging-guided administration of a wide range of pharmaceuticals, such as pharmacological agents, protein therapeutics, and nucleic acid treatments.

Feng and et al. have reported the development of a biomimetic nanomotor, mPPy@COF-Por, which comprises a polypyrrole (PPy) core, a porphyrin-based covalent organic framework (COF) shell, and a coating of HCT116 cancer cell membrane (Fig. 3.20) (Feng et al. 2023). This nanomotor, camouflaged by the cancer cell membrane, is capable of targeted recognition of cancer cells and demonstrates efficient self-propulsion through thermophoresis upon near-infrared (NIR) light irradiation. In vitro and in vivo experiments have confirmed that this COF-based nanomotor possesses trimodal imaging guidance (infrared reflectance imaging/photoacoustic imaging/fluorescence imaging) and can perform combined photothermal therapy (PTT) and photodynamic therapy (PDT) against HCT116 cancer cells. The COF nanomotor, functioning without external fuel, exemplifies phototheranostics by merging imaging and therapeutic modalities into a singular multifunctional platform for targeted tumor treatment. This novel multifunctional therapeutic approach considerably expands the applicability of COF-based materials in biological fields. It is important to highlight that the majority of COF-based nanomotors documented to date depend on external stimuli, primarily light, for their propulsion. Nonetheless, given the distinctive physiological traits of tumor microenvironments, abundant in glutathione and hydrogen peroxide, there exists a significant opportunity to engineer COF-based nanomotors that may harness these endogenous fuels for propulsion.

A units resulted in the most extensive response region and the most effective photodynamic therapy (PDT) effect, outperforming analogs with one or three groups (Fig. 3.21) (Liu et al. 2024b). DAC and the high-performance aggregation-induced emission (AIE) generator were enclosed in a hypoxia-responsive COF based on azo compounds. This COF can degrade in the presence of azo reductases under hypoxic conditions, hence enabling the regulated release of the DAC medication. The AIE generator's PDT effect significantly exacerbated hypoxia, thereby expediting the drug release process. Interestingly, the azo-COF was able to suppress the emission of the AIE generator through a PET mechanism, resulting in a significant increase in both fluorescence and PDT capabilities following hypoxia-induced COF degradation. The nanoagent was essential for the precise identification of tumors in vivo, as it possessed an in situ activatable fluorescence signature. The nanoplatform induced a robust PDT effect and effective pyroptosis in tumor cells upon infiltration of the tumor site, which was attributed to the synergistic influence of regulated drug release and augmented PDT efficacy. The nanoagent not only suppressed the growth of the primary tumor but also impeded the progression of distant tumors in 4T1 tumor-bearing mice, as a result of its exceptional generation of reactive oxygen species (ROS) and the self-amplifying release of the pyroptosis inducer. This research provides valuable insights into the development of TME-activatable phototheranostic and chemotherapeutic systems, thereby improving self-synergistic cancer immunotherapy.

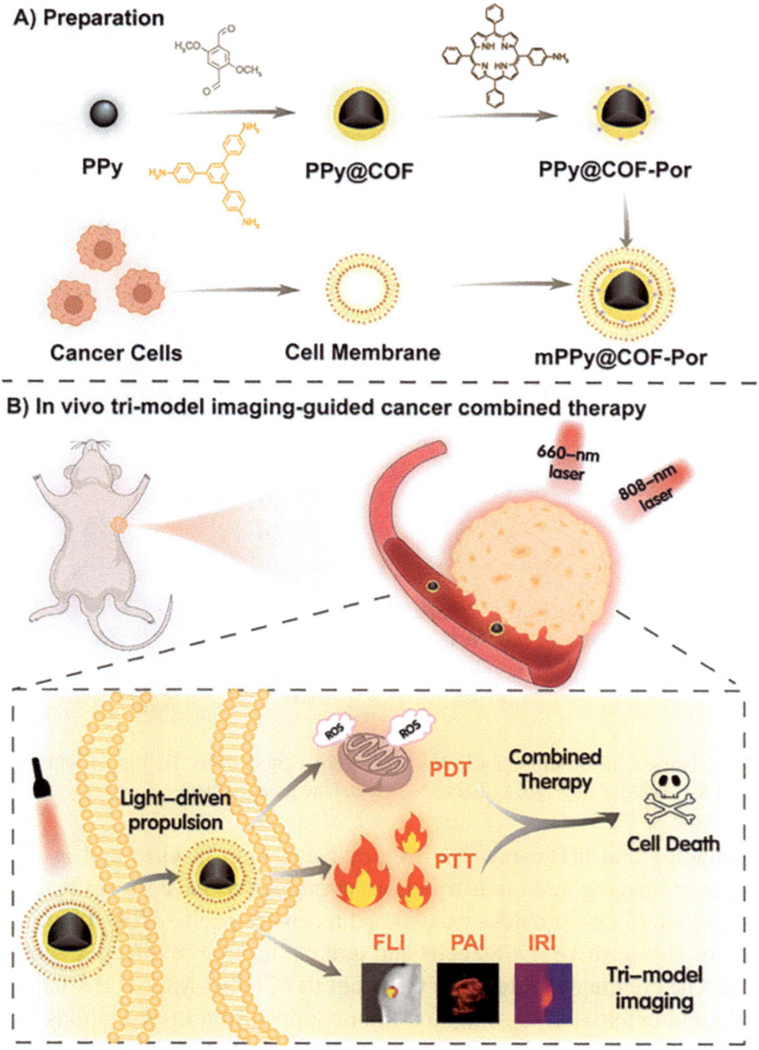

Fig. 3.20 Schematic illustration of the multifunctional COF-based nanomotor for synergistic tumor therapy. (**a**) The preparation of mPPy@COF-Por. (**b**) in vivo tri-model imaging-guided cancer therapy. Reproduced from Feng et al. (2023) with permission. Copyright © 2023 Chen et al. Advanced Science published by Wiley-VCH GmbH

A novel neodymium-MOF crystal exhibiting near-infrared II (NIR-II) fluorescence at 1067 nm, substantial Stokes shifts, and a high quantum yield of 8.91% has been synthesized using the solvothermal method (Fig. 3.22) (Cui et al. 2021). Furthermore, CQ/Nd-MOF@HA nanohybrids with a regulated particle size of 250 nm have been effectively used to create a multipurpose drug delivery system. This is the inaugural application of Ln-MOF-based nanoparticles for targeted

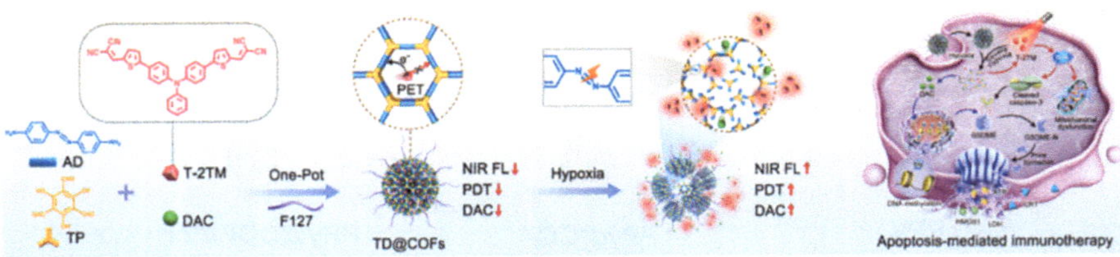

Fig. 3.21 Schematic illustration of TD@COFs. Reproduced from Liu et al. (2024b) with permission. Copyright © 2024 Elsevier Ltd.

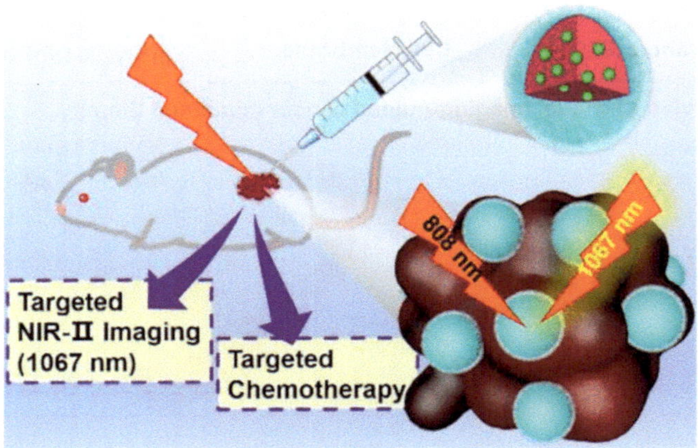

Fig. 3.22 Schematic illustration of CQ/Nd-MOF@HA nanohybrids. Reprinted with permission from Cui et al. (2021). Copyright © 2021, American Chemical Society

NIR-II imaging and pH-responsive drug delivery, both in vitro and in vivo. NIR luminescence imaging and biodistribution investigations have shown that this drug delivery system (DDS) provides exceptional receptor-specific NIR-II luminescence imaging for HA-positive A2780 cells and tumors, leading to effective tumor accumulation. The continuous release of CQ from the CQ/Nd-MOF@HA nanohybrids has facilitated cytotoxicity testing and tumor suppression investigations in tumor-bearing animals. The experimental group treated with CQ/Nd-MOF@HA demonstrated considerable reduction of cancer cell and tumor growth in comparison to the control group. This study presents a DDS using Ln-MOFs that incorporates pH-responsive release, targeted cancer treatment, and NIR-II imaging, establishing a basis for efficient tumor-targeted detection and therapy in vivo, and is expected to facilitate other clinical uses of MOF-based materials.

Copper-based single-atom SAzymes (FNUC) are meticulously engineered and synthesized to enhance chemodynamic therapy (CDT) upon ultrasound (US) excitation (Fig. 3.23) (Wang et al. 2024). The coordinated Cu^{2+} ions within FNUC function as glutathione oxidase (GSHOx)-mimetic SAzymes, depleting intracellular glutathione (GSH) and thereby boosting the generation of reactive oxygen species (ROS). Subsequently, the reduced FNUC:Cu^{2+} reacts with residual hydrogen

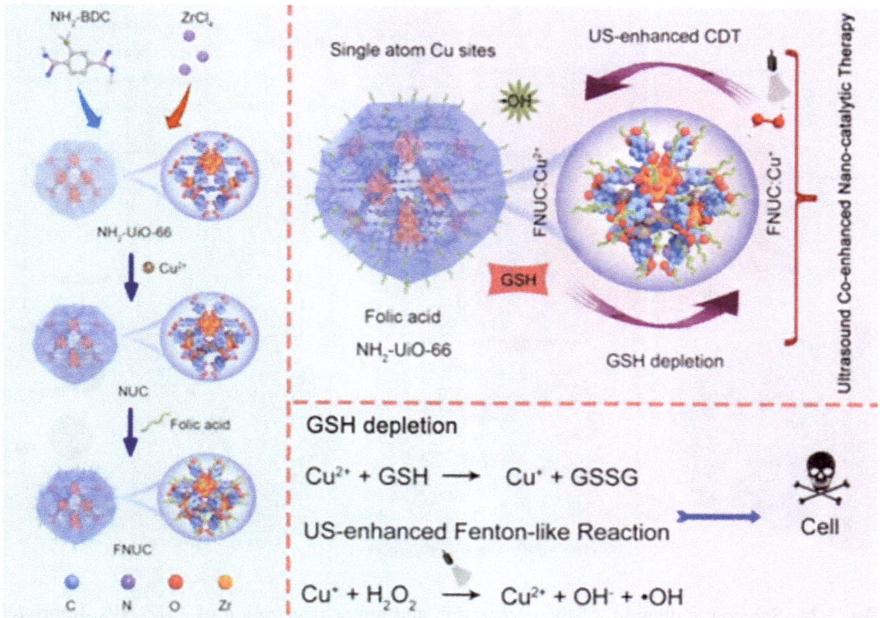

Fig. 3.23 Schematic diagram of the primary synthesis process and antitumor mechanism of FNUC. Reprinted with permission from Wang et al. (2024). Copyright © 2024, American Chemical Society

peroxide (H_2O_2) to produce hydroxyl radicals (•OH) via peroxidase (POD)-like activity. In comparison to CDT nanomedicine based on copper clusters, FNUC with its single-atom sites demonstrates superior POD-like activity and achieves a rapid •OH generation rate (Vm = 8.39×10^{-8} m s^{-1}), a low Michaelis constant ($K_m = 14.13$ mM), and a high •OH concentration ($M = 0.104$ mM) under US irradiation, benefits derived from an efficient cavitation effect. Furthermore, FNUC exhibits minimal side effects and robust tumor growth inhibition both in vitro and in vivo, suggesting its considerable potential for a wide array of clinical therapeutic applications. Notably, the labeling of FNUC with IR-1061 dye facilitates precise guidance for its US-enhanced nanocatalytic therapy. Nevertheless, there remain challenges to be addressed, including the lack of imaging properties and the relatively low copper content. In summary, this category of SAzymes warrants further investigation for its application in cancer treatment.

By reducing the size and loosening the extracellular matrix (ECM), Li and et al. have effectively developed a two-pronged strategy for deep-tumor photothermal therapy that is guided by NIR-II multimodal imaging (Fig. 3.24) (Li et al. 2022b). The NIR-II nanocomposite (AHZ NPs) suppressed hyaluronidase (HAase) activity and preserved an optimal size at physiological pH, enabling the accurate delivery and release of native HAase and tiny Ag_2S nanoparticles in the extracellular milieu of tumor tissues. Consequently, the active HAase facilitates the disaggregation of the HA-associated ECM, thereby improving the distribution and infiltration of

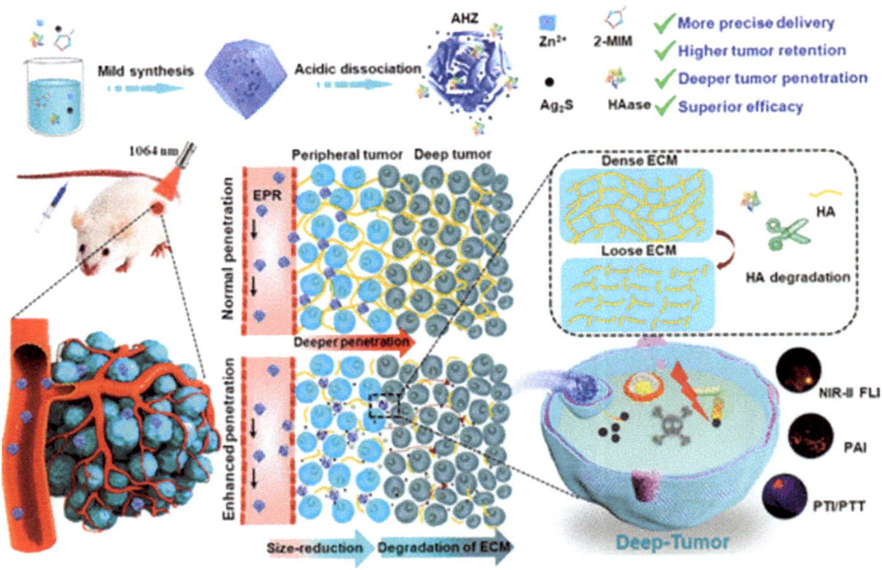

Fig. 3.24 Schematic diagram of the preparation and tumor penetration of AHZ NPs. Reprinted with permission from Li et al. (2022b). Copyright © 2022, American Chemical Society

ultrasmall Ag_2S nanodots into deep solid tumors. Importantly, these Ag_2S nanodots exhibit remarkable fluorescence and exceptional photoacoustic (PA) signals, facilitating accurate tumor treatment guidance and leading to photothermal therapy that effectively eradicates practically the whole tumor. This study introduces a systematic approach for the simultaneous delivery of functional enzymes and photothermal agents (PTAs) by a one-pot method, facilitating very effective NIR-II imaging and therapy for deep-seated tumors, hence providing new insights into the clinical management of solid tumors.

Zhang and coworkers have introduced the pioneering example of a NIR-II emissive benzobisthiadiazole-based COF-980 (Fig. 3.25) (Zhang et al. 2024). COF-980's architecture is capable of more effectively reducing the energy gap (ΔES1-T1) between the excited state and the triplet state in comparison to its ligands, thereby increasing the efficacy of photodynamic treatment. COF-980 exhibits extraordinary anti-diffusion properties, substantial photostability, superior efficacy in the production of ROS, compelling imaging potential, and ROS formation in deep tissue (approximately 8 mm). It is remarkable that COF-980, when used in conjunction with laser irradiation, can elicit a significant increase in intracellular ROS, which leads to highly effective apoptosis in cancer cells. PDT directed by NIR-II fluorescence imaging is facilitated by COF-980 nanoparticles, which effectively suppress the proliferation of 4T1 tumors with minimal adverse effects. This study introduces a comprehensive methodology for the development of long-wavelength emissive COFs and examines their potential applications in biomedicine.

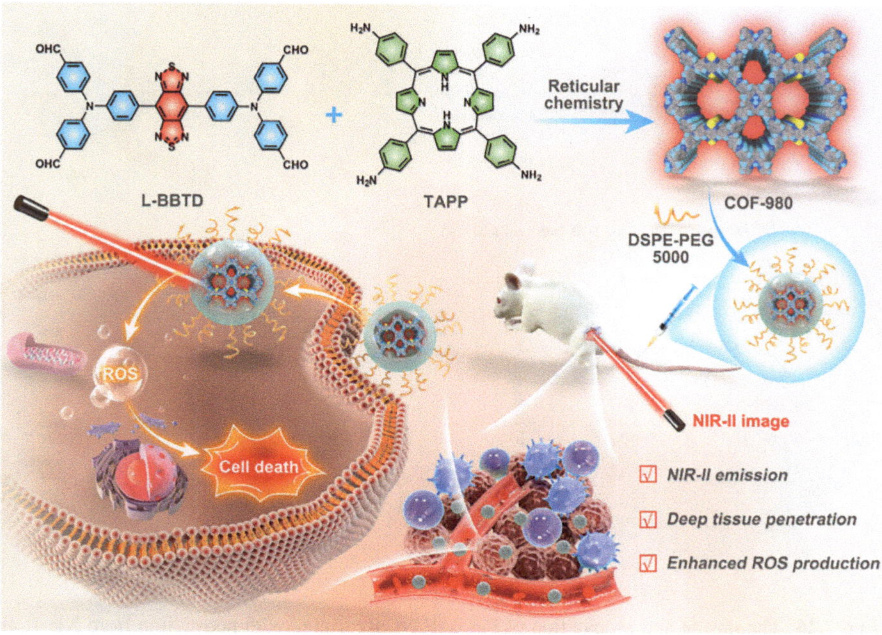

Fig. 3.25 The design and preparation of COF-980, and its NIR-II fluorescence imaging-guided in vivo phototherapy. Reprinted with permission from Zhang et al. (2024). Copyright © 2024 Wiley-VCH GmbH

3.3.3 Multimodal Imaging-Based Tumor Diagnosis

Phototheranostic metal-phenolic networks (PFG MPNs) have been successfully engineered to facilitate advanced NIR II fluorescence and photoacoustic imaging-guided precise photothermal therapy, as well as to enhance anti-exosomal PD-L1-mediated ferroptosis (Fig. 3.26) (Xie et al. 2022). The PFG MPNs significantly mitigate the suppression of exosomes on dendritic cell maturation and T cell function. Remarkably, the downregulation of exosomal PD-L1 correlates positively with the augmentation of ferroptosis, as evidenced by the elevated secretion of IFN-γ from rejuvenated T cells. Consequently, this work combined therapeutic strategy offers new perspectives on the potentiated antitumor immune response, achieved by synergistically enhancing photothermal therapy and targeted ferroptosis through the inhibition of exosomal PD-L1. This approach holds promise for the development of innovative exosomal immune therapeutics aimed at combating malignant metastatic tumors.

An alkyl radical generator, specifically 2,2-azobis[2-(2-imidazolin-2-yl)propane] dihydrochloride (AIPH), has been integrated onto a zirconium metal-organic framework (Zr-MOF) to create a dual-sonosensitizer nanoplatform (Fig. 3.27) (Zhang et al. 2022b). The resultant Zr-MOF@AIPH nanoparticles (NPs) are capable of generating singlet oxygen, which effectively eradicates tumor cells under

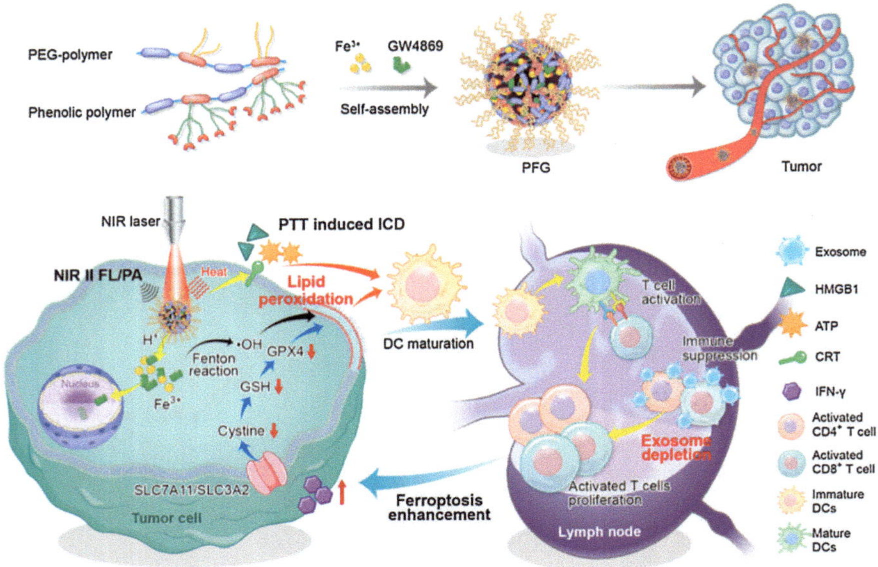

Fig. 3.26 The design and preparation of PFG MPNs. Reprinted with permission from Xie et al. (2022). Copyright © 2022, American Chemical Society

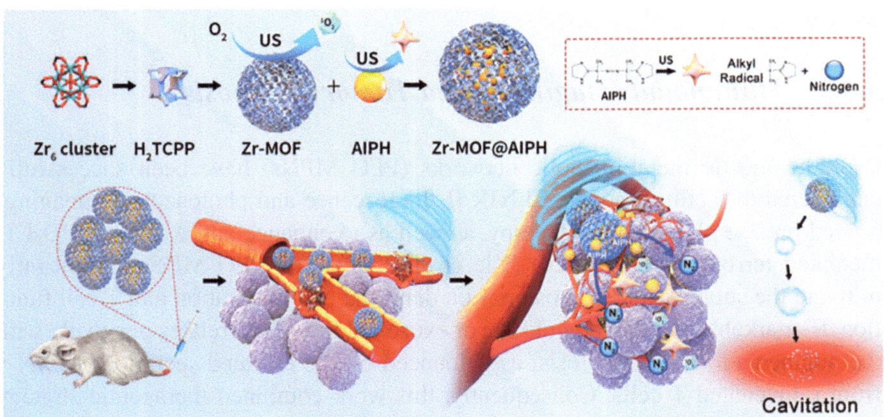

Fig. 3.27 Schematic illustration of the Zr-MOF@AIPH nanoplatform for cancer treatment. Reprinted with permission from Zhang et al. (2022b). Copyright © 2021 Wiley-VCH GmbH

normoxic conditions, and alkyl radicals, which are potent against tumor cells in both normoxic and hypoxic environments. The synergistic effect of these free radicals significantly boosts the efficiency of sonodynamic therapy (SDT). Additionally, the nitrogen released during AIPH decomposition not only lowers the cavitation threshold but also enhances the acoustic cavitation effect, thereby facilitating the penetration of NPs into the tumor site. Zr-MOF@AIPH NPs also demonstrate exceptional photoacoustic, fluorescence, and ultrasound imaging capabilities, attributed to their porphyrin-based structure and the nitrogen produced. These imaging properties facilitate the identification of the optimal therapeutic time window and remote control of NP delivery, thereby maximizing the efficiency of SDT. The superior antitumor efficacy, outstanding biocompatibility, and favorable imaging properties of Zr-MOF@AIPH have been confirmed by in vitro and in vivo evaluations. This research is a pioneer in the advancement of SDT efficacy in hypoxic conditions by utilizing a combination of complementary sonosensitizers.

Zhang and coworkers have engineered an intelligent oxygen (O_2)-evolving photodynamic therapy (PDT) nanoparticle that features homologous cancer cell targeting and dual-mode cancer cell imaging (MRI/fluorescence) capabilities (Fig. 3.28) (Zhang et al. 2019a). The nanoparticle's core structure is composed of a porphyrin-based metal-organic framework (MOF), which is encapsulated within a thin layer of MnO_2 nanosheets (referred to as MMNPs). This hybrid nanostructure exhibits a sensitive response to H_2O_2 in acidic environments, capable of generating O_2 to enhance the efficiency of O_2-mediated PDT. Concurrently, the Mn^{2+} ions released into the solution demonstrate exceptional MRI contrast enhancement, as evidenced by T1-weighted MRI images. The nanoparticles, which are referred to as CM-MMNPs, are endowed with cancer cell targeting capabilities as a result of the use of cell membranes to encapsulate them. Microscopic imaging studies suggest that CM-MMNPs have strong homologous cancer cell targeting capabilities and extraordinary colloidal stability in biological buffers. Red fluorescence is observed in the nanostructure when it is exposed to laser irradiation in cellular fluorescence imaging experiments. This is due to the effective ablation of the MnO_2 layer

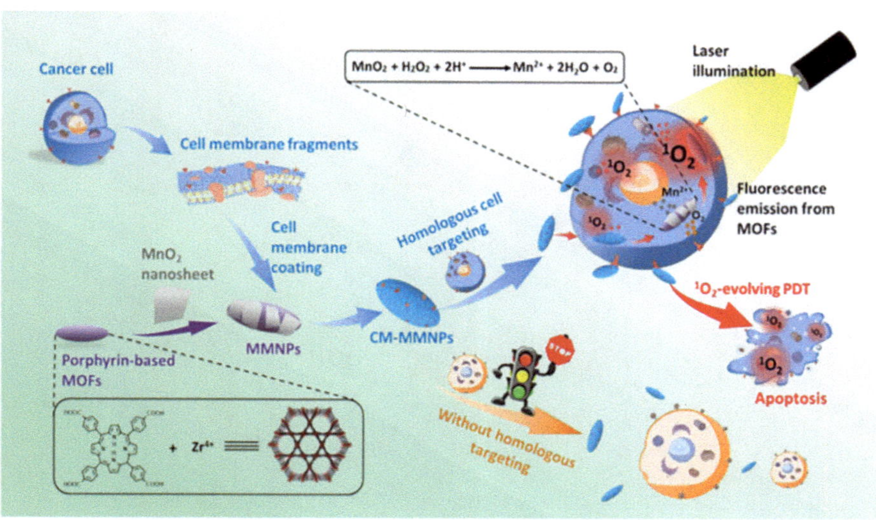

Fig. 3.28 Schematic illustration of CM-MMNPs for homologous targeting, MRI/fluorescence dual-mode imaging and PDT toward cancer cells. Reprinted with permission from Zhang et al. (2019a). Copyright © 2019, American Chemical Society

following endocytosis. The Mn^{2+} and singlet oxygen ($^1O_2$) that are produced are demonstrated to be advantageous for the PDT and dual-mode imaging of cancer cells. The nanoparticle described here has the potential to reduce hypoxia in cancer cells and provides a more sophisticated and efficient approach to cancer cell imaging and PDT.

Cun and et al. have developed a Fe_3O_4@MIL-100 nanotherapeutic agent loaded with IR780 for enhanced photodynamic nanocatalytic tumor therapy (Fig. 3.29) (Cun et al. 2022a, b). Compared with MIL-100 and Fe_3O_4 nanoparticles, IFM exhibits higher peroxidase-like activity. The catalytic production of hydroxyl radicals (•OH) by IFM is further augmented by increasing the concentration of H_2O_2 and the ambient temperature. β-lapachone pretreatment considerably elevates H_2O_2 levels in cancer cells and tumor tissues, which, when combined with near-infrared laser therapy, markedly amplifies the cytotoxic efficacy of IFM both in vitro and in vivo against cancer cells. IFM results in the generation of excessive extremely toxic •OH inside cells and significantly depletes glutathione (GSH), leading to intracellular redox imbalance and thus initiating cell death via apoptosis and ferroptosis. In vivo

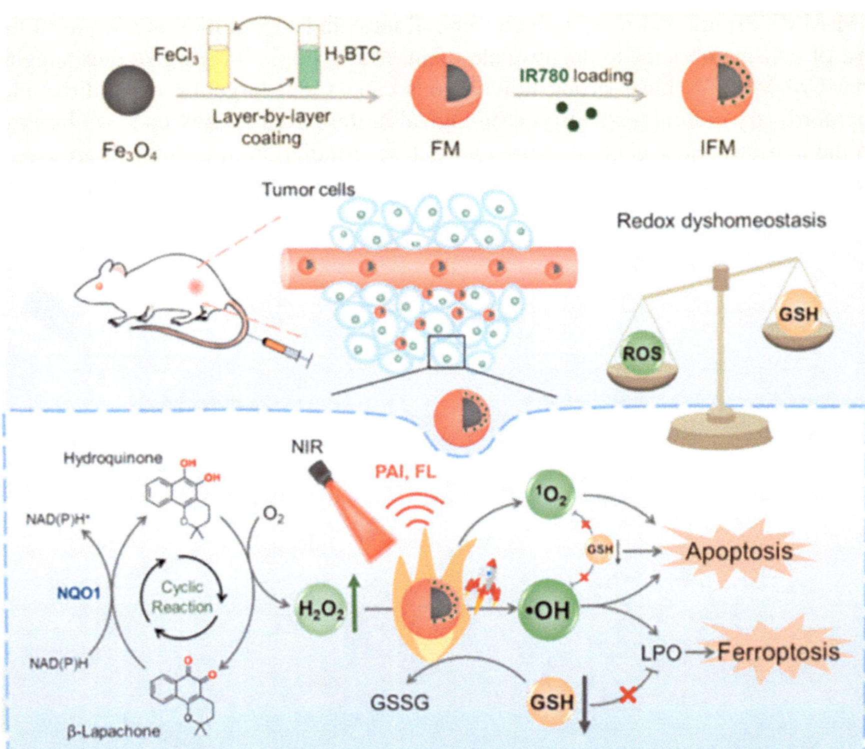

Fig. 3.29 Illustration of IR780-loaded Fe_3O_4@MIL-100 nanocomposite (IFM) for photo-enhanced upcycling H_2O_2 into highly cytotoxic•OH for intense nanocatalytic tumor therapy. Reprinted with permission from Cun et al. (2022a, b). Copyright © 2022 Elsevier Ltd

photoacoustic imaging (PA) and fluorescence imaging demonstrate that IFM attains peak accumulation in tumors 24 h postinjection. In vivo anticancer assessment demonstrates that IFM has a synergistic tumor reduction effect when augmented by laser treatment, accompanied by little systemic toxicity. These findings combined illustrate the possibility of increased endogenous H_2O_2 levels and localized photothermal therapy in improving nanocatalytic tumor treatment.

Yue and coworkers have developed an integrated DNA nanodevice featuring a dual-key-and-lock mechanism, designed for precise diagnosis and combined therapy of cancers (Fig. 3.30) (Yue et al. 2024). This device has the following benefits: (1) On-demand activation: By including endogenous pH/ATP-responsive locks, the DNA nanodevice's spatial specificity has been enhanced, reducing nonspecific activation during transport and manufacturing. (2) Using FL/SERS dual-mode imaging of intracellular microRNA, the device leverages the complementary advantages of both imaging modalities to provide an accurate diagnosis via in situ multimodal imaging for cancer detection. Signal-to-background ratios and detection sensitivity are improved by the spatial selectivity provided by endogenous activation and the CHA-mediated signal amplification. Furthermore, when combined with optical imaging modalities, Mn^{2+} generated by DKL-DND acts as a T1-weighted contrast agent for MRI, making it easier to get multiscale tumor-related diagnostic data. (3) Integrated treatment and reduced off-site delivery: This DNA nanodevice enhances antitumor effectiveness in vitro and in vivo by delivering efficient chemotherapy and gene silencing effects within specific tumor microenvironments. Additionally, TA is a highly effective small molecule inhibitor in the fight against metastasis. To sum up, this work proposed DNA nanodevice offers a novel approach to nucleic acid-based precision medicine and customized theragnostic systems by presenting

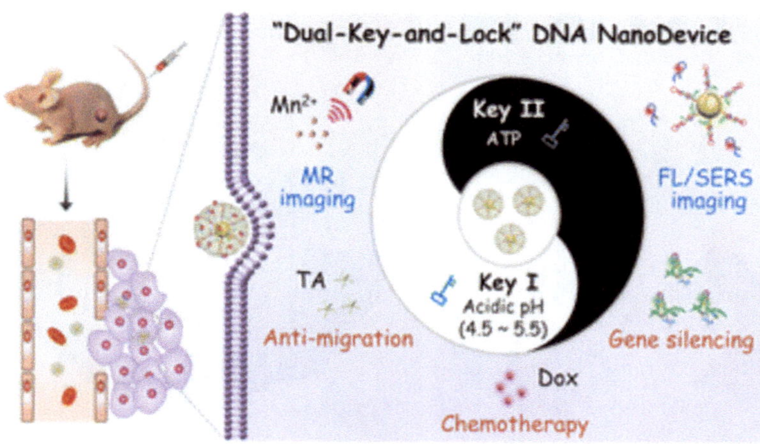

Fig. 3.30 Schematic illustration of the dual-key-and-lock DNA nanodevice (DKL-DND) in response to acidic pH and ATP for on-demand activation and multimodal imaging-guided combined tumor therapy. Reproduced from Yue et al. (2024) with permission from the Royal Society of Chemistry

a viable nanoplatform for integrated diagnosis and therapy with improved anti-interference properties.

You and colleagues have successfully engineered and synthesized ICG-PtMGs@HGd nanoparticles, which function as a multifaceted nanotheranostic agent (Fig. 3.31) (You et al. 2020). This agent is capable of modulating the TME and offers multimodal imaging guidance for enhanced photodynamic therapy (PDT) and photothermal therapy (PTT). The synthesized ICG-PtMGs@HGd nanoplatforms continuously deplete endogenous H_2O_2 within the solid tumor microenvironment using the embedded Pt nanozyme. This process locally increases oxygen levels, significantly mitigating tumor hypoxia and thereby enhancing the efficacy of PDT. The nanoplatform exhibits a significant photothermal therapy impact with elevated photothermal conversion efficiency, owing to its novel octahedral gold nanoshells. The integration of HGd hybrids endows the nanocomposites with enhanced biocompatibility and remarkable MRI functionality. In vitro and in vivo studies demonstrate that the nanoplatform has significant biosafety and an effective therapeutic impact, including oncolysis and metastasis suppression in breast cancer. Thus, this biomimetic, multifunctional nanoplatform, which integrates synergistic treatments, imaging, and ongoing modulation of the hypoxic tumor

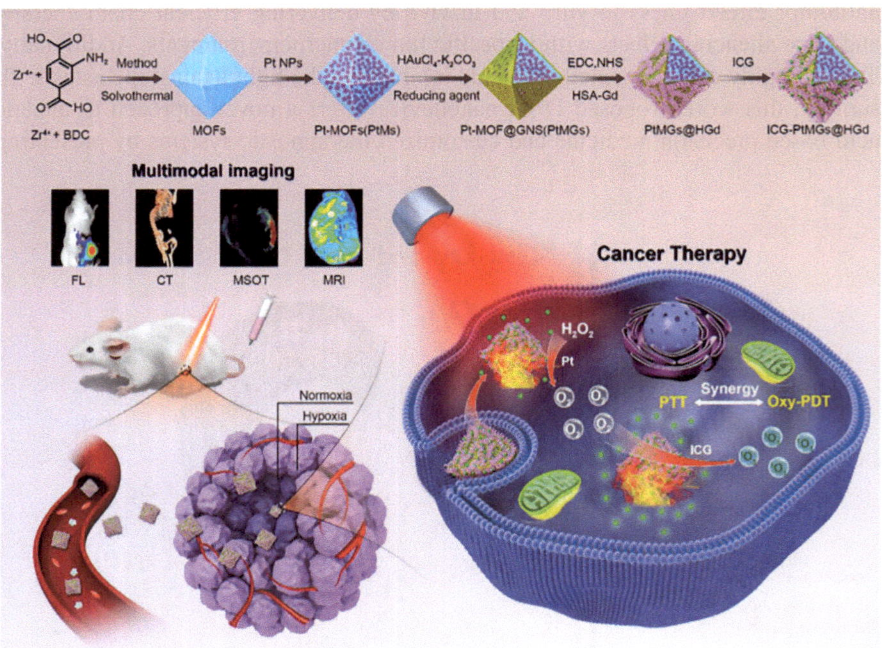

Fig. 3.31 Schematic illustration of the ICG-PtMGs@HGd nanoplatforms as H_2O_2-driven oxygenator for FL/MOST/CT/MRI multimodal imaging-guided enhanced PDT and PTT synergistic therapy in a solid tumor. Reproduced from You et al. (2020) with permission. Copyright© 2020 Wiley-VCH GmbH

microenvironment, significantly improves the precision and efficacy of cancer therapy while facilitating the use of metal-based nanozymes in biomedicine.

3.3.4 Electrochemical Biosensing of MOFs in Tumor Diagnosis

MOFs, as highly attractive materials for electrochemical sensors, possess the following potential significant advantages (Luo et al. 2019; Li et al. 2020b). Firstly, they can serve as porous frameworks, thereby increasing the surface area for analyte adsorption and transportation. Secondly, their pores and channels have a high degree of variability in terms of size and shape. This property can bring a size effect to specific analytes and confer strong selectivity. Furthermore, they can provide a large number of convenient functions and accessible sites (including metal centers, linkers, and active guests in the pore channels), thereby achieving the purposes of analysis activation and adsorption, and can also directly transfer electrons from compounds to redox-active regions. In addition, MOF frameworks have a high degree of orderliness, a clear and definite definition, and can continuously change, which is extremely beneficial for studying the sensing mechanism and the connection between structure and properties.

Because MOFs themselves contain a variety of functional groups, MOFs and biometric probes can be combined in a variety of ways, such as hydrogen bonding, electrostatic force, and π-π stacking. Thus, it can link a variety of probe molecules and is widely used in many aspects of electrochemical sensors. In Fig. 3.32, Liu et al. used Zr-MOFs to strongly bind to phosphate groups by forming Zr-O-P bonds (Liu et al. 2021). The external layer of the exosome's phospholipid bilayer is abundant in hydrophilic phosphate heads. Therefore, they developed a recognition system using Zr-MOFs and aptamers for selective electrochemical quantitative analysis of exosomes. At the same time, it is combined with a wire mesh electrode to create a labeling free, paper-based electrochemical biosensor. This biosensor utilizes Zr-MOFs modified on the paper's surface to collect exosomes via coordination interactions, demonstrating the strong binding affinity of Zr-MOFs with biomolecules.

In addition, Li et al. found through research that MOF materials with structures similar to Zr-MOFs were successfully synthesized by replacing different monomers (Chang et al. 2019), for example, UiO-66, UIO-66-NH_2, UIO-66-$2NH_2$, etc. The MOF material can fix the aptamer modified by phosphate group on the Zr (IV) cluster, so as to construct an electrochemical aptamer sensor. More importantly, electrochemical biosensors based on UIO-66-$2NH_2$ exhibit better electrical activity and stronger affinity for aptamers through electrostatic interactions, π-π stacking, and Zr-O-P coordination bonds, thus enhancing the immobilization of aptamers with unique biosensing properties. Chang et al. used porous UIO-66-NHUIO-66-$2NH_2$ to prepare functionalized MOFs and loaded them with different kinds of

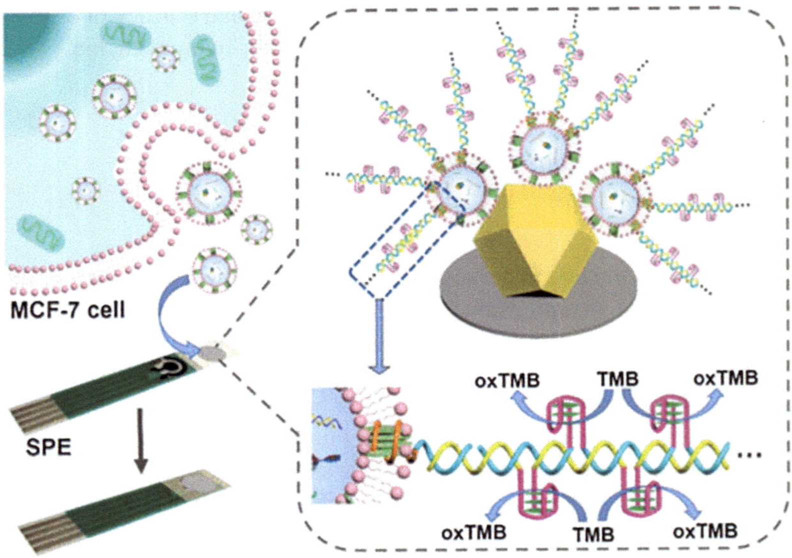

Fig. 3.32 A simple and cost-efficient electrochemical biosensor by combining a MOF-functionalized paper and a screen-printed electrode (SPE) for portable, ultrasensitive, and quantitative determination of cancer-derived exosomes was developed. Reprinted with permission from Liu et al. (2021). Copyright © 2021, American Chemical Society

electrochemically active substances as nanocontainer. The outer layer of UIO-66-NHUIO-66-2NH_2 is loaded with a large amount of single-stranded DNA, which acts as a recognition probe.

Compared with monomeric MOFs materials, bimetallic MOFs with different metal valence states have attracted more attention in the design and preparation of electrochemical biosensors. This is because the bimetallic synthesis of MOFs often produces synergistic effects, which greatly promote the transfer of electrons. For example, Hu et al. synthesized a novel bimetallic CoNi-MOF electrochemical biosensor for trace detection of the tumor marker miRNA-126 (Fig. 3.33) (Hu et al. 2021). First of all, the unique structure of CoNi-MOF enables miRNA-126 identification probes to be stably bound to the surface of CoNi-MOF by various interaction forces such as hydrogen bonds, π-π stacking, coordination bonds, and van der Waals forces. Secondly, CoNi-MOFs multiple metals work together to improve the sensing performance of the entire sensor. The sensor achieves trace detection of miRNA-126 with a detection limit of 0.14 fM. Therefore, the sensing performance of bimetallic MOFs materials is better than that of mono-metallic MOFs materials. Tian and his team prepared bimetallic ZnNi-MOFs microspheres with different properties by optimizing the molar ratio of Zn (II) and Ni (II) (He et al. 2019). Among them, the biosensor prepared by ZnNi-MOFs (1:2) showed more outstanding sensing performance for adenosine. In this biosensor, the bimetallic ZnNi-MOFs are used as a carrier to support a large number of aptamers, so ZnNi-MOFs (1:2) is more conducive to binding aptamers and showing good sensing

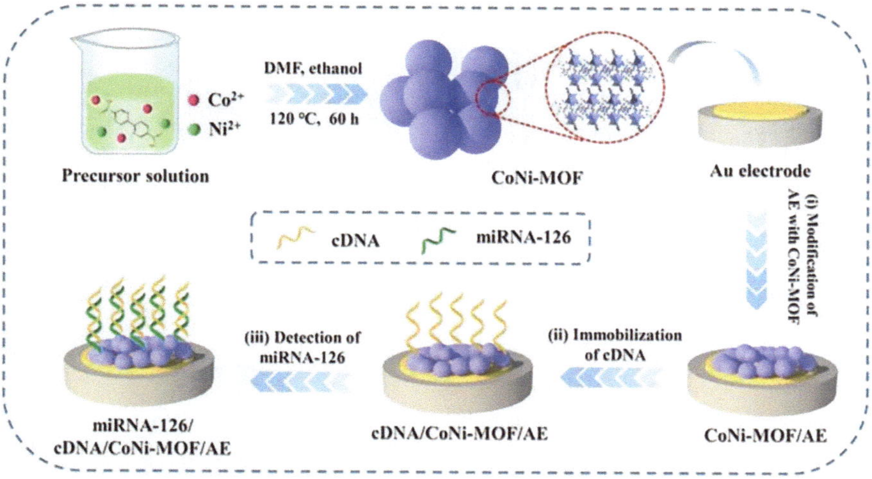

Fig. 3.33 A bimetallic CoNi-MOF with 2,2′-bipyridine-5,5′-dicarboxylic acid ligand was prepared, which can be used as a high-efficiency platform for anchoring the complementary DNA probe (cDNA) to develop the electrochemical biosensor for miRNA-126 detection. Reprinted with permission from Hu et al. (2021). Copyright © 2020 Elsevier B.V. All rights reserved

performance. In addition, Huang et al. studied the sensing properties of bimetallic MOF materials synthesized from different metals (Fig. 3.32) (Huang et al. 2021b). Using Ti-MOFs as the precursor, they prepared different kinds of bimetallic MOFs by composite with Co (II), Fe (II), and Ni (II) metals, respectively. Among them, Co (II) doped Ti-MOFs showed the best BPA sensing performance, with LOD as low as 0.09 fg mL^{-1}. In summary, bimetallic MOFs show excellent selectivity, high stability and good reproducibility, and are more suitable for the preparation of efficient electrochemical biosensors than mono-metallic MOFs.

3.3.5 Surface-Enhanced Raman-Based MOFs in Tumor Diagnosis

Bifunctional Au NRs/Fe-MOF were synthesized by Zhao et al. for photo-enhanced peroxidase-like catalysis and SERS substrates for the detection and degradation of MB (Fig. 3.34) (Zhang et al. 2022a, b). The Au 4f XPS peak exhibited a positive shift of 0.26 eV, while the Au absorption peak in Au NRs/Fe-MOF had an approximate 12 nm red shift, indicating electron transfer from Au NRs to Fe^{3+} in Fe-MOF. Under light irradiation at a wavelength corresponding to the LSPR band of Au NRs, an increased transfer of photo-generated electrons occurred from Au NRs to Fe-MOF, resulting in the reduction of Fe(III) on the surface of Fe-MOF. The resultant Fe(II) facilitated the breakdown of H_2O_2, producing •OH radicals for the

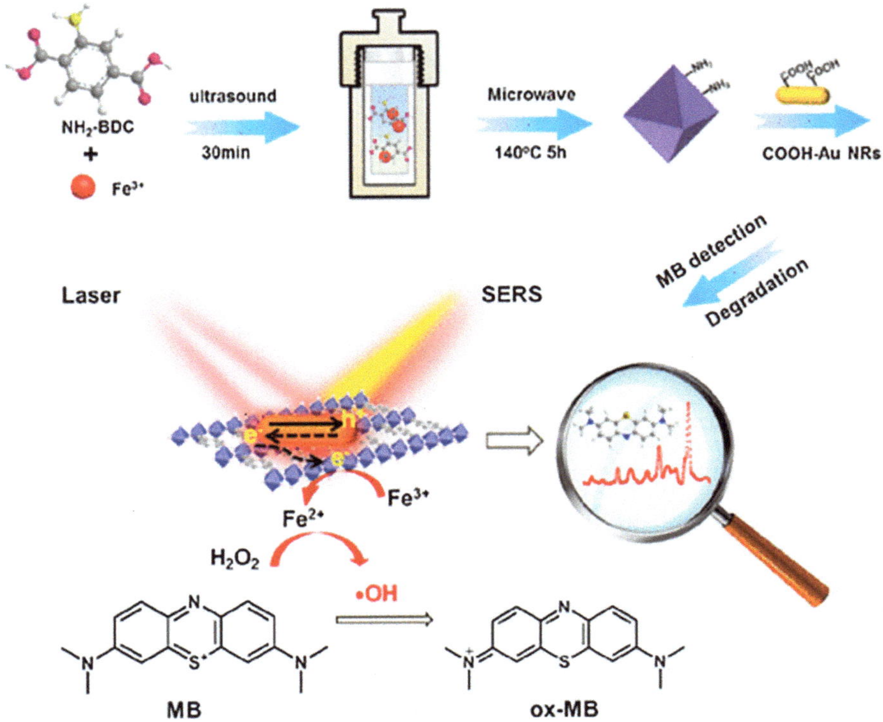

Fig. 3.34 Schematic diagram of the preparation of Au NRs/Fe-MOFs for methylene blue in situ catalytic degradation and SERS detection. Reprinted with permission from Zhang et al. (2022a, b). Copyright © 2022 American Chemical Society

elimination of MB by an augmented Fenton-like reaction. Furthermore, Fe-MOF demonstrated semiconducting characteristics and received electrons from Au NRs during LSPR excitation. Consequently, Fe-MOF impeded the electron-hole recombination on Au NRs, whereas Au NRs/Fe-MOFs promoted the electromagnetic and chemical amplification of Raman signals. Due to their superior SERS activity, Au NRs/Fe-MOF hybrids have shown significant promise for detecting harmful chemical and biological compounds, including MB, with a detection limit of 9.3×10^{-12}M.

Qi and colleagues have effectively created a durable and practical micron-nano composite probe, utilizing plasmonic AuNPs and micron-sized MOF particles (AuNPs@ZIF-8), and have accomplished label-free single-particle SERS sensing of phosphatidylserine (PS) externalization on the cell membrane during apoptosis triggered by electrical stimulation (Fig. 3.35) (Qi et al. 2021). The AuNPs@ZIF-8 demonstrates superior biocompatibility and several "hot spots," improving the sensitivity for PS detection. The adjustable dimensions of ZIF-8 enhance biomolecule concentration and enable the identification of individual cells at the single-particle scale. The sensing technology shows potential for tracking PS externalization on the cell membrane owing to its long-term stability in serum.

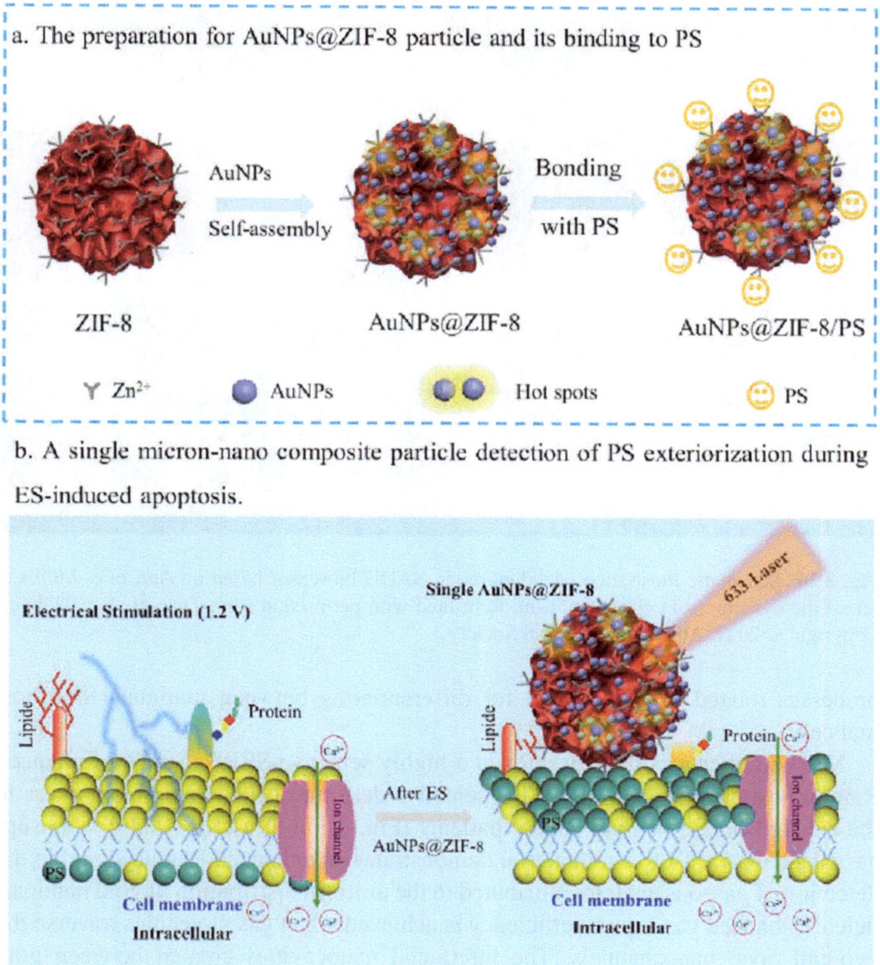

Fig. 3.35 Schematic of the preparation of AuNPs@ZIF-8 particles and their binding to PS. (**a**) The self-assembly of AuNPs with ZIF-8, forming AuNPs@ZIF-8, which then binds with PS. (**b**) The detection of PS exteriorization during ES-induced apoptosis. Reprinted with permission from (Qi et al. 2021).Copyright © 2021, American Chemical Society

A unique dual-mode sensor, using Au/$CoFe_2$-MOFs, has been developed that demonstrates significant Raman enhancement and a colorimetric response specific to NADH (Fig. 3.36) (Zhao et al. 2022b, c). NADH may be visually identified in the colorimetric manner within a linear range of 1.93×10^{-8} to 1.93×10^{-10} M, with a limit of detection (LOD) of 2.7×10^{-9} M. The sensor has remarkable sensitivity in SERS detection, providing an extensive linear range from 3.86×10^{-10} M to 3.86×10^{-3} M, with a limit of detection of 2.8×10^{-11} M. This new method has shown the intracellular distribution of NADH and its dynamic behavior during glycometabolism. It serves as a distinctive tool for clarifying dynamic cellular

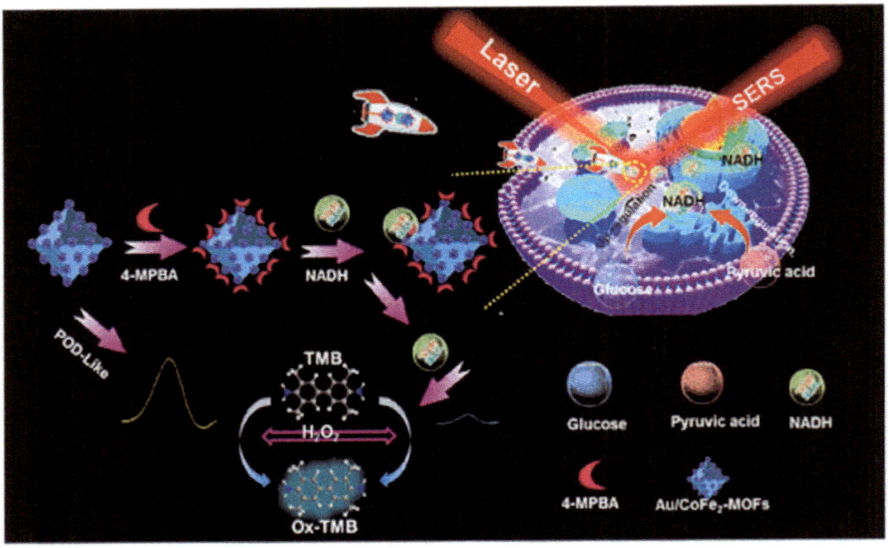

Fig. 3.36 Schematic illustration of a dual-mode NADH biosensor based on Au/$CoFe_2$-MOFs to reveal the dynamics of cell metabolism. Reprinted with permission from Zhao et al. (2022b, c). Copyright © 2022, American Chemical Society

processes related to NADH and for differentiating between malignant and normal cells.

Xu and colleagues have developed a highly sensitive SERS (Surface-Enhanced Raman Scattering) substrate for the selective detection of aldehyde biomarkers in the exhaled breath of lung cancer patients (Fig. 3.37) (Xu et al. 2023). The suggested ZIF-8/Au/TiO_2 nanomaterial demonstrates an enhanced Raman signal in the detection of gaseous analytes, attributed to the uniform distribution of gold nanoparticles. Enhanced gas capture efficiency is achieved when gas molecules traverse the two-end open nanochannels. The interfacial nanocavities created between gold nanoparticles and metal-organic frameworks serve as gas confinement chambers, successfully addressing the challenges of gas sensing due to the difficult absorption of gas molecules onto solid substrates. The sensor attains parts-per-billion (ppb) sensitivity by a Schiff base reaction involving pre-grafted Raman-active 4-ATP and the target aldehydes. Statistical study of Raman spectra, in conjunction with Principal Component study (PCA) and Machine Learning (ML), was used to simulate the aldehydes found in exhaled breath. This study highlights the considerable potential of the sensor for biomarker screening assays in lung cancer detection.

Qin and et al. have introduce an innovative buoyant MOF corona-driven assay, inspired by the structure of barnacles, designed to streamline the rapid isolation and ultrasensitive profiling of circulating extracellular vesicles (EVs) for cancer diagnosis and postoperative assessment (Fig. 3.38) (Qin et al. 2024). This assay harnesses the power of on-bubble coordinative recognition and multiplexed suspension surface-enhanced Raman scattering (SERS) readout. In comparison to previous EV

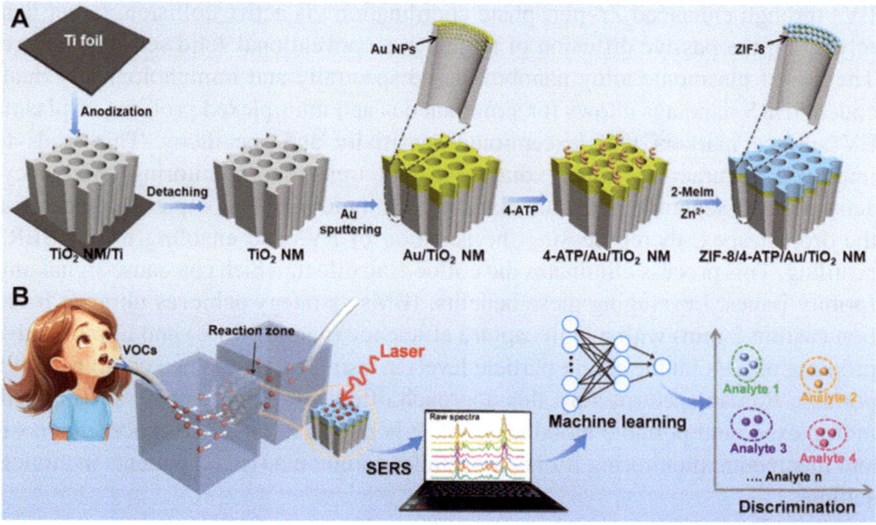

Fig. 3.37 Schematic of the fabrication of ZIF-8/4-ATP/Au/TiO_2 NM-based nanochannels SERS sensor. (**a**) The fabrication of M. (**b**) A laser is applied for SERS analysis. The resulting spectra are processed by machine learning. Reprinted with permission from Xu et al. (2023). Copyright © 2023, American Chemical Society

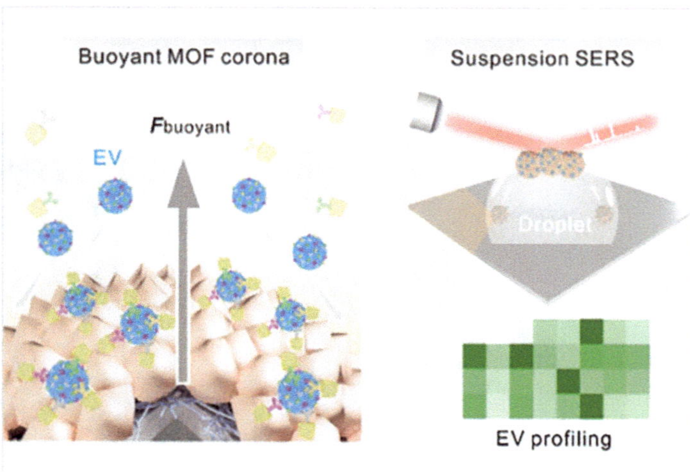

Fig. 3.38 Overview of IBMsv for EV isolation, surface profiling, and cancer diagnosis. Reprinted with permission from Qin et al. (2024). Copyright © 2024, American Chemical Society

assays, the integrated bubble-MOF-SERS EV assay (IBMsv) technology boasts several distinct advantages: The biomimetic adhesion design facilitates the high-density assembly of MOF nanoparticles on the bubble surface, creating a self-powered high-curvature coordinative corona. This design accelerates the capture of

EVs through enhanced Zr–phosphate coordination via active collision, rather than relying on the passive diffusion of targets at a conventional solid sensor interface. The use of plasmonic alloy nanobox-based spectrally and immunologically dual-coded SERS nanotags allows for simultaneous and multiplexed profiling of plasma EV surface markers with exceptional sensitivity and specificity. This leads to improved accuracy in cancer diagnosis and treatment monitoring. Buoyancy-activated self-separation ensures that the plasmonic bubble complex aggregates at the droplet apex, thereby easing the isolation of EVs and enabling in situ SERS profiling. This process eliminates the coffee-ring effect, which can cause signal uniformity issues. Leveraging these benefits, IBMsv strategy achieves ultrafast isolation (within 2 min) with a high capture efficiency (nearing 90%) and ultrasensitive profiling of EVs (at the single-particle level) in a single assay. When combined with portable Raman spectroscopy, this approach offers a rapid, sensitive, streamlined, multiplexed, and portable blood test tool. It is poised to facilitate cancer diagnosis and therapeutic monitoring for hepatocellular carcinoma (HCC) patients in clinical settings.

The yolk-shell hierarchical pore structure of MesoAu@ZIF-8 has been innovatively developed by Shi and colleagues, and introduced as an exceptional SERS-active substrate, providing superior sensitivity and enabling swift detection of gaseous aldehydes (Fig. 3.39) (Shi and Fang 2024). Leveraging the distinctive structural benefits of MesoAu and the inherent microporosity of the ZIF-8 shell, the MesoAu@ZIF-8 yolk–shell architecture has effectively realized rapid enrichment and sequestration of gaseous aldehydes within 'hot spot' areas. This breakthrough

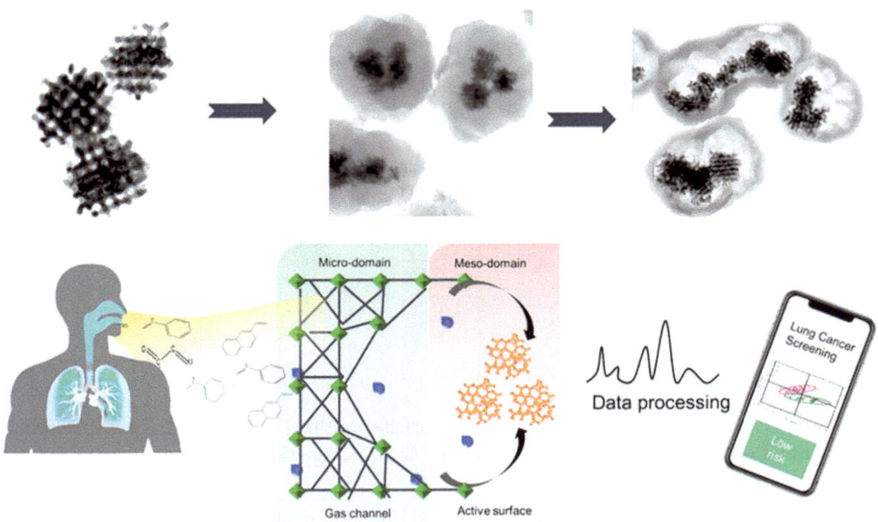

Fig. 3.39 Schematic of the MesoAu@ZIF-8. Reprinted with permission from Shi and Fang (2024). Copyright © 2024, American Chemical Society

surmounts long-persisting obstacles in the SERS detection of gaseous analytes. The MesoAu@ZIF-8 yolk–shell substrate has demonstrated exceptional SERS efficacy, with a detection threshold as low as 0.32 parts per billion (ppb) for benzaldehyde (BA), a biomarker commonly associated with the early stages of lung malignancy. Consequently, this newly developed SERS sensor possesses significant potential for the noninvasive detection of lung malignancies in practical scenarios.

Li and et al. have developed a flexible, paper-based plasmonic MOF film, which is composed of silver (Ag) nanowires encapsulated within a ZIF-8 shell, for the purpose of Surface-Enhanced Raman Scattering (SERS) detection of gaseous molecules (Fig. 3.40) (Li et al. 2024). The film's ability to adsorb gases is attributed to the micronanopores created by the interlacing network of nanowires and the ZIF-8 coating. Additionally, the paper's microfiber network forms natural microchannels that facilitate gas transport, complementing the gas adsorption properties of the Ag nanowires@ZIF-8 material. The combination of this cost-effective, flexible paper-based film with deep learning algorithms presents a noninvasive diagnostic approach for colorectal cancer by analyzing the gaseous biomarkers in patients' breath. This SERS-active substrate not only demonstrates excellent SERS performance but also provides highly sensitive responses, ensuring the retention of distinct Raman signals over a period of 30 days during repeated measurements. The substrate has successfully collected 1000 SERS spectra from the exhaled gaseous analytes of both simulated healthy individuals and those with colorectal cancer (CRC). With the aid of the AAN algorithm, the substrate has achieved a high accuracy of 93.7% in identifying benzaldehyde in simulated breath samples, showcasing significant clinical promise for the early diagnosis of CRC. This research delves into the utilization of SERS in the detection of gaseous analytes and reveals its extensive potential applications across clinical medicine, food safety, and environmental monitoring.

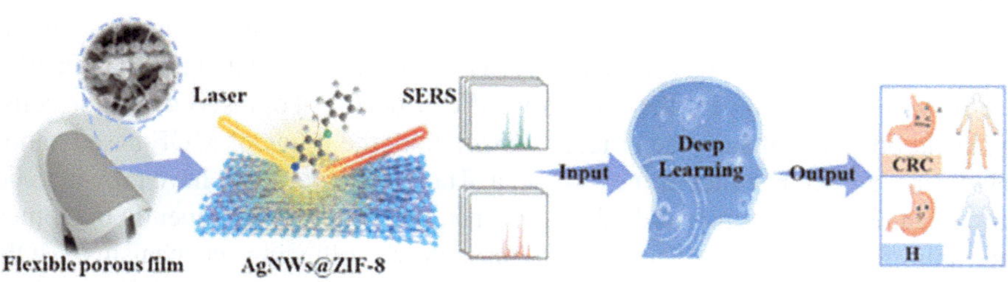

Fig. 3.40 Schematic of the fabrication of the flexible porous film. Reprinted with permission from Li et al. (2024). Copyright © 2024, American Chemical Society

3.4 The MOFs Materials in Treatment of Tumors

MOFs have a wide range of uses in biological detection, imaging, and drug loading thanks to their distinctive structure and characteristics (Hou et al. 2024; Vodyashkin Andrey et al. 2023). With additional study and development, it is anticipated that it will assume a more significant position in the future of the biomedical area.

3.4.1 Radiation Therapy

Radiation therapy (RT) is a conventional and efficient way for treating tumors, known for its ability to deeply penetrate tissues and effectively eliminate malignant cells. It employs a high dosage of live radiation to specifically eradicate cancer cells in a localized area and impede the proliferation of tumor cells. Nevertheless, the therapeutic efficacy of radiation in clinical practice is limited by its inability to selectively eliminate tumor cells. Simultaneously, the low rate at which radiation is absorbed by tumor tissue necessitates greater radiation doses to effectively impede tumor growth, hence elevating the likelihood of systemic harmful side effects. Prolonged exposure to radiation therapy can lead to the development of vascular thickening, which in turn can result in the occurrence of atherosclerosis, thrombosis, and other related medical conditions. Moreover, repeated radiation therapy might trigger the development of resistance in tumor cells against radiation, hence diminishing the efficacy of radiation therapy in treating tumors. Typically, a nanoplatform made of metal elements with a high atomic number or metal element nanoparticles functionalized with MOF (Metal-Organic Framework) has a strong ability to attenuate X-rays and a high X-ray absorption coefficient. When exposed to X-ray radiation, it can emit secondary electrons, such as Compton electrons, optoelectrons, and electrons. This emission can affect intracellular components, water, and electricity, leading to radiological sensitization. Simultaneously, they have the ability to interact with ionizing radiation, causing the conversion of intracellular O_2 into reactive oxygen species (ROS). This process enhances the sensitivity of tumors and triggers the programmed cell death (apoptosis) of tumor cells. Thus, in order to enhance the efficacy of radiation, the usage of MOF-based nanoplatforms has become increasingly popular. These nanoplatforms have proven to be good candidates for tumor radiotherapy, since they improve therapeutic outcomes and minimize systemic adverse effects through the following mechanisms. MOF, such as Hf-MOF, is made up of metal ions or clusters that contain high Z elements. These high Z metal atoms have a strong ability to attenuate X-rays, making MOF a useful radiosensitizer for enhancing the effectiveness of radiation therapy. Inorganic nanoparticles that also contain high Z elements can be added to MOF to create MOF-based nanocomposites. This can further enhance the therapeutic effect on tumors by improving radiosensitization. MOF is widely regarded as an excellent nanocarrier due to its exceptional porosity, optimal pore size, and excellent biological compatibility. It is particularly well-suited for loading radiosensitizing compounds, hence enhancing radiological efficiency.

Chen et al. developed a versatile nanocomposite using Zr-MOF to treat BALB/c nude mice with A549 tumors by the combined use of dual sensitization of RT (as shown in Fig. 3.41 (Ma et al. 2019). Zr-MOF was synthesized using Zr4 as the metal node and 1,4-phenyldicarboxylic acid (BDC) as the organic linker. The resultant Zr-MOF was further altered using bovine serum albumin (BSA) to produce Zr-MOF-QU, which exhibited an average diameter of 62 nm. The experimental findings demonstrate that Zr-MOF-QU has the ability to bind the Zn^{2+} ions of CAIX in the tumor microenvironment (TME) by decomposing BDC in Zr-MOF-QU. This binding effectively suppresses the catalytic activity of carbonic anhydrase IX (CAIX), which is advantageous in reducing hypoxia-induced drug resistance at the tumor site and enhancing the efficacy of radiotherapy on tumors. Furthermore, the nanocomposites generated with Zr-MOF-QU demonstrated exceptional radiosensitization properties in both laboratory settings (in vitro) and living organisms (in vivo) by effectively enhancing the aggregation of γ -H_2AX after DNA breakage. This is because of the exceptional X-ray attenuation properties of QU in Zr-MOF-QU. Consequently, Zr-MOF-QU exhibits remarkable biodegradability, a dual sensitization effect, and little side effects. It also has the capacity to alter the resistance of A549 tumor-bearing mice when administered with radiation.

Tang et al. fabricated a catalase-like MOF-based radiosensitizer, MnTCPP-Hf-FA NPs (Chen et al. 2019a, b). TCPP-Hf MOF was constructed via solvothermal

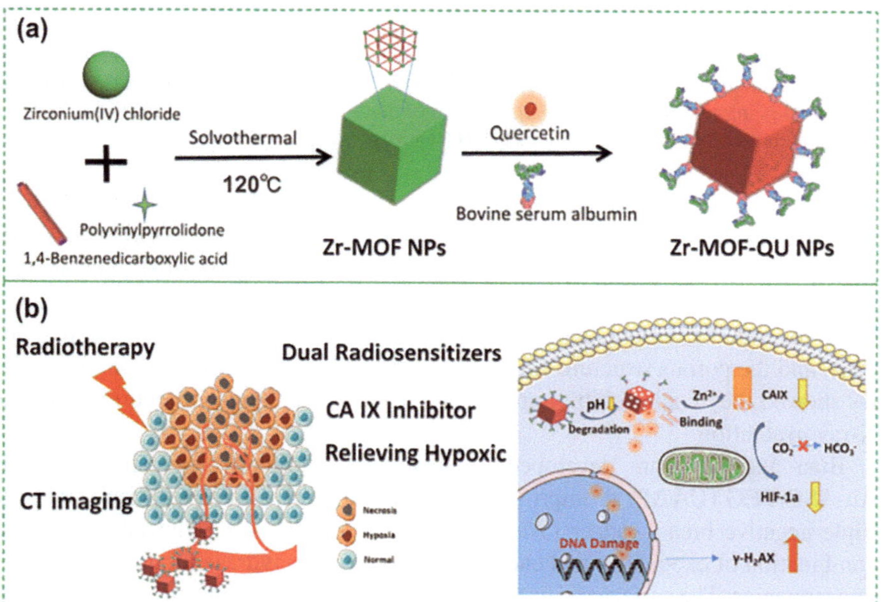

Fig. 3.41 Schematic diagram of the preparation of Zr-MOF-QU and its application in RT. Reprinted with permission from Ma et al. (2019). Copyright © 2019, American Chemical Society

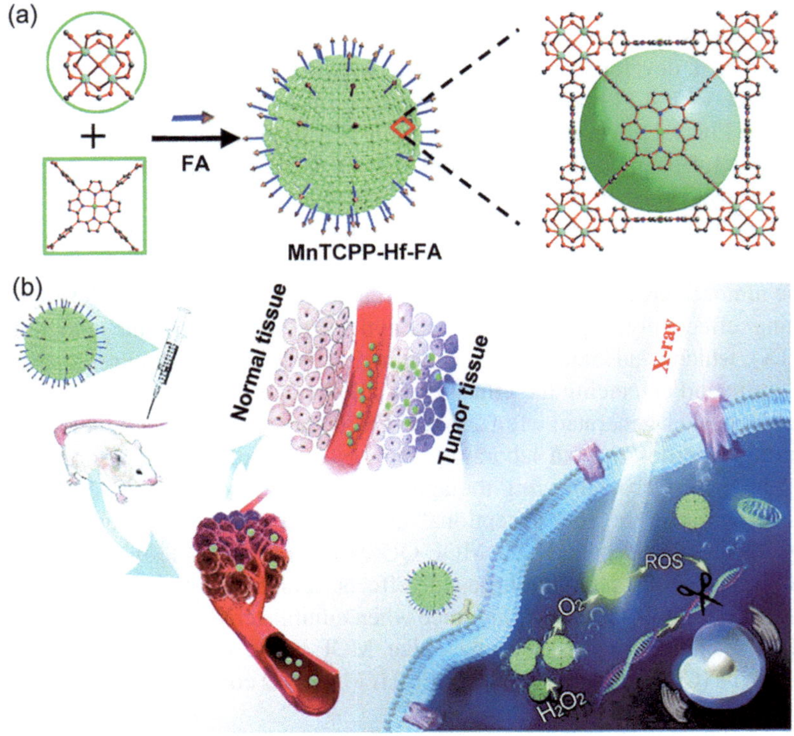

Fig. 3.42 Schematic representation of MnTCPP-Hf-FA used to enhance hypoxia-induced radiation resistance. (**a**) The formation of a compound labeled MnTCPP-Hf-FA, with molecular structures combining to form a larger spherical structure. (**b**) The application process, where the compound is injected into a mouse, targeting tumor tissue. Reproduced from Chen et al. (2019a, b) with permission from the Royal Society of Chemistry

method using Hf clusters and TCPP as organic ligands (Fig. 3.42). Then, Mn ions and FA were integrated into it through chelation and coordination to form MnTCPP-Hf-FA NPs with good biocompatibility and tumor targeting. With Hf clusters and Mn ions, these NPs could decompose endogenous H_2O_2 into O_2 to enhance RT effect for hypoxic tumors, convert O_2 and H_2O into ROS to induce tumor cell apoptosis, and the porous structure allowed O_2 and ROS to diffuse rapidly. In vivo studies showed they could inhibit hypoxic tumor growth and prevent recurrence under X-ray irradiation.

Bian and his team developed a biodegradable nanohybrid called UiO-66/Au-ASO/PEG (UAAP) to improve the effectiveness of radiation therapy (RT) for triple-negative breast cancer (Wang et al. 2021c). This was achieved by using a combination of exogenous and endogenous carbonic anhydrase IX (CA IX) inhibition strategies. The surface of UiO-66 was changed with the Zr 4 node and p-phthalic acid (PTA) ligand to create Au NPs, which act as a radiosensitizer. The production of biodegradable UAAP was achieved through the modification of thified CAIX antisense oligonucleotides (ASO) and polyethylene glycol (PEG) using an Au-S link. During the targeted molecular event (TME), the degradation of phosphotungstic acid (PTA) by UiO-66 might form a bond with zinc 2 (Zn 2) of carbonic anhydrase IX (CAIX). This bond inhibits the activity of CAIX, hence reducing the resistance caused by tumor hypoxia and enhancing the radiation therapy (RT) effect. Meanwhile, CA IX ASO, which is regarded as an endogenous inhibitory method, has the ability to suppress the expression of CAIX. Significantly, besides serving as binding sites for thiolated ASO and PEG, Au NPs also function as a radiosensitizer to augment the RT effect during X-ray irradiation. This nanohybrid has shown a strong ability to be compatible with living organisms and effectively suppressed tumor growth by alleviating resistance caused by low oxygen levels and enhancing the sensitivity of tumor cells to radiation through the use of gold nanoparticles in mice with MDA-MB-231 tumors.

In a recent study, Cheng and his team described a method for treating esophageal cancer using radiotherapy. The methodology involves employing a nanoscale UiO-66-NH2 (Hf) material, which is prepared using a straightforward one-step process (Zhou et al. 2022). This is illustrated in Fig. 3.40d of their publication. The metal node used for synthesis was Hf, whereas the chemical ligand was NH2-BDC, also known as 2-amino terephthalic acid. Based on the nitrogen adsorption and desorption data, UiO-66-NH2 (Hf) has a specific surface area of 415.5 $m^2 g^{-1}$ and a pore diameter of 0.6 nm. This indicates that UiO-66-NH2 (Hf) possesses porous characteristics and a persistent pore structure. The remarkable stability of UiO-66-NH2 (Hf) in phosphate-buffered saline (PBS) and Dulbecco-modified Eagle medium (DMEM) makes it suitable for applications in biomedicine. UiO-66-NH2 (Hf) has been demonstrated in laboratory experiments to efficiently enhance X-ray absorption when exposed to X-ray radiation. This leads to the induction of DNA breakage and apoptosis in tumor cells, hence augmenting the effectiveness of radiation therapy (RT). Furthermore, the in vivo experimental results further confirmed the substantial impact of UiO-66-NH2 (Hf) as a radiosensitizer in tumor treatment, offering a novel method for developing MOF-based radiosensitizers to enhance radiotherapy by a straightforward procedure.

3.4.2 *Chemical Therapy*

As of now, chemotherapy is one of the most effective and often used methods for tumor treatments. Chemotherapeutic agents, such as doxorubicin (DOX), 5-fluorouracil (5-Fu), oxaliplatin (oxPt), and paracetamol (APAP), are often used in the treatment of various solid and nonsolid tumors (Yang et al. 2021; Horcajada et al. 2010). Nevertheless, the prompt utilization of chemotherapeutic medications is hindered by the drawbacks of limited bioavailability, inadequate tumor permeability, susceptibility to multidrug resistance, and uncontrollable side effects, all of which significantly diminish the efficacy of treatment. Nanoscale MOF-based nanoplatforms can effectively encapsulate and anchor anticancer medicines or prodrugs, enhancing their accumulation at tumor locations by leveraging the enhanced permeability and retention (EPR) effect. Furthermore, the surface of MOFs can be altered via covalent and non-covalent interactions, resulting in enhanced activity targeting and lengthy cycling capacity. This modification is advantageous for improving chemotherapeutic efficacy and minimizing adverse effects (Begum et al. 2019; Cohen 2012; Zhang et al. 2016a, b).

Based on the University of Michigan crystal material (UMCM) -1-NH2 and carboxycolumn aromatics (CP5), CP5, UMCM-1-NH-Py, a dual-stimulation-responsive nanoplatform was created for controlling cargo release and tumor therapy (Fig. 3.43; Tan et al. 2015). Rhodamine 6G or DOX was encapsulated using UMCM-1-NH-Py, which is made up of Zn^{2+} and NH2-BDC/4,4′,4′-benzene-1,3,5-triyl-tribenzoic acid (BTB) ligands. The CP5 nanovalve and the Py stem on the UMMM-1-NH-Py surface exhibit tuneable host–guest interactions that result in pH-responsive and competitive agent-responsive release behavior. Additionally, the nanoplatform exhibits favorable biodegradability and biocompatibility, little cytotoxicity, and minimal premature release, which qualifies it as an improved drug delivery system for oncological therapy and cellular imaging. For the on-demand delivery of 5-Fu to aid in tumor treatment, a Zn^{2+} and thermoresponsive nanocomposite comprising negatively charged CP5 supramolecular nanovalves and positively charged quaternary ammonium (Q)-modified UiO-66-NH_2 was created. With its large surface area and notable stability, UiO-66-NH_2, which is composed of Zr^{6+} clusters and NH_2-BDC ligands, is well suited for encapsulating 5-Fu. Increased Zn^{2+} under pathological conditions reduced the binding affinity between CP5 and Q, which caused CP5 to separate from Q and promote drug release. A temperature-responsive drug release may be made possible by concurrently lowering the binding affinity between CP5 and Q by external heating. For the treatment of bone tumors, CP5-gated, modified Py-functionalized Zr-MOF was created because bone tumor cells had higher intracellular Ca^{2+} levels. The results of the drug release study showed that 5-Fu release could be precisely controlled by high-temperature

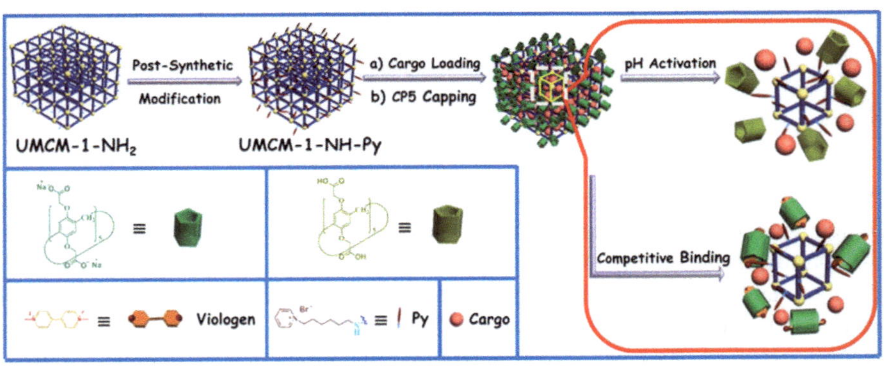

Fig. 3.43 Schematic diagram of the synthesis process and discharging behavior of UMCM-1-NH-Py of cargo CP5 sealing for chemotherapy. Reproduced from Tan et al. (2015) with permission from the Royal Society of Chemistry

stimulation, low pH, and increased Ca^{2+} concentration, which would make treating diseases linked to bone tumors easier.

Zhou and his team created a tyrosinase (TYR)-MOF using NPCN-333 (Al) as a powerful prodrug activator (Lian et al. 2018). This was done to cure cancer by tumor-specific xenoactivation in mice with HeLa tumors. A 100 nm diameter NPN-333 (Al) was synthesized using the trimer-oxygen cluster as the core and 4,4′,4′-s-triazine-2,4,6-tribenzoyl acid (TATB) as the ligand. TYR-NPN-333 (Al) is formed by encapsulating and protecting TYR in its skeleton, due to the existence of size 5.5 nm in NPN-333 (Fig. 3.41b). This compound exhibits good stability under acidic microenvironment and proteolytic circumstances. TYR-NPCN-333 (Al) successfully triggers the conversion of the prodrug APAP into a poisonous substance called 4-acetyl-O-benzoquinone by generating ROS and depleting glutathione (GSH). This process hinders cell growth and leads to programmed cell death or tissue death at the tumor location. Hence, the TYR-NPCN-333 (Al) nanoreactor exhibits significant promise in the biomedical and biotechnology sectors due to its exceptional enzyme decapping capacity, excellent chemical stability, and convenient post-functionalization capabilities.

Sun et al. developed an approach using MOF-based intelligent therapeutic diagnostic nanoplatform to target DOX release and guide tumor precision chemotherapy through imaging (Fig. 3.44) (Zhang et al. 2019a, b). Gadolinium (Gd) and Ytterbium (Yb) metal-organic frameworks (MOFs) were synthesized using Gd and Yb as the metal nodes and 5-borene-1,3-dicarboxylic acid (BBDC) as the ligands connecting the metal nodes. Following the encapsulation of DOX, D-glucose was utilized as a gatekeeper that responds to changes in pH. It was coated on the outer surface of the nanoplatform by forming reversible diol-borate condensation between D-glucose and borate groups in BBDC. This resulted in the development of DOX @ MOFs-Glu, which exhibits excellent biocompatibility and specifically targets tumors. The results from both laboratory experiments (in vitro) and live animal studies (in vivo) showed that DOX @ MOFs-Glu has the ability to release the drug DOX in response to changes in pH. It also displayed strong compatibility with biological systems, the ability to specifically target tumors, and the capability to guide tumor therapy using imaging. Moreover, this efficient nanoplatform preparation method may be readily utilized for the development and fabrication of other nanoplatforms, including Yb-MOF-Glu for gastrointestinal imaging. This opens up possibilities for precise image-guided chemotherapy employing a MOF-based intelligent nanoplatform.

Yang, Wang, Xia, and their colleagues developed a biomimetic MOF nanoplatform called ZIF-8-DOX-LY-RM, which is hidden by the RBC membrane (Qiao et al. 2022). This nanoplatform was designed to address the issue of hypoxia-induced chemoresistance in mice with 4T1 tumors. ZIF-8, consisting of Zn2 ions and 2-methimidazole (2-MIM) ligands, acts as a container. It was secured to concurrently deliver the anticancer agent DOX and the type 1 transforming growth factor β receptor inhibitor LY364947 (LY) with good efficiency. It was subsequently applied to the outer layer of a red blood cell membrane to enhance the stability and accumulation of the nanoplatform in tumors. The synthesized ZIF-8-DOX-LY-RM, with a diameter of 125 nm, exhibited pH-dependent drug release properties and

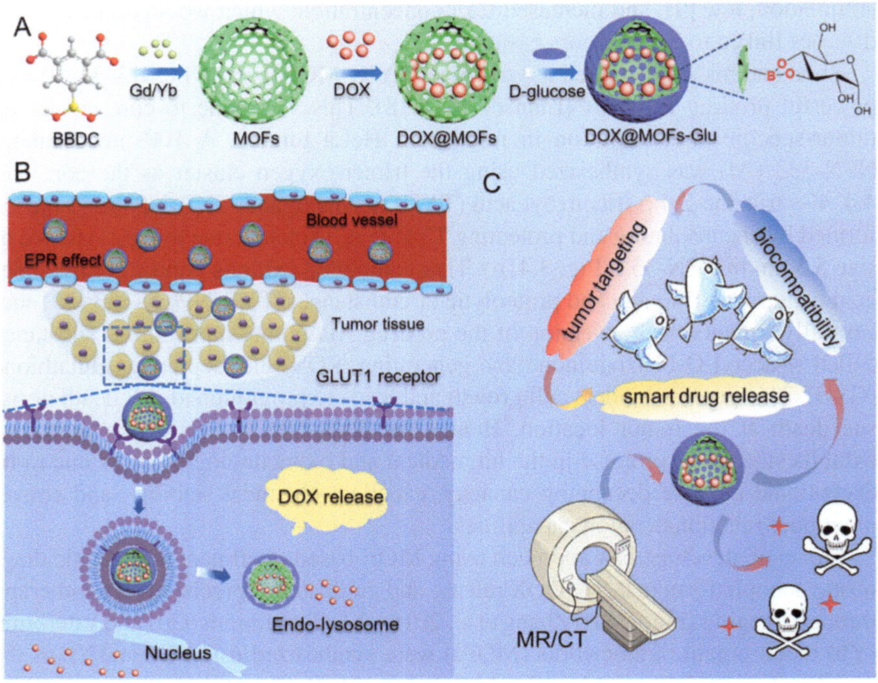

Fig. 3.44 Schematic Interpretation of the design of smart MOF-based nanoplatforms for imaging-guided precise chemotherapy. (**a**) The synthesis process of DOX@MOFs-Glu with D-glucose. (**b**) The delivery mechanism of MOFs to blood vessel and target tumor tissue via the EPR effect, bind to GLUT1 receptors, and release DOX in endo-lysosomes near the nucleus. (**c**) Tumor targeting, biocompatibility, smart drug release, and imaging via MR/CT. Reprinted with permission from Zhang et al. (2019a, b). Copyright © 2019, American Chemical Society

remarkable stability in both PBS and full cell culture media. These properties make it very suitable for future utilization in cancer therapy. ZIF-8-DOX-LY-RM, used in the TME, is capable of efficiently eliminating the extracellular matrix by releasing LY in a regulated manner. This leads to enhanced penetration of nanoplatforms and increased generation of ROS, ultimately reducing the chemoresistance caused by hypoxia for DOX (Fig. 3.41d). ZIF-8-DOX-LY-RM demonstrated excellent biocompatibility and better biodistribution, resulting in significant tumor reduction in vivo. This breakthrough paves the way for enhanced effectiveness of chemotherapy in treating solid tumors.

3.4.3 Chemodynamic Therapy

Chemodynamic therapy (CDT) has since been extensively utilized in the treatment of various diseases, including tumor therapy, after being inspired by Fenton responses and Fenton-like responses (Zhang et al. 2016a, b; Chen et al. 2019b). The function of CDT is to accelerate the conversion of the naturally occurring chemical H_2O_2 into highly poisonous OH using metal ions such as Fe^{3+}, Cu^{2+}, and Mn^{2+}. This process subsequently disturbs the internal redox equilibrium of cells, leading to the death of tumor cells. It is important to observe that the TME has a low level of acidity and a high concentration of H_2O_2, which creates an appropriate environment for the tumor therapy mediated by CDT. CDT offers the benefits of not relying on oxygen, utilizing external energy and equipment, exhibiting great selectivity, and causing minimal harm to normal tissues. Nevertheless, there is inadequate correlation between the H_2O_2 concentration, which could potentially diminish the effectiveness. Remarkably, MOFs made up of metal ions/clusters and organic ligands possess adjustable openings, straightforward customization, compatibility with living organisms, and the capacity to break down over time. These characteristics can significantly alleviate the obstacles encountered in CDT. The nanoplatform based on MOFs improves the efficacy of CDT on tumors by effectively incorporating metal ions/NPs, generating metal ions by self-degradation, and augmenting the H_2O_2 concentration at the tumor location (Zhong et al. 2020). Furthermore, the presence of metal ions such as Fe^{3+}, Cu^{2+}, and Mn^{2+} in MOFs imparts catalytic qualities, GSH consumption, and imaging capabilities to MOF-based nanoplatforms. This is advantageous for enhancing the efficacy of tumor treatment and enabling the diagnosis of cancer therapy.

Zhao and his team developed a nanocarrier using ZIF-8 and amphiphilic poloxamer 188 (F68) to create ClO@MOF/68 for treating breast cancer using a $^1O_2$-based CDT (Fig. 3.42a) (Chen et al. 2019b). By utilizing ZIF-8 to load NaClO, it is possible to induce a reaction between NaClO and ascorbic acid, resulting in the production of H_2O_2. This process enhances the presence of $^1O_2$ specifically at the tumor site. Finally, the end is sealed with a layer of F68 to produce a spherical ClO@MOF/68. The experimental findings demonstrate that ClO@MOF/68 exhibits excellent biocompatibility, a high ClO load, and pH/ROS-triggered degradability. Moreover, it can passively accumulate at tumor sites following intravenous administration, and has the capability to generate $^1O_2$ through ascorbic acid injection. Furthermore, it effectively treats tumors through $^1O_2$-based tumor treatment using CDT in 4T1 tumor-bearing xenograft mice, with minimal adverse effects. This study focuses on the challenges associated with the transportation of ClO in living organisms. It aims to overcome these challenges by developing nanocarriers made of MOF/polymer materials. The research findings uncover a new approach to tumor-specific cancer therapy known as ROS-based CDT.

Li et al. described the use of a cascade cobalt-ferrocene (Fc) NMOF to stimulate CDT in mice with 4 T1 tumors (Fig. 3.45) (Fang et al. 2020). A nanomaterial with a diameter of 80 nm was created by incorporating glucose oxidase (GOx) into a cobalt-ferrocene-based nanomolecular organic framework (NMOF). The NMOF was made using cobalt(II) and ferrocene carboxylic acid as metal sources and organic ligands. The addition of Fe^{2+} in the ferrocene ligand in Co-Fc@GOx results in an increased Fenton response capability. Through the process of TME, the enzyme glucose oxidase (GOx) encapsulated in Co-Fc @ GOx may effectively catalyze the conversion of endogenous glucose and O_2 into gluconic acid and H_2O_2

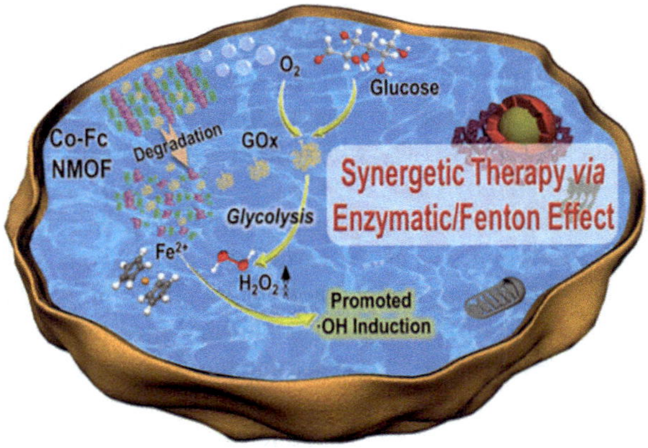

Fig. 3.45 Schematic illustration of Co-Fc@GOx as a cascade enzymatic/Fenton reaction platform for promoted •OH induction. Reprinted with permission from Fang et al. (2020). Copyright © 2020 WILEY-VCH Verlag GmbH & Co. KGaA, Weinheim

using the glycolytic pathway. This leads to an increase in acidity and the production of H_2O_2. Simultaneously, the H_2O_2 produced was transformed into a harmful OH by Co-Fc @ GOx via the Fenton reaction, thereby significantly amplifying the efficacy of CDT. The experimental findings demonstrate that Co-Fc @ GOx possesses favorable degradability, notable glucose utilization, and a substantial capacity for generating H_2O_2/•OH. As a result, it serves as an exceptional nanoplatform for efficiently suppressing tumor growth by CDT.

Ren et al. developed a novel agent called PZIF-67-AT, which is based on a MOF (Metal-Organic Framework) and is designed to destroy H_2O_2 in a steady-state manner (Sang et al. 2020). The agent works by increasing the amounts of H_2O_2 inside cells through promoting its creation and preventing its clearance. This ultimately leads to the targeted killing of tumors, as shown in Fig. 3.42c. ZIF-67, which consists of cobalt ions and 2-MIM, underwent a post-synthesis modification process using the small molecule inhibitor 3-amino-1,2,4-triazole (3-AT), achieving a loading efficiency of around 28.5%. Subsequently, the PEG undergoes additional alterations on its external surface to enhance its compatibility with biological systems and its stability within physiological conditions. Through the use of flow cytometry and confocal laser scanning microscopy, it was determined that PZIF-67-AT was effectively taken up by cells and accumulated in acidic lysosomes, which facilitated the release of 3-AT. In addition, PZIF-67-AT can increase the buildup of H_2O_2 in cancer cells by facilitating the transformation of superoxide anion into H_2O_2, suppressing catalase activity, and reducing the consumption of GSH. Significantly, elevated levels of H_2O_2 can be transformed into harmful •OH, leading to the initiation of intracellular oxidative stress and consequently amplifying the effectiveness of tumor CDT. This phenomenon was further confirmed by experiments conducted in both laboratory settings and living organisms.

Wen and colleagues introduced a nanoreactor called DHA @ MIL-101, which is based on MIL-101 and functions as a peroxidase-type catalyst (Lu et al. 2014). This nanoreactor was developed for the treatment of Lewis lung cancer using a highly efficient CDT. Dihydroartemisinin (DHA) is a substance derived from *Artemisia annua*. It is extracted and embedded in the pore of MIL-101, which is made from Fe3 and NH_2-BDC ligand. This results in the formation of DHA @ MIL-101, which has improved water solubility and biological compatibility. The loading efficiency of DHA in this molecule is 20 wt%. Within this system, Fe3, which is produced by DHA @ MIL-101 in TME (tumor microenvironment), can catalyze the conversion of endogenous H_2O_2 into •OH by the Fenton reaction. Furthermore, because of the unstable inner peroxide bridge (–O–O–) in the structure, the released DHA not only enhances the activation of iron ions to produce hazardous •OH, but also induces the influx of iron ions, leading to the generation of ROS. In addition, DHA can directly suppress the activity of GSH peroxidase 4, resulting in the programmed cell death of tumor cells. Significantly, both laboratory experiments (in vitro) and experiments conducted on living organisms (in vivo) have shown that DHA @ MIL-101 has a beneficial impact on inhibiting the growth of cancer cells.

3.4.4 Photodynamic Therapy

Photodynamic therapy (PDT) is an encouraging treatment approach that employs photosensitizers to trigger programmed cell death (apoptosis) and tissue injury when exposed to light and oxygen (Yan et al. 2020; Qin et al. 2021). There are two categories of photodynamic therapy (PDT) mechanisms: type I PDT, which generates superoxide anion and reactive oxygen species (ROS) such hydroxyl radical (•OH) and hydrogen peroxide (H_2O_2), and type II PDT, which produces highly lethal singlet oxygen ($^1O_2$) (Huang et al. 2021c). During PDT, the photosensitizer undergoes a transition from its initial state (S0) to either the first excited state (S1) or the second excited state (S2) upon absorbing photon energy at the specific wavelength of light. S2 undergoes rapid degradation to S1 via internal conversion (Chen et al. 2020). Conversely, the S1, which is in an unstable state, can undergo a transformation into a more stable excited triplet state (T1). T1 exhibits a notable ability to interact with cell substrates by means of an electron transfer mechanism, resulting in the production of free radicals. These free radicals then react with O_2 or H_2O to form reactive oxygen species (ROS), including OH and O_2^-. This process ultimately leads to the induction of apoptosis, namely by type I PDT. Furthermore, the energy of T1 can be transferred to three molecules of O_2 to produce one molecule of singlet oxygen (referred to as a type PDT process), which exhibits potent cytotoxic effects (Zhou et al. 2016). When compared to standard chemotherapy and radiotherapy, PDT offers several benefits such as minimal side effects, noninvasiveness, and high selectivity (Wei et al. 2022). As a result, PDT has found extensive use in the clinical treatment of pancreatic cancer, skin cancer, and head and neck cancer. Despite significant advancements in tumor treatment, PDT still faces challenges such as the inadequate stability of photosensitizers, limited penetration depth of irradiation, and a high reliance on oxygen concentration. Due to its high porosity and ease of functionalization, the MOF-based nanoplatform can serve as a nanocarrier for delivering photosensitizer molecules or nanoparticles to tumor sites. This nanoplatform has a high loading capacity, which greatly enhances the stability and targeting of the photosensitizer, leading to improved therapeutic efficacy. In addition, the intelligent MOF made from porphyrin-based organic ligands can be utilized as a photosensitizer for PDT, offering a hopeful alternative for anticancer PDT.

Lin and colleagues published a study on a nanoscale Hf-porphyrin MOF called DBP-UiO, which was used for PDT to treat head and neck cancer that is resistant to other treatments (Lu et al. 2014). The DBP-UiO material, created by combining the 5,15-di-(benzoyl) porphyrin (H2DBP) ligand with the Hf4 node by a solvent heat reaction, has a porous structure and functions as a very efficient photosensitizer. It has an average diameter of 76.3 nm. DBP-UiO has exceptional effectiveness in generating $^1O_2$ due to the sequestration of the site and the inherent porous structure of the H2DBP ligand. Moreover, in vivo experiments conducted with DBP-UiO material have shown significant inhibition of tumor growth and minimal harm following photodynamic therapy (PDT) in SQ20B subcutaneous xenograft mice. This suggests a highly encouraging potential for the use of DBP-UiO material in anticancer treatment. The group also enhanced a MOF by incorporating a dichloride compound, resulting in improved properties for light absorption and higher efficiency in generating $^1O_2$ (Lu et al. 2015). This MOF, called DBC-UiO, consists of Hf 4 metal ions and 5,15-di(p-benzoato) chlorin (H2DBC) ligands. Notably, DBC-UiO exhibited significant PDT effects in both CT26 and HT 29 mouse models of colorectal adenocarcinoma.

Zhou and his colleagues successfully created a size-controlled Zr (IV) group porphyrin MOF (Park et al. 2016). They utilized FA modification as the active targeting method for PDT, as shown in Fig. 3.46. The PCN-224 is a spherical structure composed of a Zr 6 metal cluster and four (4-carboxenyl) porphyrin (H2TCPP) ligands, created via a bottom-up method. By adjusting the concentration of benzoic acid and H2TCPP, the size of PCN-224 may be controlled. Among the many sizes tested, 90 nm-PCN-224 had the greatest uptake in HeLa cells and had a notable influence on photodynamic therapy (PDT). In addition, the surface of PCN-224 was altered by coordinating FA with the carboxylate groups of the Zr cluster. This modification enhances the active targeting of the MOF and improves the effectiveness of PDT.

Zhang and their colleagues developed a biomimetic and multifunctional nanoparticle called aMMTm, based on MOFs (Min et al. 2019). They used this nanoparticle to effectively eliminate GSH (glutathione) and provide antiangiogenic treatment in mice with 4T1 tumors by PDT (Photodynamic Therapy). The Zr-MOF (PCN-224), a photosensitive porphyrin, was loaded with apatinib, an FDA-approved inhibitor of vascular endothelial growth factor receptor 2 (VEGFR2). It was then covered with a MnO_2 shell and the cell membrane retrieved from the 4 T1 layer of a mouse breast cancer cell line. The encapsulation effectiveness of this process was greater than 98.3%. Within the tumor microenvironment (TME), the surplus of glutathione

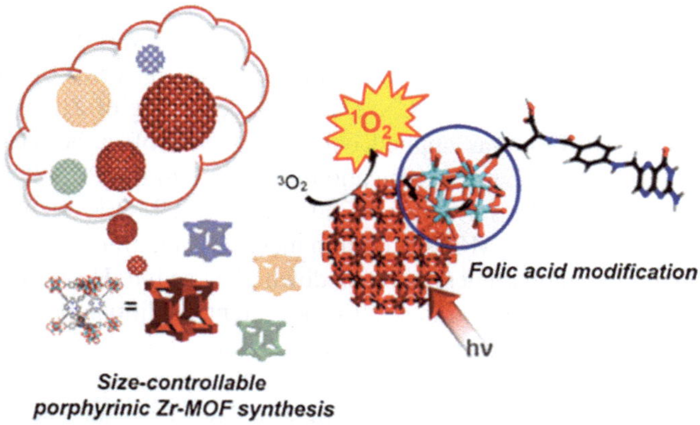

Fig. 3.46 Targeted photodynamic therapy (PDT) with Zr(IV)-based porphyrinic metal–organic framework (MOF) nanoparticles. Reprinted with permission from Park et al. (2016). Copyright © 2016, American Chemical Society

(GSH) can be counteracted by the manganese dioxide (MnO_2) coating on the surface of the aMMTm. This coating inhibits the process of angiogenesis generated by photodynamic therapy (PDT) and facilitates the effective release of apatinib. Furthermore, Mn^{2+} can serve as a contrast agent for in vivo magnetic resonance imaging (MRI) of tumors. Significantly, aMMTm demonstrated robust homologous targeting capability, substantial anticancer effectiveness in both laboratory and animal models, and favorable biocompatibility attributed to alterations in the cell membrane.

Zhao, Guo, and their team developed a flexible MOF-based nanoformulation called OxgeMCC-rSAE using a one-pot multicomponent self-assembly approach (Wang et al. 2020). This formulation is aimed to enhance the production of oxygen at solid tumor locations for the treatment of cancer using PDT. The Mn 3 [Co (CN) 6] 2 MOF consists of manganese (Mn) atoms and organic ligands [Co (CN) 6]. It has single-atomic ruthenium (Ru) atoms attached through coordination, as well as hydrophobic and hydrophobic interactions. The MOF also contains the hydrophobic photosensitizer chlorin e6. The Ru loading weight ratio is approximately 2.23 wt% and the Chlorin e6 loading efficiency is 75.8%. During the tumor microenvironment (TME), the presence of Ru in OxgeMCC-rSAE serves as a very efficient catalytic site for the quick conversion of endogenous H_2O_2 to O_2. This process effectively improves tumor hypoxia without depleting itself or without external activation. Consequently, it leads to increased levels of reactive oxygen species (ROS) and enhances the efficiency of photodynamic therapy (PDT). Additionally, the OxgeMCC-rSAE demonstrated a remarkable T1-weighted MRI impact attributed to the existence of Mn 2. OxgeMCC-rSAE demonstrated remarkable tumor suppression, excellent biocompatibility, and significant accumulation at the tumor site in vivo. This further demonstrates the effectiveness of self-assembled nanoagents for cancer treatment using photodynamic therapy (PDT).

An efficient nuclear-shell nanoreactor based on MOFs, supersmall Au NPs, and upconversion nanoparticles (UCNPs), UMOFs @ Au NPs, was described by Chen and colleagues for the use of PDT in the treatment of cancer in U87MG tumor-bearing mice (He et al. 2020). A PCN-222 MOF, consisting of Zr clusters and Fe (III) ligands (4-TCP-Fe), was applied onto the surface of carboxylate-modified UCNPs. Subsequently, a 2 nm modification was made to the PCN-222 by in situ reduction, resulting in the formation of nuclear-shell UMOFs @ Au NPs. This design involves the use of UMOFs (ultrathin metal-organic frameworks) combined with Au NPs (gold nanoparticles) to break down glucose into H_2O_2 (hydrogen peroxide) when exposed to near-infrared (NIR) laser light. Additionally, the UMOFs @ Au NPs act as catalysts to facilitate the production of O_2 (oxygen) from the H_2O_2 specifically at the tumor site. Furthermore, the UCNP core can transform near-infrared laser light into visible light, which then stimulates the nanoreactor to consistently generate $^1O_2$. This process enables a series of PDT reactions to occur. This study offers a novel approach to creating MOF-based cascade catalytic nanoreactors that can be used for efficient tumor therapy by PDT.

Numerous nanoplatforms based on MOFs have been created and developed to address the issues of unstable photosensitizer, shallow penetration of irradiation, and excessive reliance on oxygen levels in photodynamic therapy for tumors. It is important to understand that the effectiveness of PDT can be enhanced by controlling the size, porosity, and colloidal stability of the MOF-based nanoplatform, as this can help to minimize the diffusion distance of ROS. Additionally, there is a growing method of creating degradable hollow nanoplatforms using MOFs (Metal-Organic Frameworks) to effectively load the photosensitizer and retain a high oxygen production. Additionally, the NIR-II imaging technique, which has the ability to deeply penetrate tissues, was employed. Using lasers as a light source is an alternative method to enhance the effectiveness of PDT. It is crucial to take into account both the biosafety and biodegradability of the MOF-based nanoplatforms at the same time.

3.4.5 Photothermal Therapy

Photothermal therapy is a technique used in the treatment of malignant tumors (Zheng et al. 2021; Zhou et al. 2021; Li et al. 2021). It involves turning photoenergy into thermotherapy utilizing photothermal agents to promote thermal ablation and necrosis of tumor cells. Typically, the light source used for photothermal therapy (PTT) contains the first near-infrared (NIR) window (750–1000 nm) and the second NIR window (1000–1700 nm). This is because NIR lasers may penetrate deep into tissues and are not strongly absorbed by biological cells and tissues. Furthermore, cancer cells have a higher susceptibility to hyperthermia in comparison to normal cells. This characteristic is advantageous in enhancing the effectiveness of anticancer treatments and minimizing undesirable side effects (Sun et al. 2021a, b). PTT is extensively utilized in cancer therapy because of its noninvasive nature, high level of control, capacity to function without oxygen, and few adverse effects. Nevertheless, the lack of tumor selectivity and limited tissue penetration hindered the progress of photothermal therapy (PTT). By employing logical planning and combining different elements, several nanomaterials have been developed to improve the effectiveness of photothermal therapy (PTT) (Yang et al. 2022). The MOF-based nanoplatform plays a crucial role in the photothermal therapy (PTT) process because of its exceptional capacity for loading and its effortless ability to attach to photothermal agents.

Liu and his team synthesized a versatile nanocomposite using MIL-100 (Fe) and hyaluronic acid (HA), called MOF @ HA @ ICG NPs (Cai et al. 2017). This nanocomposite was used for multimodal imaging-guided PTT in mice with MCF-7 tumors. MIL-100 (Fe) is synthesized using a hydrothermal process employing Fe^{3+} nodes and 1,3,5-phenyltriphenylpropionic acid (BTC) as organic ligands. This material has pore sizes of 2.5 nm and 2.9 nm, making it ideal for cargo encapsulation. The integration of MIL-100 (Fe) with HA to improve targeting efficacy for CD44-overexpressing tumors resulted in the creation of MOF @ HA @ ICG nanoparticles, including the water-soluble near-infrared organic dye Indocyanine Green (ICG) at a maximum loading capacity of 40%. The experimental results indicate that PTT effectively inhibited MOF @ HA @ ICG NPs, exhibiting good biocompatibility, efficient cellular uptake, and stable photothermal characteristics, as shown by fluorescence imaging, photoacoustic imaging (PAI), and T2-weighted MRI.

Xie and colleagues synthesized a nanocomposite of MOFs known as UiO-66@CyP using a straightforward and widely applicable method to perform PTT for cancer treatment (Wang et al. 2017a, b). The surface of UiO-66 was modified by incorporating a cyanine-containing polymer (CyP) using a multicomponent Passerini reaction. This led to the creation of octahedral UiO-66@CyP nanoparticles with a consistent size of 100 nm and superior dispersion. The photothermal conversion efficiency of UiO-66@CyP was ascertained to be 27.3% according to the experimental findings. The amalgamation of UiO-66 and CyP, in conjunction with an 808-nm laser, effectively obliterated tumor cells and inhibited tumor proliferation by photothermal therapy, without inducing any discernible damage in mice with CT26 tumors. This research lays the groundwork for the development of nanocomposites including metal-organic frameworks (MOFs) and polymers, applicable for highly effective photothermal therapy (PTT) in cancers. Following that, the research group created a different type of nanocomposite called Cy@ZIF-8 NPs, which is based on MOF. This nanocomposite was developed to contain carboxy anthocyanins (Cy) with an NIR dye, and it was used for the PTT of cervical cancer. This information is illustrated in and can be found in reference (Li et al. 2018). Rhomboid dodecahedral Cy@ZIF-8 nanoparticles, with an average size of 118.7 nm, were synthesized utilizing straightforward nanoprecipitation techniques. These nanoparticles exhibit pH responsiveness and excellent water solubility. The presence of Cy loading resulted in the ZIF-8 NPs exhibiting exceptional photothermal conversion efficiency (33.2%) and notable photostability when exposed to 808 nm laser irradiation. Significantly, both in vitro and in vivo studies demonstrate that Cy@ZIF-8 NPs had a remarkable capacity to suppress tumor growth, with few side effects associated with PTT. This highlights the extensive potential of MOF in PTT.

Zeng, Wu, Lu, and their colleagues successfully created a sophisticated nuclear-shell nanoplatform called AuNS-@MOF-ZD 2, using gold nanostar (AuNS) and MIL-101-NH2 (Fe). This nanoplatform was designed for T1 weighted MRI and targeted treatment of triple-negative breast cancer using photothermal therapy (PTT), as shown in Fig. 3.47 (Zhang et al. 2018a, b). AuNS (gold nanoparticles) modified with polyvinyl pyrrolidone (PVP) were encapsulated in a MIL-101-NH2 shell for four cycles. The resulting structure was then connected with NHS-PEG-COOH and a short peptide (CTVRTSADC, ZD 2). This led to the development of AuNS @ MOF-ZD 2, which has precise targeting ability and excellent biological compatibility. The AuNS core provides the AuNS @ MOF-ZD 2 with a highly effective T1-weighted MRI effect and a photothermal conversion efficiency of 40.5%. Furthermore, the alteration of ZD 2 on the exterior of AuNS @ MOF-ZD 2 exhibited selective targeting of different subtypes of MDA-MB-231 cells while excluding breast cancer cells. This supports the molecular categorization and accurate therapy of breast cancer. AuNS@MOF-ZD 2 had exceptional biocompatibility and exhibited remarkable photothermal therapy (PTT) efficacy, as well as T1-weighted MRI capability and selective in vivo targeting. These findings offer novel insights into the potential use of visual therapeutic diagnostics for various subtypes of breast cancer.

Yang, Lin, and their colleagues have developed a straightforward PTT nanoplatform using MOF-808 and polyaniline (PANI), specifically called Gd-DTPA-MOF-808 @ PANI. This platform was used for MRI-guided PTT in mice with 4T1 tumors (Jia et al. 2021). The construction of a stable Gd-DTPA-MOF-808 @ PANI with high MRI relaxation and 30.6% photothermal conversion efficiency involves

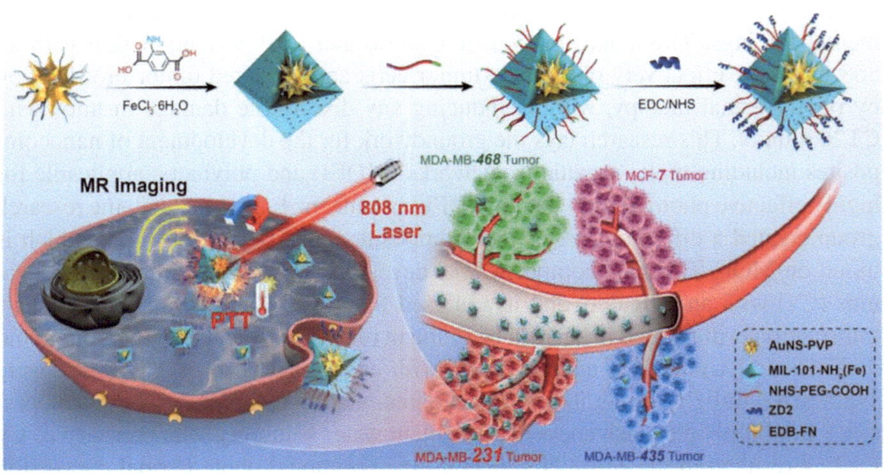

Fig. 3.47 Schematic illustration of the synthesis of AuNS@MOF-ZD2 nanoprobes. Reproduced from Zhang et al. (2018a, b) with permission. Copyright © 2018 WILEY-VCH Verlag GmbH & Co. KGaA, Weinheim

using $Zr_6O_4^-$ $(OH)_4$ $(–CO_2)_6$ $(HCOO)_6$ as a cluster, BTC as a ligand, and then functionalizing it to the surface with PANI. According to laboratory and animal testing, Gd-DTPA-MOF-808 @ PANI demonstrated a powerful T1-weighted MRI effect (5.4 times stronger than the commercial contrast agent Magnevist), significant photothermal therapy (PTT) capacity when exposed to an 808 nm laser, and minimal negative effects in mice with 4T1 tumors. These findings suggest that MOF-based nanocomposites have potential as a suitable option for cancer therapy diagnostics.

3.4.6 *Hunger Therapy*

Tumor cells require substantial quantities of nutrients and energy to sustain their life and growth, in contrast to regular tissues, owing to their disrupted metabolic pathways (Vander et al. 2009). As per the Warburg effect, tumor cells primarily rely on the glycolytic pathway as their main energy source. This allows tumor cells to consume a greater amount of glucose compared to normal cells. Therefore, when the supply of glucose is terminated, the cancer cells are deprived of essential nutrients and ultimately perish. Starvation therapy is now acknowledged as a viable method to decrease crucial elements necessary for tumor growth. Glucose oxidase (GOx) is a common enzyme that naturally consumes glucose (Wang et al. 2019a). It is extensively utilized in the treatment of tumor starvation (Wang et al. 2019b). Nevertheless, GOx has drawbacks such as inadequate stability, limited environmental resilience, and susceptibility to degradation, resulting in diminished effectiveness of fasting therapy. Encapsulating or anchoring glucose oxidase (GOx) in nanomaterials is anticipated to address the aforementioned limitations (Ge et al. 2022). Recently, there has been significant interest in MOF-based nanocomposites, which are considered to be highly important nanomaterials (Yu et al. 2021). These nanocomposites have been found to enhance the development of versatile nanocarriers and improve the effectiveness of starvation therapy.

Qu, Ren, and their colleagues developed a biomimetic MOF-based nanocomposite called TGZ @ eM (Zhang et al. 2018a). This nanocomposite was designed to simultaneously carry the prodrugs telapazine (TPZ) and GOx for the treatment of starvation-activated colon cancer in CT26 tumor-bearing mice. A ZIF-8 material, composed of Zn^{2+} ions and 2-MIM ligands, was synthesized and used to encapsulate both GOx and TPZ. This resulted in the formation of ZIF-8 (TGZ) loaded with GOx and TPZ. To enhance its immune evasion and prolong its presence in the bloodstream, the TGZ material was further coated with red blood cell membrane, forming spherical TGZ @ eM. During the tumor microenvironment (TME), the levels of naturally occurring glucose and oxygen can be reduced due to the presence of glucose oxidase (GOx) released by thiazolidinedione (TGZ) at the extracellular matrix (eM). This depletion of glucose and oxygen can lead to hunger and worsen the condition of tumor hypoxia. Additionally, it aids in the efficient activation of the prodrug TPZ into extremely harmful free radicals, which ultimately triggers apoptosis. The TGZ compound exhibited enhanced anticancer efficacy in CT26 cells and CT26 tumor-bearing mice, both in laboratory settings and in live subjects.

Zhang and colleagues developed a pH-responsive degradable nanoplatform called HZ @ GD, which is based on ZIF-8 (Wu et al. 2022). This nanoplatform was specifically designed for treating melanoma in B16-F10 tumor-bearing C57BL/6 mice. This information is illustrated in Fig. 3.44e. The glucose transporter 1 (GLUT 1) mRNA cleaving DNA enzyme (GD) was efficiently loaded into ZIF-8 with a loading efficiency of around 81.5%. Subsequently, the ZIF-8 was modified with HA to improve tumor targeting of the resulting HZ@GD formulation. ZIF-8, which consists of Zn2 nodes and 2-MIM ligands, can release Zn2 and GD into the cytoplasm because to its degradability in the acidic tumor environment. As a result, the release of Zn 2 can both decrease NAD and GAPDH inactivation, effectively inhibiting glycolysis, and activate GD to downregulate the expression of GLUT 1, which helps inhibit the glycolysis process. This leads to the energy consumption of the tumor site and provides a new approach for MOF-based nanocarriers to inhibit tumor growth through starvation therapy.

3.4.7 Immunotherapy

Immunotherapy is a specialized treatment that harnesses the patient's immune system to activate or strengthen the body's fight against tumor cells (Huang et al. 2021a). Immunotherapy, in contrast to traditional therapies, can direct the immune system to specifically attack cancer cells, thereby minimizing adverse effects. Simultaneously, immunotherapy offers the benefits of broad adaptability and robust memory impact. Macrophages, natural killer cells, and T cells, along with hydrolysis enzymes, defensins, and cytokines such as interleukin-1 (IL-1), interleukin-6 (IL-6), interferon-γ (IFN-γ), and tumor necrosis factor-α (TNF-α), play a role in the process of immunotherapy (Sang et al. 2019). Immunotherapy presents a hopeful approach to treating malignancies, and notable progress has been achieved. Nevertheless, the use of these treatments in oncology is hindered by various drawbacks, including the occurrence of an inflammatory response, the presence of a complex immunosuppressive tumor microenvironment, and an unusual clinical response rate. These limitations restrict their potential for further application in cancer treatment. MOFs are utilized as nanocarriers in immunotherapy due to their high loading capacity, appropriate size, varied composition, adjustable pore size, and EPR impact (Liang et al. 2021). They are employed to boost the therapeutic effect by delivering substances such as antigens, adjuvants, and immunomodulators.

Ren and his team developed a nanoscale vaccine called OVA @ ZIF-8-CpG, which is built on a Metal-Organic Framework (MOF) (Zhang et al. 2016a, b). This vaccine is designed to enhance cancer immunotherapy by stimulating robust humoral and cellular immune responses, as well as promoting strong immunological memory. Ovalbumin (OVA) was incorporated into ZIF-8 to create OVA @ ZIF-8, which had a diameter of 200 nm and demonstrated efficient cellular uptake.

Thereafter, negatively charged unmethylated cytosine-phosphate-guanine oligodeoxy nucleotides (CpG ODNs) were affixed to the surface of OVA @ ZIF-8 via electrostatic interactions, leading to the creation of OVA @ ZIF-8-CpG. This composite demonstrated pH-responsive breakdown, exceptional biological compatibility, and robust immunogenicity. Both in vitro and in vivo studies have significantly shown that this dual loading strategy allows the OVA @ ZIF-8-CpG to deliver both CpG ODN and OVA antigen to the same antigen-presenting cells and release them in acidic environments, thereby enhancing the generation of robust immune responses, including humoral and cellular immunity, as well as a strong immunological memory upon re-exposure to the same antigen.

Li and his colleagues created a simple nanoplatform that reacts to pH variations and enhances cancer treatment in the B16-OVA melanoma model (Duan et al. 2017). The OVA antigen was integrated into the MOF using Eu3 as the metal node and guanine monophosphate (GMP) as the linker. It was then included onto the outside of the CPG by Watson–Crick base pairing. The experimental results indicate that the biodegradable nanocomposite shows a good encapsulation effectiveness for the OVA antigen (55%, w/w). Furthermore, it augments the capacity to evade endosomes/lysosomes, expedites the activation of CD8 cytotoxic T cells, and raises the anticancer efficacy (with a survival rate nearing 100%). These findings facilitate the initiation of a vigorous immune response in cancer treatment.

Wang and his colleagues developed a core-shell nanoparticle using mesoporous silica and a metal-organic framework (MOF) to enclose OVA and a substance that boosts the immune system (polyinosine-polycytidine, polyIC) (Li et al. 2020a, b). They used a self-assembly method that can be applied universally. These nanoparticles were then used in the immunotherapy of mice with G7-OVA tumors. Experimental studies have shown that nuclear-shell nanoparticles display robust immune responses that are specific to antigens and have long-lasting tumor-suppressing effects.

Zhang and colleagues have developed a nanoplatform called LYS-NPs, based on ZIF-8, which specifically targets lysosomes (Zhao et al. 2021a). They used this nanoplatform for tumor treatment in mice with 4 T1 tumors by immunotherapy. Therapeutic proteins such as perforin and granzyme B are enclosed within acid-degraded ZIF-8 in this system. Subsequently, calcium ions (Ca^{2+}) were introduced to the surface or interior of ZIF-8 to improve the biocompatibility of the nanoplatform by mineralization and to augment the efficacy of perforin and granzyme B. The CD63-aptamer was modified on the outermost layer of the nanocarrier to produce LYS-NPs with enhanced targeting of T cells to lysosomes. LYS-NPs degrade in the acidic milieu of T cell lysosomes, therefore releasing perforin, granzyme B, and Ca^{2+} ions stored inside the lysosomes of T cells. Notably, when the primary tissue compatibility complex of tumor cells activates the T cell receptor, the chemicals contained inside T cell lysosomes are rapidly discharged into the immunological synapse, as shown by laboratory and animal studies.

3.5 Multimodal Therapy Based on MOFs

Lin and his colleagues developed a straightforward nanoplatform using Cu-MOF for photoacoustic imaging (PAI) and tumor therapy through chemotherapy, chemodynamic therapy (CDT), and photothermal therapy (PTT) (Shen et al. 2021). HKUST-1 is a nanocarrier composed of Cu2 nodes and BTC organic ligand. It serves as a multifunctional platform for delivering the FDA-approved antitumor drug disulfiram (DSF). The platform is further modified by functionalizing the PVP polymer, resulting in the creation of a spherical nanoplatform called DSF @ HKUST-1. This nanoplatform has a DSF load of 12.60% and exhibits a photothermal conversion efficiency of 26.69% in response to the tumor microenvironment (TME). Through the process of Targeted Molecular Elimination (TME), DSF (Disulfiram) in combination with HKUST-1 can break down and release Copper (Cu2) and DSF. This leads to the activation of Copper (Cu2)-mediated chemodynamic therapy (CDT) through the Fenton response, as well as DSF-induced chemotherapy. Lian, Chen, and their colleagues also documented the development of a smart nanocomposite using CuS NPs, Cu-MOF, and PEG shells. This nanocomposite, called DOX @ CuS @ Cu-MOF/PEG, has the ability to encapsulate the anticancer drug DOX for combined chemotherapy, CDT, and PTT treatments. Based on laboratory experiments conducted in test tubes and living organisms, it has been found that DOX @ CuS @ Cu-MOF/PEG is capable of efficiently suppressing the growth of tumors through chemotherapy, combined chemo-photothermal therapy (CDT), and photothermal therapy (PTT). This treatment exhibits excellent compatibility with biological systems and expands the range of options for creating uncomplicated nanocomposites based on metal-organic frameworks (MOFs) for the treatment of tumors using several modes of action.

Li and colleagues have developed a smart nanocomposite called Mil-101 (Fe) @ GOx/DOX @ TPP/FA (MGDFT NPs) that exhibits phosphate liability (Peng et al. 2021). This nanocomposite has been used for synergistic chemotherapy, combined with CDT and starvation treatment, in 4T1 tumor-bearing mice. MIL-101 (Fe) was employed for the modification of chemotherapeutic drugs DOX and GOx, as well as the coating of tumor target molecules GA-PEG and mitochondrial target molecules (3-carboxyl) triphenylphosphine bromide. This resulted in the creation of polyfaceted MGDFT NPs with an average size of around 150 nm and excellent targeting capability. Upon reaching the tumor cells, the MGDFT NPs undergo degradation due to the presence of elevated phosphate concentrations. Subsequently, the liberated GOx enzyme can accelerate the conversion of glucose into H_2O_2, so inducing tumor starvation. In addition, the degradation of MGDFT NPs produces Fe3, which can catalyze the conversion of H_2O_2 to OH, hence enhancing the effectiveness of CDT. In addition, effective tumor targeting enables MGDFT NPs to be efficiently taken up by tumor cells, resulting in improved treatment outcomes through chemotherapy/CDT/starvation therapy while causing little systemic damage.

Tian and his colleagues developed a biomimetic hollow MOF nanoplatform for imaging-guided synergistic chemotherapy, photothermal therapy (PTT), and photodynamic therapy (PDT) (Sun et al. 2021a, b). The surface of PCN-222, which consists of Zr 4 nodes and TCPC organic ligands, was coated with ZIF-8 to create a nuclear shell MOF. The ZIF-8 template was then eliminated using a self-sacrificing template approach, resulting in the formation of a hollow MOF skeleton. Following the encapsulation of the anticancer medication DOX and photosensitizer ICG, they are applied as a coating on the cancer cell membrane. This results in the formation of DIHPm NPs, which exhibit notable tumor targeting capabilities, a high drug load, and the capacity to evade the immune system. The DIHPm NPs, which were produced, exhibited a significant photothermal conversion capability and photodynamic therapy (PDT) effect when exposed to laser irradiation. Meanwhile, DOX confers DIHPm NPs with significant chemotherapeutic efficacy. Following the intravenous administration of hollow DIHPm NPs in mice with 4T1 tumors, the NPs exhibited remarkable biocompatibility, targeted the tumor cells well, released the drug DOX in response to changes in pH and photothermal stimuli, and demonstrated synergistic effects in chemotherapy, photothermal therapy (PTT), and photodynamic therapy (PDT) guided by photothermal and fluorescence (FL) imaging.

Chen and colleagues later described a sophisticated core-shell nanocomposite consisting of MOF and Au nanostar AuNS @ ZrTCPP-GA@LP (AZGL) (Li et al. 2022c). This nanocomposite was developed for the treatment of breast cancer using chemotherapy, photothermal therapy (PTT), and photodynamic therapy (PDT) (Fig. 3.48). The approach involved the introduction of a heat shock protein 90 (HSP 90) inhibitor and a photothermal agent called Au nanostar into MOF reservoirs using a straightforward one-pot strategy. These reservoirs were then coated with PEGylated liposomes to create an AZGL. The resulting AZGL exhibited excellent biocompatibility and an encapsulation rate of approximately 38.4%. Based on the experimental findings, AZGL demonstrated drug release that is dependent on pH, as well as effective accumulation in tumors. Additionally, it was used for combination chemotherapy, photothermal therapy, and photodynamic therapy, guided by FL imaging.

Xu and his colleagues developed a versatile nanoplatform based on MOFs (metal-organic frameworks) for the treatment of malignancies using chemical, photothermal, and immunotherapy approaches (Liu et al. 2022). MIL-100 was produced using a microwave technique to efficiently load the NIR dye ICG and the anticancer medication OxPt with a high loading efficiency. Afterward, the HA was altered with the emergence of MIL-100 in order to produce OIMH NPs that have a prolonged cycle capacity. ICG and OxPt provide OIMH NPs with photothermal treatment (PTT) and chemotherapeutic capabilities, respectively. On the other hand, OIMH NPs demonstrate highly responsive PAI effects that can be utilized for tumor therapy guided by imaging. OxPt and ICG in OIMH NPs have the ability to induce immunogenic cell death, enhance the infiltration of immune cells, and raise the population of effector memory T cells in the spleen. These effects are advantageous for the precise treatment of tumor recurrence and suppression. Furthermore, this nanoplatform demonstrated the capacity to stimulate strong anticancer immune responses by inhibiting immunological escape and activating the immune system in

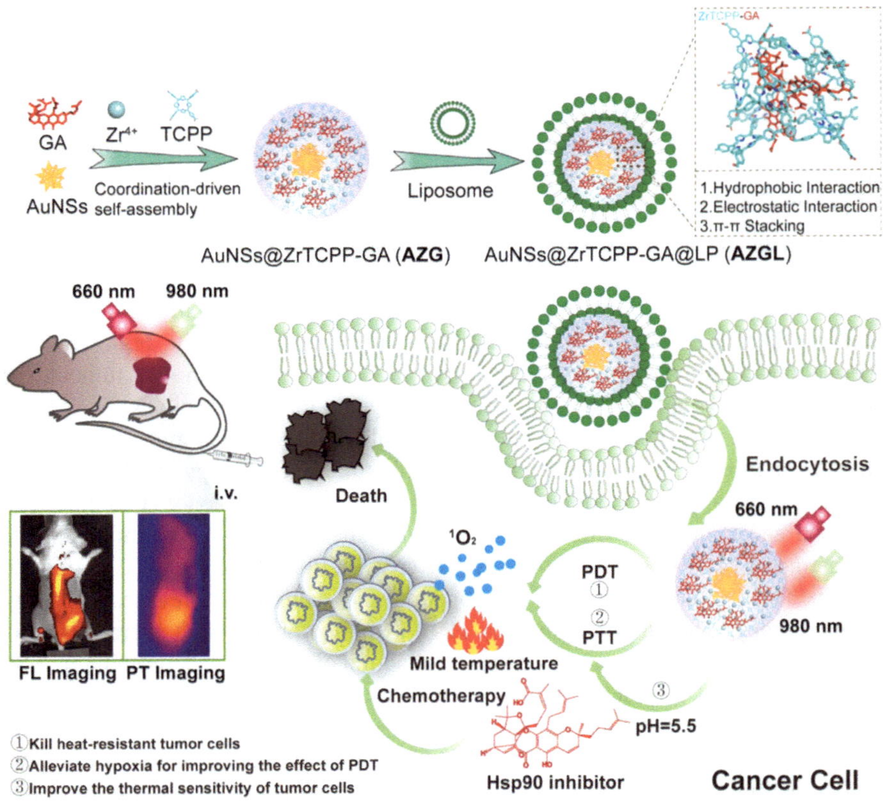

Fig. 3.48 The synthetic illustration of AZGL nanocomposite formation and its synergistic PTT-PDT-chemotherapy for breast cancer (Li et al. 2022c). Reprinted with permission from Springer Nature. Copyright © 2022 Rong-Tian Li et al.

a mouse model with bilateral CT26 tumors. This offers a promising alternative for effectively treating colon cancer.

Li and his team created a sophisticated core-shell nanocomposite using porphyrin MOFs and UCNPs (Shao et al. 2020). This nanocomposite was designed to treat hypoxic cancers by combining chemotherapy, photodynamic therapy (PDT), and immunotherapy. The design involved utilizing porphyrin PN-224 to encapsulate the core of UCNPs, and constructing the TPZ/UCSs nanocomposite by including the hypoxia-activated prodrug TPZ. When exposed to a laser that emits near-infrared light, the UCNP core in TPZ/UCSs can transmit this light energy to the porphyrin molecules in the MOF shell. This process generates $^1O_2$ and enables effective laser-induced photodynamic therapy (PDT) for malignancies. Furthermore, as a result of the hypoxic-activating features of TPZ, the TPZ/UCSs that were created demonstrated improved chemotherapeutic effects specifically at the location of the tumor. In addition, the concurrent use of chemotherapy/PDT and immune checkpoint blockade therapy can successfully inhibit the spread of tumors throughout the body by inducing the production of targeted cytotoxic T cells that infiltrate the tumor cells in a mouse model with CT26 tumors. This study offers a potential solution for the efficient treatment of systemic malignancies by utilizing multifunctional MOFs.

Yin and his team created a nanocomposite material called hybrid porous Cu/Zn-MOFICG@Mn/Cu/Zn-MOF MnO_2 (Cheng et al. 2021). This material was designed for multimodal imaging-guided CDT/PTT/PDT in mice with U87 tumors. The system involved the self-assembly of Cu^{2+}, Zn^{2+}, and imidazole-2-carboxyaldehyde (2-ICA) at room temperature to create a mixed Cu/Zn MOF. Ostwald Following maturation, a layer of MnO_2 was applied to the surface of Cu/Zn MOF, resulting in the formation of ICG @ Mn/Cu/Zn-MOF @ MnO_2 with the ability to load ICG. The nanocomposite released ICG, which demonstrated fluorescence and photothermal imaging, as well as PTT effects within the tumor microenvironment (TME). Conversely, elevated levels of H2O2 can accelerate the production of O_2 and harmful OH through the Fenton reaction, hence intensifying the effects of CDT and PDT. In addition, the nanocomposite's surface is coated with a MnO_2 shell that may eat GSH and produce Mn^{2+}, allowing for MRI to be performed at the location of the tumor. Significantly, utilizing FL/MRI/photothermal imaging, the Cu/Zn-MOF mixture, containing a substantial amount of ICG, exhibits no adverse effects and amplifies the combined effects of CDT/PTT/PDT. This breakthrough paves the path for the application of MOF-based nanocomposites in multimodal cancer detection and therapy.

Zhang and his colleagues published a study describing a nanoplatform based on MOFs that use multimodal imaging to effectively treat tumors in mice with 4T1 tumors (Fan et al. 2021). The treatment involves a combination of PTT, PDT, and immunotherapy. A multifunctional ICG-CpG @ MOF was created by covalently attaching and loading the immunoadjuvant CpG onto the core carrier, octahedral MIL-101-NH2. The resulting MOF has a diameter of approximately 150 nm and a CpG loading efficiency of around 76%. The ICG-CpG @ MOF, once manufactured, can be directed to tumor locations through passive targeting facilitated by the EPR effects, which are a result of its appropriate diameter. Furthermore, the activation of the immune system by CpG and the secreted tumor-associated antigens can transform tumor cells from a state of low immune response (cold) to a state of heightened immunological response (hot), hence enhancing the effectiveness of tumor therapy. The experimental findings demonstrate that the biocompatible ICG-CpG @ MOF not only displays several imaging characteristics such as fluorescence (FL), photoacoustic (PA), and magnetic resonance imaging (MRI), but also effectively treats tumors through photothermal therapy (PTT), photodynamic therapy (PDT), and immunotherapy using 808 nm laser irradiation. This system holds great promise for the diagnosis and treatment of cancer.

3.6 Conclusions and Perspectives

In conclusion, while MOF-based nanoplatforms in cancer treatment still encounter some obstacles, notable progress achieved thus far enables their continued utilization. With the rapid advancement of various technologies that span multiple disciplines in recent decades, there will be further progress in the development of multifunctional nanoplatforms based on MOFs (Metal-Organic Frameworks) for the treatment of cancer. These nanoplatforms will utilize a range of therapies, from single treatments to combinations, in order to enhance the effectiveness of anticancer treatments and improve the quality of life for patients.

MOFs have shown great potential in the field of tumor diagnosis and treatment due to their unique porous structure, high specific surface area, and tunable chemical properties. Future research may focus on the following aspects: Firstly, the porous structure of MOFs enables them to effectively load drug molecules, contrast agents, or other therapeutic agents, thus playing the role of a drug carrier in tumor treatment. Researchers can explore the drug-loading efficiency and release kinetics of different MOFs materials to optimize drug delivery systems. Secondly, the surface of MOFs can be functionalized to improve their stability and targeting ability in the body. By introducing specific ligands or antibodies, MOFs can be designed to target tumor cells, reducing toxicity to normal cells and improving treatment outcomes. Thirdly, the fluorescent properties and magnetic characteristics of MOFs make them promising for tumor imaging diagnosis. Researchers can develop MOFs-based contrast agents with high contrast and resolution for MRI, CT, PET, or fluorescence imaging to achieve early diagnosis and treatment monitoring of tumors. In addition, the photothermal conversion and photodynamic therapy properties of MOFs have also attracted attention. By designing photo-responsive MOFs materials, heat or reactive oxygen species can be generated under specific wavelengths of light irradiation, thus achieving the destruction of tumor cells. The optimization of the synthesis and post-processing methods of MOFs materials is also a focus of future research. By precisely controlling the size, morphology, and pore structure of MOFs, their application performance in tumor diagnosis and treatment can be further improved.

References

Bag PP, Xu-Sheng W, Rong C (2015) Microwave-assisted large scale synthesis of lanthanide metal–organic frameworks (ln-MOFs), having a preferred conformation and photoluminescence properties. Dalton Trans 44(26):11954–11962

Bavykina A, Nikita K, Son KI et al (2020) Metal–organic frameworks in heterogeneous catalysis: recent progress, new trends, and future perspectives. Chem Rev 120(16):8468–8535

Begum S, Hassan Z, Brase S et al (2019) Metal-organic framework-templated biomaterials: recent progress in synthesis, functionalization, and applications. Acc Chem Res 52(6):1598–1610

Beldon PJ, László F, Stein Robin S et al (2010) Rapid room-temperature synthesis of zeolitic imidazolate frameworks by using mechanochemistry. Angew Chem 50(122):9834–9837

Bumstead AM, Ignas P, Richards Kieran D et al (2022) Post-synthetic modification of a metal–organic framework glass. Chem Mater 34(5):2187–2196

Cai W, Gao H, Chu C et al (2017) Engineering Phototheranostic nanoscale metal-organic frameworks for multimodal imaging-guided cancer therapy. ACS Appl Mater Interfaces 9(3):2040–2051

Chang J, Wang X, Wang J et al (2019) Nucleic acid-functionalized metal-organic framework-based homogeneous electrochemical biosensor for simultaneous detection of multiple tumor biomarkers. Anal Chem 91(5):3604–3610

Chen Y, Deng J, Liu F et al (2019a) Energy-free, singlet oxygen-based chemodynamic therapy for selective tumor treatment without dark toxicity. Adv Healthc Mater 8(18):e1900366

Chen Y, Zhong H, Wang J et al (2019b) Catalase-like metal-organic framework nanoparticles to enhance radiotherapy in hypoxic cancer and prevent cancer recurrence. Chem Sci 10(22):5773–5778

Chen J, Fan T, Xie Z et al (2020) Advances in nanomaterials for photodynamic therapy applications: status and challenges. Biomaterials 237:119827

Chen Y, Lyu R, Wang J et al (2023) Metal-organic frameworks nucleated by silk fibroin and modified with tumor-targeting peptides for targeted multimodal cancer therapy. Adv Sci 10(28):e2302700

Cheng Y, Wen C, Sun YQ et al (2021) Mixed-metal MOF-derived hollow porous nanocomposite for trimodality imaging guided reactive oxygen species-augmented synergistic therapy. Adv Funct Mater 31:37

Choi J-S, Won-Jin S, Jahoen K et al (2008) Metal–organic framework MOF-5 prepared by microwave heating: factors to be considered. Microporous Mesoporous Mater 116(1-3):727–731

Cohen SM (2012) Postsynthetic methods for the functionalization of metal-organic frameworks. Chem Rev 112(2):970–1000

Cui R, Sun W, Liu M et al (2021) Near-infrared emissive lanthanide metal-organic frameworks for targeted biological imaging and pH-controlled chemotherapy. ACS Appl Mater Interfaces 13(49):59164–59173

Cun JE, Fan X, Pan Q et al (2022a) Copper-based metal-organic frameworks for biomedical applications. Adv Colloid Interf Sci 305:102686

Cun JE, Pan Y, Zhang Z et al (2022b) Photo-enhanced upcycling H(2)O(2) into hydroxyl radicals by IR780-embedded Fe(3)O(4)@MIL-100 for intense nanocatalytic tumor therapy. Biomaterials 287:121687

Daglar H, Can GH, Gokay A et al (2021) Effect of metal–organic framework (MOF) database selection on the assessment of gas storage and separation potentials of MOFs. Angew Chem Int Ed 60(14):7828–7837

Dong MJ, Li W, Xiang Q et al (2022) Engineering metal-organic framework hybrid AIEgens with tumor-activated accumulation and emission for the image-guided GSH depletion ROS therapy. ACS Appl Mater Interfaces 14(26):29599–29612

Duan F, Feng XC, Yang XJ et al (2017) A simple and powerful co-delivery system based on pH-responsive metal-organic frameworks for enhanced cancer immunotherapy. Biomaterials 122:23–33

Echaide-Górriz C, Coralie C, Fernando C-B et al (2018) New strategies based on microfluidics for the synthesis of metal–organic frameworks and their membranes. J Mater Chem A 6(14):5485–5506

Espallargas GM, Eugenio C (2018) Magnetic functionalities in MOFs: from the framework to the pore. Chem Soc Rev 47(2):533–557

Fan Z, Liu H, Xue Y et al (2021) Reversing cold tumors to hot: an immunoadjuvant-functionalized metal-organic framework for multimodal imaging-guided synergistic photo-immunotherapy. Bioact Mater 6(2):312–325

Fang C, Deng Z, Cao GD et al (2020) Co-ferrocene MOF/glucose oxidase as cascade nanozyme for effective tumor therapy. Adv Funct Mater 30(16):1910085

Feng J, Yang SP, Shao YQ et al (2023) Covalent organic framework-based Nanomotor for multimodal cancer photo-theranostics. Adv Healthc Mater 12(30):e2301645
Gao P, Shen X, Liu X et al (2021a) Covalent organic framework-derived Carbonous Nanoprobes for cancer cell imaging. ACS Appl Mater Interfaces 13(35):41498–41506
Gao P, Shen X, Liu X et al (2021b) Nucleic acid-gated covalent organic frameworks for cancer-specific imaging and drug release. Anal Chem 93(34):11751–11757
Ge X, Wong R, Anisa A et al (2022) Recent development of metal-organic framework nanocomposites for biomedical applications. Biomaterials 281:121322
Głowniak S, Barbara S, Jerzy C et al (2021) Mechanochemistry: toward green synthesis of metal–organic frameworks. Mater Today 46:109–124
Głowniak S, Barbara S, Jerzy C et al (2023) Recent developments in sonochemical synthesis of nanoporous materials. Molecules 28(6):2639
Hafizovic J, Morten B, Unni O et al (2007) The inconsistency in adsorption properties and powder XRD data of MOF-5 is rationalized by framework interpenetration and the presence of organic and inorganic species in the nanocavities. J Am Chem Soc 129(12):3612–3620
He LH, Li ZZ, Guo CP et al (2019) Bifunctional bioplatform based on NiCo Prussian blue analogue: Label-free impedimetric aptasensor for the early detection of carcino-embryonic antigen and living cancer cells. Sens Actuators B Chem 298:126852
He L, Ni Q, Mu J et al (2020) Solvent-assisted self-assembly of a metal-organic framework based biocatalyst for Cascade reaction driven photodynamic therapy. J Am Chem Soc 142(14):6822–6832
Horcajada P, Chalati T, Serre C et al (2010) Porous metal-organic-framework nanoscale carriers as a potential platform for drug delivery and imaging. Nat Mater 9(2):172–178
Hou Y, Can Z, Ge B et al (2024) Advancements and challenges in the application of metal-organic framework (MOF) nanocomposites for tumor diagnosis and treatment. Int J Nanomedicine 19:6295–6317
Hu X, Lu Y, Zhou L et al (2020) Post-synthesis strategy to integrate porphyrinic metal-organic frameworks with CuS NPs for synergistic enhanced photo-therapy. J Mater Chem B 8(5):935–944
Hu M, Zhu L, Li ZZ et al (2021) CoNi bimetallic metal-organic framework as an efficient biosensing platform for miRNA 126 detection. Appl Surf Sci 542:148586
Huang SJ, Chen RY, Zhao SR et al (2021a) Diverse metal ions-doped titanium-based metal-organic frameworks as novel bioplatforms for sensitively detecting bisphenol A. Electrochim Acta 384:138403
Huang J, Yang B, Peng Y et al (2021b) Nanomedicine-boosting tumor immunogenicity for enhanced immunotherapy. Adv Funct Mater 31(21):2011171
Huang L, Zhao SJ, Wu JS et al (2021c) Photodynamic therapy for hypoxic tumors: advances and perspectives. Coord Chem Rev 438:213888
Jia M, Yang X, Chen Y et al (2021) Grafting of Gd-DTPA onto MOF-808 to enhance MRI performance for guiding photothermal therapy. J Mater Chem B 9(41):8631–8638
Kalaj M, Cohen Seth M (2020) Postsynthetic modification: an enabling technology for the advancement of metal–organic frameworks. ACS Central Sci 6(7):1046–1057
Khan NA, Hwa JS (2015) Synthesis of metal-organic frameworks (MOFs) with microwave or ultrasound: rapid reaction, phase-selectivity, and size reduction. Coord Chem Rev 285:11–23
Klimakow M, Peter K, Thunemann Andreas F et al (2010) Mechanochemical synthesis of metal-organic frameworks: a fast and facile approach toward quantitative yields and high specific surface areas. Chem Mater 22(18):5216–5221
Klinowski J, Almeida PF, Patrícia S et al (2011) Microwave-assisted synthesis of metal–organic frameworks. Dalton Trans 40(2):321–330
Koryakina IG, Bachinin Semyon V, Gerasimova Elena N et al (2023) Microfluidic synthesis of metal-organic framework crystals with surface defects for enhanced molecular loading. Chem Eng J 452:139450

Kou W-T, Cheng-Xiong Y, Xiu-Ping Y (2018) Post-synthetic modification of metal–organic frameworks for chiral gas chromatography. J Mater Chem A 6(37):17861–17866

Krusenbaum A, Sven G, Tamiru TG et al (2022) The mechanochemical synthesis of polymers. Chem Soc Rev 51(7):2873–2905

Kuppler RJ, Timmons Daren J, Qian-Rong F et al (2009) Potential applications of metal-organic frameworks. Coord Chem Rev 253(23-24):3042–3066

Kwon HT, Ky VT, Jeong-Hoon K et al (2020) Microwave-assisted continuous flow synthesis of mesoporous metal-organic framework MIL-100 (Fe) and its application to cu (I)-loaded adsorbent for CO/CO2 separation. Mater Chem Phys 253:123278

Li Z-Q, Ling-Guang Q, Tao X et al (2009) Ultrasonic synthesis of the microporous metal–organic framework Cu3 (BTC) 2 at ambient temperature and pressure: an efficient and environmentally friendly method. Mater Lett 63(1):78–80

Li Y, Xu N, Zhou J et al (2018) Facile synthesis of a metal-organic framework nanocarrier for NIR imaging-guided photothermal therapy. Biomater Sci 6(11):2918–2924

Li X, Wang XP, Ito A et al (2020a) A nanoscale metal organic frameworks-based vaccine synergises with PD-1 blockade to potentiate anti-tumour immunity. Nat Commun 11(1):3858

Li HY, Zhao SN, Zang SQ et al (2020b) Functional metal-organic frameworks as effective sensors of gases and volatile compounds. Chem Soc Rev 49(17):6364–6401

Li J, Zhang W, Ji W et al (2021) Near infrared photothermal conversion materials: mechanism, preparation, and photothermal cancer therapy applications. J Mater Chem B 9(38):7909–7926

Li H, Huang M, Wei Z et al (2022a) Hydrogen sulfide activatable metal-organic frameworks for fluorescence imaging-guided photodynamic therapy of colorectal cancer. Front Bioeng Biotechnol 10:1032571

Li S, Ma Z, Zhang K et al (2022b) A two-pronged strategy for enhanced deep-tumor penetration and NIR-II multimodal imaging-monitored photothermal therapy. ACS Appl Mater Interfaces 14(37):41684–41694

Li RT, Zhu YD, Li WY et al (2022c) Synergistic photothermal-photodynamic-chemotherapy toward breast cancer based on a liposome-coated core-shell AuNS@NMOFs nanocomposite encapsulated with gambogic acid. J Nanobiotechnol 20(1):212

Li M, He X, Wu C et al (2024) Deep learning enabled SERS identification of gaseous molecules on flexible plasmonic MOF nanowire films. ACS Sens 9(2):979–987

Lian X, Huang Y, Zhu Y et al (2018) Enzyme-MOF nanoreactor activates nontoxic paracetamol for cancer therapy. Angew Chem Int Ed Engl 57(20):5725–5730

Liang JL, Luo GF, Chen WH et al (2021) Recent advances in engineered materials for immunotherapy-involved combination cancer therapy. Adv Mater 33(31):e2007630

Liang R, Li F, Chen X et al (2023) Multimodal imaging-guided strategy for developing (177) Lu-labeled metal-organic framework nanomedicine with potential in cancer therapy. ACS Appl Mater Interfaces 15(39):45713–45724

Liu X, Gao X, Yang L et al (2021) Metal-organic framework-functionalized paper-based electrochemical biosensor for ultrasensitive exosome assay. Anal Chem 93(34):11792–11799

Liu HS, Xu CN, Meng M et al (2022) Metal-organic framework-mediated multifunctional nanoparticles for combined chemo-photothermal therapy and enhanced immunotherapy against colorectal cancer. Acta Biomater 144:132–141

Liu D, Liang M, Tao Y et al (2024a) Hypoxia-accelerating pyroptosis nanoinducers for promoting image-guided cancer immunotherapy. Biomaterials 309:122610

Liu Y, Xia S, Xiao M et al (2024b) Synthesis of a metal-organic framework cu-mi-UiO-66-based fluorescent nanoprobe for the simultaneous sensing and intracellular imaging of GSH and ATP. Nanoscale 16(31):14831–14843

Lu K, He C, Lin W (2014) Nanoscale metal-organic framework for highly effective photodynamic therapy of resistant head and neck cancer. J Am Chem Soc 136(48):16712–16715

Lu K, He C, Lin W (2015) A Chlorin-based nanoscale metal-organic framework for photodynamic therapy of colon cancers. J Am Chem Soc 137(24):7600–7603

Luo HB, Ren Q, Wang P et al (2019) High proton conductivity achieved by encapsulation of imidazole molecules into proton-conducting MOF-808. ACS Appl Mater Interfaces 11(9):9164–9171

Ma T, Liu Y, Wu Q et al (2019) Quercetin-modified metal-organic frameworks for dual sensitization of radiotherapy in tumor tissues by inhibiting the carbonic anhydrase IX. ACS Nano 13(4):4209–4219

Mandal S, Srinivasan N, Prabu M et al (2021) Post-synthetic modification of metal–organic frameworks toward applications. Adv Funct Mater 31(4):2006291

Mao C, Shuchen W, Jiankai L et al (2023) Metal–organic frameworks in microfluidics enable fast encapsulation/extraction of DNA for automated and integrated data storage. ACS Nano 17(3):2840–2850

Martinez JA, Jana J-A, Pablo S-C et al (2012) Electrochemical synthesis of some archetypical Zn^{2+}, Cu^{2+}, and Al^{3+} metal organic frameworks. Cryst Growth Des 12(7):3489–3498

Meng X, Zhang K, Yang F et al (2020) Biodegradable metal-organic frameworks power DNAzyme for in vivo temporal-spatial control fluorescence imaging of aberrant MicroRNA and hypoxic tumor. Anal Chem 92(12):8333–8339

Min H, Wang J, Qi Y et al (2019) Biomimetic metal-organic framework nanoparticles for cooperative combination of Antiangiogenesis and photodynamic therapy for enhanced efficacy. Adv Mater 31(15):e1808200

Mohan B, Sandeep K, Hui X et al (2022) Fabricated metal-organic frameworks (MOFs) as luminescent and electrochemical biosensors for cancer biomarkers detection. Biosens Bioelectron 197:113738

Novaković SB, Bogdanović Goran A, Christian H et al (2015) Charge-density distribution and electrostatic flexibility of ZIF-8 based on high-resolution X-ray diffraction data and periodic calculations. Inorg Chem 54(6):2660–2670

Park J, Jiang Q, Feng D et al (2016) Size-controlled synthesis of Porphyrinic metal-organic framework and functionalization for targeted photodynamic therapy. J Am Chem Soc 138(10):3518–3525

Peng H, Qin YT, Feng YS et al (2021) Phosphate-degradable nanoparticles based on metal-organic frameworks for chemo-starvation-Chemodynamic synergistic antitumor therapy. Acs Appl Mater Inter 13(31):37713–37723

Qi G, Wang J, Ma K et al (2021) Label-free single-particle surface-enhanced Raman spectroscopy detection of phosphatidylserine externalization on cell membranes with multifunctional micron-nano composite probes. Anal Chem 93(4):2183–2190

Qian J, Fuan S, Lizhen Q (2012) Hydrothermal synthesis of zeolitic imidazolate framework-67 (ZIF-67) nanocrystals. Mater Lett 82:220–223

Qiao CQ, Wang XF, Liu GH et al (2022) Erythrocyte membrane camouflaged metal-organic framework nanodrugs for remodeled tumor microenvironment and enhanced tumor chemotherapy. Adv Funct Mater 32(6):2107791

Qin YT, Feng YS, Ma YJ et al (2020) Tumor-sensitive biodegradable nanoparticles of molecularly imprinted polymer-stabilized fluorescent Zeolitic Imidazolate Framework-8 for targeted imaging and drug delivery. ACS Appl Mater Interfaces 12(22):24585–24598

Qin S, Xu Y, Li H et al (2021) Recent advances in in situ oxygen-generating and oxygen-replenishing strategies for hypoxic-enhanced photodynamic therapy. Biomater Sci 10(1):51–84

Qin X, Xiang Y, Mao L et al (2024) Buoyant metal-organic framework Corona-driven fast isolation and ultrasensitive profiling of circulating extracellular vesicles. ACS Nano 18(22):14569–14582

Qiu L-G, Zong-Qun L, Yun W et al (2008) Facile synthesis of nanocrystals of a microporous metal–organic framework by an ultrasonic method and selective sensing of organoamines. Chem Commun 31:3642–3644

Railey P, Song Y, Tianyu L et al (2017) Metal organic frameworks with immobilized nanoparticles: synthesis and applications in photocatalytic hydrogen generation and energy storage. Mater Res Bull 96:385–394

Sachdeva S, Alexey P, Ernst S et al (2016) Control of interpenetration of copper-based MOFs on supported surfaces by electrochemical synthesis. CrystEngComm 18(22):4018–4022

Sang W, Zhang Z, Dai YL et al (2019) Recent advances in nanomaterial-based synergistic combination cancer immunotherapy. Chem Soc Rev 48(14):3771–3810
Sang Y, Cao F, Li W et al (2020) Bioinspired construction of a nanozyme-based H(2)O(2) homeostasis disruptor for intensive Chemodynamic therapy. J Am Chem Soc 142(11):5177–5183
Shang Y, Zhang H, Cheng Y et al (2022) Fluorescent imaging-guided chemo- and photodynamic therapy of hepatocellular carcinoma with HCPT@NMOFs-RGD nanocomposites. Int J Nanomedicine 17:1381–1395
Shao YL, Liu B, Di ZH et al (2020) Engineering of upconverted metal-organic frameworks for near-infrared light-triggered combinational photodynamic/chemo-/immunotherapy against hypoxic tumors. J Am Chem Soc 142(8):3939–3946
Shen J, Zhou W, Jia M et al (2021) Tumor microenvironment-responsive reagent DFS@HKUST-1 for photoacoustic imaging-guided multimethod therapy. ACS Appl Bio Mater 4(7):5753–5764
Shi Y, Fang J (2024) Yolk-Shell hierarchical pore au@MOF nanostructures: efficient gas capture and enrichment for advanced breath analysis. Nano Lett 24(33):10139–10147
Shi J, Jianling Z, Dongxing T et al (2019) Rapid, room-temperature and template-free synthesis of metal-organic framework nanowires in alcohol. ChemCatChem 11(8):2058–2062
Shu G, Zhao H, Zhang X (2023) Persistent luminescent metal-organic framework nanocomposite enables autofluorescence-free dual modal imaging-guided drug delivery. Biomater Sci 11(5):1797–1809
Son W-J, Jun K, Jaheon K et al (2008) Sonochemical synthesis of MOF-5. Chem Commun 47:6336–6338
Sun X, He GH, Xiong CX et al (2021a) One-pot fabrication of hollow porphyrinic MOF nanoparticles with ultrahigh drug loading toward controlled delivery and synergistic cancer therapy. ACS Appl Mater Inter 13(3):3679–3693
Sun J, Xing F, Braun J et al (2021b) Progress of phototherapy applications in the treatment of bone cancer. Int J Mol Sci 22(21):11354
Sun S, Yuewu Z, Jine W et al (2022) Lanthanide-based MOFs: synthesis approaches and applications in cancer diagnosis and therapy. J Mater Chem B 10(46):9535–9564
Sun X, Yang Z, Zhang M et al (2023) A simple one step synthesis of magnetic-optical dual functional ZIF-8 in a sodalite phase for magnetically guided targeting bioimaging and drug delivery. Soft Matter 19(42):8164–8171
Tan LL, Li H, Qiu YC et al (2015) Stimuli-responsive metal-organic frameworks gated by pillar[5]arene supramolecular switches. Chem Sci 6(3):1640–1644
Tang K, Wang W, Song Z et al (2021) Multifunctional nano-biosensor based on metal-organic framework for enhanced fluorescence imaging of intracellular miRNA-122 and synergistic chemo-photothermal therapy of tumor cells. Anal Chim Acta 1176:338779
Thorarinsdottir AE, David HT (2020) Metal–organic framework magnets. Chem Rev 120(16):8716–8789
Tsai H, Shreetu S, Vilá RA et al (2021) Bright and stable light-emitting diodes made with perovskite nanocrystals stabilized in metal–organic frameworks. Nat Photonics 15(11):843–849
Van ATR, Gert D, Rob A et al (2012) Electrochemical synthesis of thin HKUST-1 layers on copper mesh. Microporous Mesoporous Mater 158:209–213
Vander HMG, Cantley LC, Thompson CB (2009) Understanding the Warburg effect: the metabolic requirements of cell proliferation. Science 324(5930):1029–1033
Vodyashkin Andrey A, Sergorodceva Antonina V, Parfait K et al (2023) Metal-organic framework (MOF)—a universal material for biomedicine. Int J Mol Sci 24(9):7819
Wan Y, Li C, Fu LH et al (2024) Erythrocyte membrane camouflaged Nanotheranostics for optical molecular imaging-escorted self-oxygenation photodynamic therapy. Small 20(31):e2309026
Wang W, Wang L, Liu S et al (2017a) Metal-organic frameworks@polymer composites containing Cyanines for near-infrared fluorescence imaging and Photothermal tumor therapy. Bioconjug Chem 28(11):2784–2793

Wang X, Wenxing C, Lei Z et al (2017b) Uncoordinated amine groups of metal–organic frameworks to anchor single Ru sites as chemoselective catalysts toward the hydrogenation of quinoline. J Am Chem Soc 139(28):9419–9422
Wang H, Chen Y, Wang H et al (2019a) DNAzyme-loaded metal-organic frameworks (MOFs) for self-sufficient gene therapy. Angew Chem Int Ed Engl 58(22):7380–7384
Wang M, Wang D, Chen Q et al (2019b) Recent advances in glucose-oxidase-based nanocomposites for tumor therapy. Small 15(51):e1903895
Wang D, Wu H, Phua SZF et al (2020) Self-assembled single-atom nanozyme for enhanced photodynamic therapy treatment of tumor. Nat Commun 11(1):357
Wang K, Ding SS, Zeng LJ et al (2021a) Antisense oligonucleotides-laden UiO-66@Au nanohybrid for enhanced radiotherapy against hypoxic tumor by dual-inhibition of carbonic anhydrase IX. Appl Mater Today 25:101201
Wang Z, Qianqian S, Bin L et al (2021b) Recent advances in porphyrin-based MOFs for cancer therapy and diagnosis therapy. Coord Chem Rev 439:213945
Wang Z, Sun QQ, Liu B et al (2021c) Recent advances in porphyrin-based MOFs for cancer therapy and diagnosis therapy. Coord Chem Rev 439:360
Wang H, Zhang Z, Wang X et al (2024) Copper single-atom-based metal-organic framework for ultrasound-enhanced nanocatalytic therapy. Nano Lett 24(31):9700–9710
Wei X, Song M, Jiang G et al (2022) Progress in advanced nanotherapeutics for enhanced photodynamic immunotherapy of tumor. Theranostics 12(12):5272–5298
Wu SX, Zhang KX, Liang Y et al (2022) Nano-enabled tumor systematic energy exhaustion via zinc (II) interference mediated glycolysis inhibition and specific GLUT1 depletion. Adv Sci 9(7):2103534
Xie LS, Grigorii S, Mircea D (2020) Electrically conductive metal–organic frameworks. Chem Rev 120(16):8536–8580
Xie L, Li J, Wang G et al (2022) Phototheranostic metal-phenolic networks with antiexosomal PD-L1 enhanced ferroptosis for synergistic immunotherapy. J Am Chem Soc 144(2):787–797
Xu J, Wang J, Ye J et al (2021) Metal-coordinated supramolecular self-assemblies for cancer theranostics. Adv Sci 8(16):e2101101
Xu J, Xu Y, Li J et al (2023) Construction of high-active SERS cavities in a TiO(2) Nanochannels-based membrane: a selective device for identifying volatile aldehyde biomarkers. ACS Sens 8(9):3487–3497
Yaghi OM, Michael O'K, Ockwig Nathan W et al (2003) Reticular synthesis and the design of new materials. Nature 423(6941):705–714
Yan K, Zhang Y, Mu C et al (2020) Versatile nanoplatforms with enhanced photodynamic therapy: designs and applications. Theranostics 10(16):7287–7318
Yang J, Dai DH, Ma LJ et al (2021) Molecular-scale drug delivery systems loaded with oxaliplatin for supramolecular chemotherapy. Chin Chem Lett 32(2):729–734
Yang K, Zhao SJ, Li BL et al (2022) Low temperature photothermal therapy: advances and perspectives. Coord Chem Rev 454:214330
Yang C, Wang K, Liang G et al (2023a) A versatile MOF-derived theranostic for dual-miRNA controlled accurate cancer cell recognition and photodynamic therapy. Talanta 265:124805
Yang C, Wang K, Tian S et al (2023b) Functionalized photosensitive metal-organic framework as a theranostic nanoplatform for turn-on detection of MicroRNA and photodynamic therapy. Anal Chim Acta 1239:340689
Yao B, Shun-Kuang L, Hui-Shi L et al (2021) Rapid ultrasound-assisted synthesis of controllable Zn/co-based zeolitic imidazolate framework nanoparticles for heterogeneous catalysis. Microporous Mesoporous Mater 314:110777
You Q, Zhang K, Liu J et al (2020) Persistent regulation of tumor hypoxia microenvironment via a bioinspired Pt-based oxygen nanogenerator for multimodal imaging-guided synergistic phototherapy. Adv Sci 7(17):1903341
Yu J, Wei Z, Li Q et al (2021) Advanced cancer starvation therapy by simultaneous deprivation of lactate and glucose using a MOF Nanoplatform. Adv Sci 8(19):e2101467

Yuan S, Weigang L, Ying-Pin C et al (2015) Sequential linker installation: precise placement of functional groups in multivariate metal–organic frameworks. J Am Chem Soc 137(9):3177–3180

Yue S, Zhan J, Xu X et al (2024) A "dual-key-and-lock" DNA nanodevice enables spatially controlled multimodal imaging and combined cancer therapy. Chem Sci 15(29):11528–11539

Yusuf VF, Malek Naved I, Kumar KS (2022) Review on metal–organic framework classification, synthetic approaches, and influencing factors: applications in energy, drug delivery, and wastewater treatment. ACS Omega 7(49):44507–44531

Zhang C, Bu W, Ni D et al (2016a) Synthesis of iron nanometallic glasses and their application in cancer therapy by a localized Fenton reaction. Angew Chem Int Ed Engl 55(6):2101–2106

Zhang Y, Wang FM, Ju EG et al (2016b) Metal-organic-framework-based vaccine platforms for enhanced systemic immune and memory response. Adv Funct Mater 26(35):6454–6461

Zhang L, Liu C, Gao Y et al (2018a) ZD2-engineered gold nanostar@metal-organic framework nanoprobes for T(1)-weighted magnetic resonance imaging and photothermal therapy specifically toward triple-negative breast cancer. Adv Healthc Mater 7(24):e1801144

Zhang L, Wang Z, Zhang Y et al (2018b) Erythrocyte membrane cloaked metal-organic framework nanoparticle as biomimetic Nanoreactor for starvation-activated colon cancer therapy. ACS Nano 12(10):10201–10211

Zhang H, Shang Y, Li YH et al (2019a) Smart metal-organic framework-based nanoplatforms for imaging-guided precise chemotherapy. ACS Appl Mater Interfaces 11(2):1886–1895

Zhang D, Ye Z, Wei L et al (2019b) Cell membrane-coated porphyrin metal-organic frameworks for cancer cell targeting and O(2)-evolving photodynamic therapy. ACS Appl Mater Interfaces 11(43):39594–39602

Zhang S, Feilong R, Chuanpan G et al (2021) Metal–organic frameworks (MOFs) based electrochemical biosensors for early cancer diagnosis in vitro. Coord Chem Rev 439:213948

Zhang P, Ouyang Y, Sohn YS et al (2022a) miRNA-guided imaging and photodynamic therapy treatment of cancer cells using Zn(II)-Protoporphyrin IX-loaded metal-organic framework nanoparticles. ACS Nano 16(2):1791–1801

Zhang C, Xin L, Li J et al (2022b) Metal-organic framework (MOF)-based ultrasound-responsive dual-sonosensitizer nanoplatform for hypoxic cancer therapy. Adv Healthc Mater 11(2):e2101946

Zhang X, Dou Y, Liu S et al (2024) Rationally designed Benzobisthiadiazole-based covalent organic framework for high-performance NIR-II fluorescence imaging-guided photodynamic therapy. Adv Healthc Mater 13(17):e2303842

Zhao Y, Heng Z, Xiao-Wei Z et al (2021a) Metal–organic frameworks as photoluminescent biosensing platforms: mechanisms and applications. Chem Soc Rev 50(7):4484–4513

Zhao Q, Gong ZJ, Li ZH et al (2021b) Target reprogramming lysosomes of CD8+T cells by a mineralized metal-organic framework for cancer immunotherapy. Adv Mater 33(17):e2100616

Zhao X, Niu R, Fan S et al (2022a) A dual-mode NADH biosensor based on gold nanostars decorated CoFe(2) metal-organic frameworks to reveal dynamics of cell metabolism. ACS Sens 7(9):2671–2679

Zhao X, Yang T, Wang D et al (2022b) Correction to gold Nanorods/metal-organic framework hybrids: photo-enhanced peroxidase-like activity and SERS performance for organic dyestuff degradation and detection. Anal Chem 94(17):6635–6636

Zhao D, Wang Z, Shuang Y et al (2022c) Application of MOF-based nanotherapeutics in light-mediated cancer diagnosis and therapy. J Nanobiotechnol 20(1):421

Zheng Q, Liu X, Zheng Y et al (2021) The recent progress on metal-organic frameworks for phototherapy. Chem Soc Rev 50(8):5086–5125

Zhong Y, Li X, Chen J et al (2020) Recent advances in MOF-based nanoplatforms generating reactive species for chemodynamic therapy. Dalton Trans 49(32):11045–11058

Zhou Z, Song J, Nie L et al (2016) Reactive oxygen species generating systems meeting challenges of photodynamic cancer therapy. Chem Soc Rev 45(23):6597–6626

Zhou M, Liu X, Chen F et al (2021) Stimuli-activatable nanomaterials for phototherapy of cancer. Biomed Mater 16(4):abfa6e

Zhou W, Liu Z, Wang N et al (2022) Hafnium-based metal-organic framework nanoparticles as a Radiosensitizer to improve radiotherapy efficacy in esophageal cancer. ACS Omega 7(14):12021–12029

Zou R, Gong Q, Shi Z et al (2020) A ZIF-90 nanoplatform loaded with an enzyme-responsive organic small-molecule probe for imaging the hypoxia status of tumor cells. Nanoscale 12(27):14870–14881

Chapter 4
Application of Nano-Drug Carrier in Tumor Diagnosis and Treatment

Huai Rong Zhang

Abstract Nano-drug carrier refers to a new class of nanomaterials with a particle size of 10–1000 nm, usually made of natural or synthetic polymer materials. Due to the advantages of improving drug stability, reducing drug toxicity, and controlled and targeted release of drugs, nanomedicine carriers have been widely used in tumor diagnosis and treatment. Some clinical trials proved that these smart drug delivery nanosystems improve the efficiency of treatment for several types of cancer compared to free drugs. The core technology of tumor diagnosis and treatment integration is mainly to integrate diagnostic agents and therapeutic agents into the same nanomaterial system, construct nanomaterial carrier diagnosis and treatment agents, and realize the organic combination of tumor diagnosis, monitoring, and treatment. Therefore, it is necessary to develop new class of nanocarriers, which can efficiently deliver specific drugs to tumor sites.

Keywords Nano-drug carrier · Nanomaterials · Properties · Tumor diagnosis · Tumor treatment

4.1 Application of Single-Atom Nanocatalysts in Tumor Diagnosis and Treatment

4.1.1 Overview of Single-Atom Nanocatalysts

Single-atom nanocatalysts (SACs) indicated that the metal is uniformly and singly supported on the carrier of metal, metal oxide, two-dimensional material and molecular sieve in the form of single atom, which is used as the catalytic active center for catalytic reaction (Fei et al. 2018; Xie and Liu 2020). As a bridge between

---

H. R. Zhang (✉)
Shandong Provincial Key Laboratory of Tumor Imaging Equipment Development and Tumor Diagnosis and Treatment Integration Technology, School of Medicine, Linyi University, Linyi, Shandong, People's Republic of China

© The Author(s), under exclusive license to Springer Nature Singapore Pte Ltd. 2026

P. Shi et al. (eds.), *Functional Organic Materials for Tumor Diagnostics and Therapy*, https://doi.org/10.1007/978-981-96-9478-5_4

homogeneous catalysis and heterogeneous catalysis, SACs provide a simple and optimal framework for designing rational nanocases at the atomic level. Because of the homogeneous active site and the inherent properties of single atom in the low coordination environment, SACs have established a suitable nanoplatform for studying the origin of enzyme-like activities.

4.1.1.1 Properties of Single-Atom Nanocatalysts

SACs have the characteristics of maximum atomic utilization and isolation of active sites, which is a very prosperous branch of catalysis research in recent years. Different from nanocatalysis and subnanocatalysis, new characteristics are caused due to the size of single atom, such as quantum size effect, sharply increasing surface free energy, metal–carrier interaction, and unsaturated coordination environment. The activity, selectivity, and stability of single-atom catalysts are different from traditional nanocatalysts in many REDOX and hydrogenation reactions, including thermal catalysis, electrocatalysis, and photocatalysis reactions (Xia et al. 2022; Yang et al. 2020). As an important component of biocatalyst, SACs have remarkable activity and specificity (Du et al. 2021; Chang et al. 2022). SACs uncover the complex mechanisms of catalytic reactions beyond the limitations of natural enzymes and nano-enzymes. Currently, SACs have shown a diverse array of enzyme properties encompassing oxidase (OXD)-like, superoxide dismutase (SOD)-like, catalase (CAT)-like, and peroxidase (POD)-like activities (Fig. 4.1). Specifically, the catalytic mechanism of OXD-like SAzymes engage O_2 as an esteemed electron acceptor, orchestrating the reduction of O_2 into both H_2O and H_2O_2, accompanied by the generation of ROS (Wu et al. 2024). SOD emerges as a catalyst par excellence, facilitating the extraordinary dismutation of superoxide into O_2 and H_2O_2 (Liu et al. 2023a). The SOD-like SAzymes represents another pivotal nanozyme, which can be widely used in catalytic therapy. CAT is an enzyme that can catalyze H_2O_2 into H_2O and O_2, mainly including the dissociation of O–O bond in H_2O_2 with the assistance of protons in solution. As an alternative to natural horseradine peroxidase, POD has the extraordinary ability to utilize hydrogen peroxide (H_2O_2) as an electron acceptor, thus coordinating the oxidation of the substrate while efficiently breaking down H_2O_2 into ROS and water molecules (Wu et al. 2022a).

4.1.1.2 Synthesis of Single-Atom Nanocatalysts

Various SACs have been developed and have shown excellent performance in many catalytic applications. The main factors affecting the enzyme-like catalytic activity of SACs are carrier morphology, synthesis strategy, active metal centers and types, numbers, and positions of coordination atoms. Therefore, it is important to establish a feasible synthesis strategy for the preparation of high-performance catalysts, deeply understand the structure of the active site and the catalytic mechanism, and

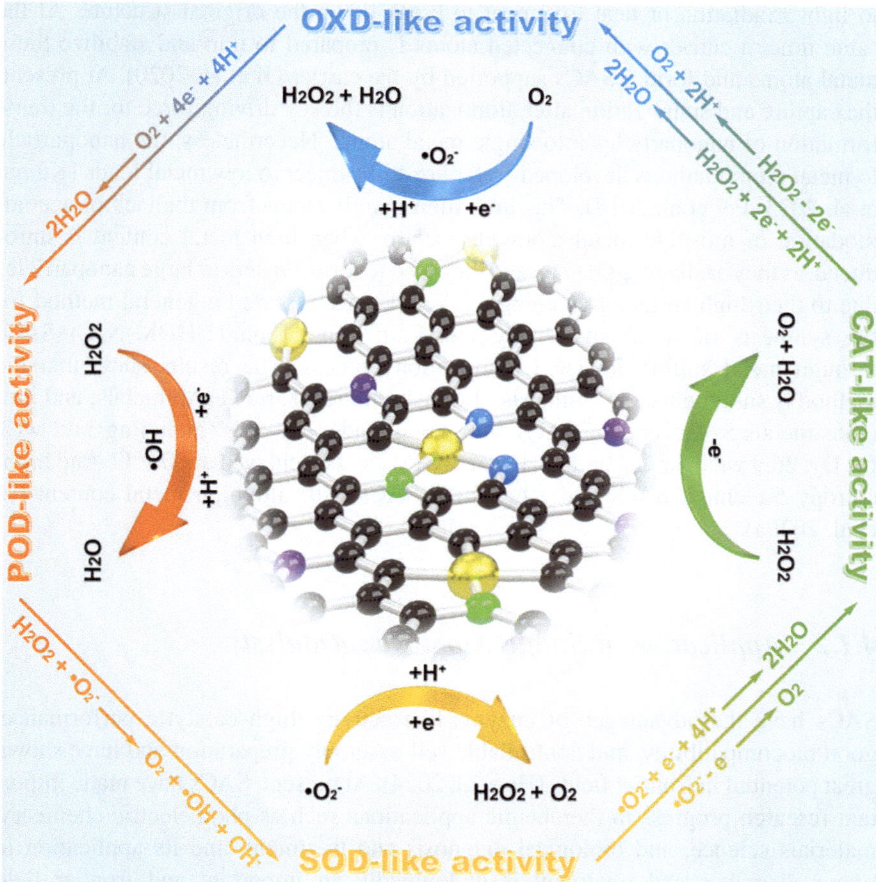

Fig. 4.1 Schematic of the main catalytic activities and mechanisms of SAzymes (Reproduced from He et al. 2024). Reproduced from He et al. (2024). Copyright 2024 John Wiley & Sons Ltd

develop catalysts with practical value (Liu et al. 2020, 2023b). In 2011, Zhang et al. synthesized an efficient Pt_1/FeO_x catalyst by co-precipitation method, and formally proposed the concept of single-atom catalysis. The selective oxidation of carbon monoxide was successfully catalyzed (Liang et al. 2022). In addition, XAS, AC-STEM, and Fourier transform infrared spectroscopy (FT-IR) were combined to clearly confirm the atomic dispersion of Pt species. Density functional theory (DFT) calculations were performed to reveal how the Pt_1 site is stabilized on FeOx and explain why Pt1/FeOx exhibits significantly enhanced CO oxidation activity.

The synthesis strategies of SACs have been developed from immersion method, mass selection soft landing method, co-precipitation method to atomic layer deposition (ALD) strategy, host-guest strategy, photochemical synthesis, etc. (Liu et al. 2021a). For most of the methods, metal precursors (such as metal chlorides, metal nitrates, and metal acetoethyl esters) are selected as starting materials and subjected to light irradiation or heat treatment to break down the original structure. At the same time, a carrier with connected atoms is prepared to trap and stabilize these metal atoms and form a SACs supported by the carrier (Ji et al. 2020). At present, the capture and stabilization after atomization is the key driving force for the transformation of nanoparticles into single metal atoms. Nevertheless, the nanoparticle to metal atom methods developed so far are still subject to low metal loads (Kumar et al. 2023; Fei et al. 2018). This limitation mainly stems from the lack of accommodation of movable metal atoms, especially when high metal content is introduced, as they easily reaggregate, and they tend to form clusters or large nanoparticles due to their high surface free energy. Lu et al. have reported a general method for the synthesis of ultrahigh density metal-nitrogen-carbon (UHDM-N-C) SACs through metal sulfide-mediated atomization process. The results show that the method is suitable for the synthesis of transition metals, rare earth metals, and precious metals SACs, and 17 SACs with metal loads >20 wt% (including 31.2 wt% for Dy, 26.9 wt% for Cu, and 33.4 wt% for Pt) can be achieved at 800 °C. And high-entropy 5-element SACs and 20-element SACs with ultrahigh metal content (Li et al. 2020a).

4.1.2 *Application of Single-Atom Nanocatalysts*

SACs have the advantages of enzyme-like activity, high catalytic performance, good biocompatibility, and controllable self-assembly preparation and have shown great potential in frontier fields (He et al. 2024). At present, SACs have made important research progress in therapeutic applications such as photoelectric chemistry, materials science, and biological diagnosis and treatment, and its application in tumor diagnosis and treatment is undoubtedly an important and frontier field (Fig. 4.2, He et al. 2024).

4.1.2.1 Single-Atom Nanocatalysts in Diagnosis of Tumor

Compared with traditional nanomaterials, SACs have precise active sites, unique geometric and electronic structures, and controllable metal–carrier interactions, which can enhance the sensitivity and selectivity of biosensors. Because of the good biocompatibility, stability, and excellent photoelectromagnetic properties, SACs haves been widely used in signal amplification strategies in biosensing. Combined with electrochemistry, fluorescence, colorimetry, photochemistry, and other technical means, SACs can be used as a carrier, catalyst, and signal source to achieve efficient signal amplification process, and show excellent sensitivity, fast response time, and good cost advantages (Jiao et al. 2020; Chellasamy et al. 2023). Researchers improve the performance of SACs by precisely regulating the size, morphology, structure, and composition of nanomaterials, thereby improving the sensitivity of biosensors. As shown in Fig. 4.3, the development of nanSACS with high catalytic

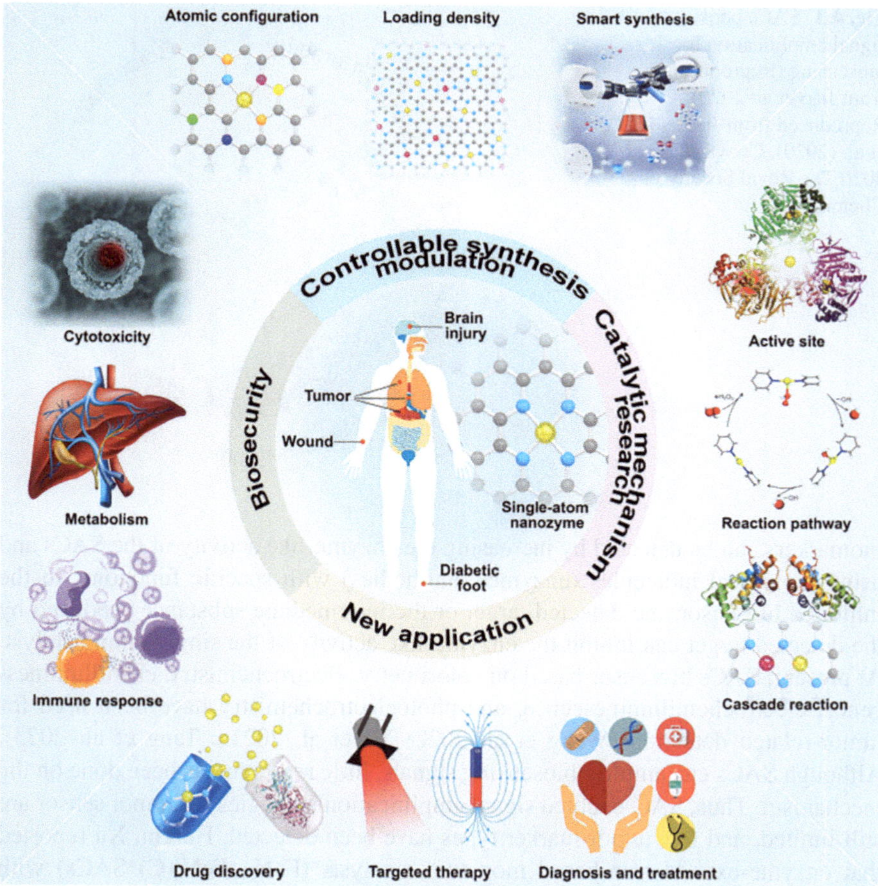

Fig. 4.2 Schematic of SAzymes for catalytic therapy, concerning clinical application (Reproduced from He et al. 2024). Reproduced from He et al. (2024). Copyright 2024 John Wiley & Sons Ltd

performance and accurate active sites is conducive to further improving the sensitivity of biosensors (Lei et al. 2021).

SACs have a well-defined active site, and with the help of advanced characterization techniques (such as in situ electron microscopy and XAFS) and theoretical calculations, its interaction with the target substrate can be easily studied, which is conducive to elucidate the catalytic mechanism and thus design signal amplification strategies with high sensitivity and selectivity (Younis et al. 2021). Based on the excellent enzyme-like properties, the SACs can catalyze a series of molecular changes to achieve signal amplification and reduce the detection limit of target objects. According to the signal amplification mechanism, the constructed tumor biosensors can be divided into enhanced biosensors and suppressed biosensors (Shen et al. 2024). In the enhanced biosensor, the concentration of the detected object is positively correlated with the detected signal. The sensitivity of tumor

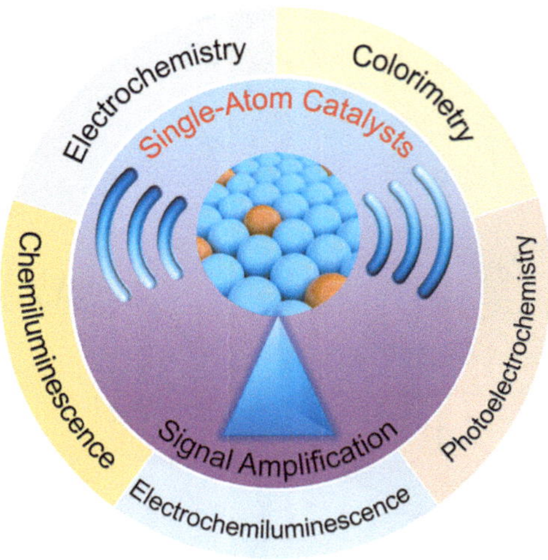

Fig. 4.3 SACs boost signal amplification for biosensing (Reproduced from Jiao et al. 2020). Reproduced from Jiao et al. (2020). Copyright 2020 The Royal Society of Chemistry

biomarkers can be detected by increasing the enzyme-like activity of the SACs and using biological molecules (enzymes, antibodies) with specific functions. In the inhibited biosensor, the detected target or the intermediate substance produced by the detected target can inhibit the enzyme-like activity of the single-atom catalyst. At present, SACs biosensor based on colorimetry, electrochemistry, chemiluminescence, electrochemiluminescence, and photoelectrochemistry have been used for tumor-related detection (Wang et al. 2022; Xu et al. 2021a; Tang et al. 2023). Although SACs can amplify biosensing signals, little research has been done on the mechanism. Thus, SAC-enabled signal amplification strategies for tumor sensor are still limited, and few tumor marker types have been detected. Herein, Xu reported that enzyme-excited iron-based monatom catalysts (FeN_5, $FeN_4(C)$ SACs) with axial N/C coordination structure can produce specific ROS for cathode/anode ECL conversion. Mechanistic studies have shown that FeN_5 sites are more inclined to produce highly active hydroxyl radicals and promote O-O bond cleavage through N-induced electron redistribution, thus providing direct cathodic luminescence. Conversely, the $FeN_4(C)$ site is prone to superoxide-free radicals, resulting in an inefficient anode ECL. Using enhanced cathode ECL, an immunosensor based on FeN_5 SACs was constructed for sensitive detection of tumor biomarkers (Xu et al. 2023a).

4.1.2.2 Single-Atom Nanocatalysts in Antitumor Therapy

Driven by the rapid development of nanotechnology, various SACs are becoming candidates for cancer-related treatment, antimicrobial therapy, biosensor, and other fields (Geng et al. 2024; Lu et al. 2021). SACs, like enzymes, exhibit a range of remarkable properties at the atomic scale, including geometric, chemical, and electronic structures. As shown in Fig. 4.4, Shi's group indicated pegylated monatomic

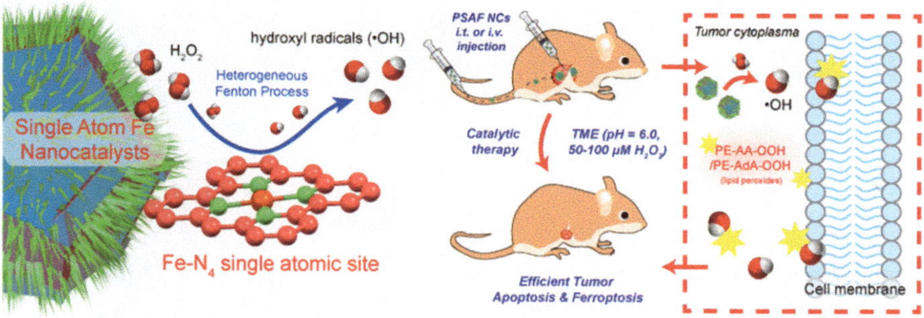

Fig. 4.4 Schematic of nanocatalytic tumor therapy by SACs (Reproduced from Huo et al. 2019). Reprinted with the permission from Huo et al. (2019). Copyright 2019 American Chemical Society

iron-containing nano-catalyst (PSAF NCs) can effectively trigger in situ tumor-specific Fenton reaction under acidic TME, which selectively produces a large number of toxic hydroxyl radicals (•OH). Based on density functional theory, it has been theoretically found that nanocatalysts can catalyze heterogeneous Fenton reaction specifically through proton-mediated H_2O_2 -hemolytic pathway. The produced free radicals can not only lead to cell death in malignant tumors, but also cause the accumulation of lipid peroxides, inducing tumor cell fatulaiton, which synergistically produce impressive tumor suppression results. At the same time, the good biocompatibility and biodegradability of PSAF NCs ensure the ideal biocompatibility in vivo and in vitro (Huo et al. 2019).

Aggregation of single atomic metals into nanoparticles is often due to the weak interactions between the carriers and metals. To conquer this challenge, it is important to select applicable carrier materials that can be used as a robust platform for the construction of SACs (Anwar et al. 2023). The support materials (MOF materials, carbon-based materials, two-dimensional materials) that have large cavities and modifiable properties can strongly trap drugs and signal tags, which endow the SACs with drug carrier properties (Bakandritsos et al. 2019; Gloag et al. 2024). The integrated nanoplatforms of diagnosis and treatment targeting mitochondria can achieve more accurate tumor treatment and detection. Therefore, various SACs are being reasonably designed and show good application prospects in a wide range of biomedical fields. Taking advantage of this, Zhang's group used Fe monatomic catalysts (SAF NPs) as nanocarrier to load with luminol and curcumin (Fig. 4.5). Besides, glycyrrhetinic acid (GA) was used as a recognition molecule to derecognize and target mitochondria. After entering the cells, GA can target and open the mitochondrial permeability transition pore (MPTP), which can promote the concentration of hydrogen peroxide (H_2O_2). After that, SAFs with fenton-like properties can catalyze H_2O_2 to •OH for chemodynamic therapy (CDT). The excess H_2O_2 and •OH can be delivered to luminol to emit chemiluminescence (CL), which can be recorded by EMCCD for specific detection of tumor cells. Meanwhile, the CL light can be transferred to Cur to produce reactive oxygen species (ROS) and realized photodynamic therapy (PDT). Moreover, Cur could not only be served as a

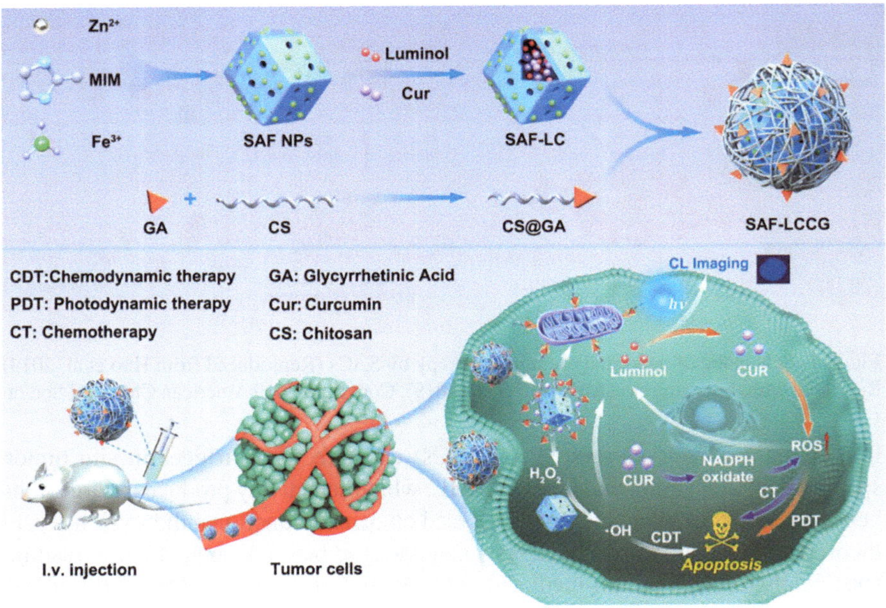

Fig. 4.5 Schematic of the SAF-LCCG nanoplatform for CL imaging and cancer therapy (Reproduced from Feng et al. 2023). Reprinted with the permission from Feng et al. (2023). Copyright 2023 American Chemical Society

chemotherapeutic (CT) drug, but also can enhance intracellular ROS for penetrating therapy. More importantly, the accumulation of ROS promotes the CL intensity, which realized the cyclic ROS amplification. This autoluminescence nanoplatform was developed for accurate cancer cells imaging and effective inhibition of tumor metastasis (Feng et al. 2023).

4.1.3 Conclusions and Perspectives

In conclusion, SACs have achieved initial success in the field of cancer diagnosis and treatment; however, both opportunities and challenges still exist. SACs with electronic space arrangements, well-designed geometries, combined with novel in situ characterization techniques, theoretical studies, and artificial intelligence techniques, will provide insights into the basic research and future clinical applications of SACs for biological fields. This will require the joint efforts of multiple fields, including nanomaterials, chemistry, computing, and biomedicine, and we look forward to more exciting research progress.

4.2 Application of Mesoporous Nanoparticles in Tumor Diagnosis and Treatment

4.2.1 Overview of Mesoporous Nanoparticles

Mesoporous nanoparticles (MNs) are a type of material with pore size between 2 nm and 50 nm. They have wide application prospects and potential use in many fields, and has become a research hotspot in the world today (Vallet-Regí et al. 2022). The physical and chemical properties of MNs, such as large specific surface area, adjustable particle size and porosity, excellent biodegradability, and biocompatibility, make them ideal platforms for drug delivery and diagnostics (Cauda and Canavese 2020; Nawaz et al. 2022). Most recent research has focused on biological applications, which involve drug delivery systems and antitumor activity. Therefore, MNs have attracted wide attention in biomedical applications such as biosensing, cell imaging, drug and gene delivery, and biomolecular separation, which become a kind of nanomaterials with great application prospects.

4.2.1.1 Classification of Mesoporous Nanoparticles

The types of MNs include dioxide silicon (Mikšík et al. 2019), carbon material (Rahman et al. 2021), metal oxide (Xie et al. 2023), metal organic framework (MOFs) (Zhang et al. 2024), etc. Mesoporous silica nanomaterials (MSNs) have regular pore structure, narrow pore size distribution, and mature technology. Silicon materials can be utilized in sensing, separation and purification, catalysis, drug embedding and slow release, and other fields (Rajendra et al. 2017). Silicon materials can be divided into two categories according to pure silicon and doped with other elements, and can be further classified according to the type of doped elements and the number of different elements (Wang et al. 2023). The introduction of heteroatoms can be regarded as the replace of heteroatoms for the original silicon atoms, and the doping of different heteroatoms will give many new properties to the nanomaterial, such as changes in hydrophilic properties, changes in stability and changes in catalytic activity (Elibol et al. 2021). Non-silicon mesoporous materials mainly include phosphates, sulfides, and transition metal oxides (Wang et al. 2019b; Baek and Joo 2022). Due to the variable valence states of those materials, it is promising to exploit new application fields for mesoporous materials, and show the application prospects that MSNs cannot reach. In addition, the charge reserve of double-layer capacitor materials prepared with mesoporous carbon is higher than that of assembled metal oxide particles, which is much higher than that of commercially available metal oxide double-layer capacitors (Dong et al. 2020; Lim et al. 2024). TiO_2-based MNs have the characteristics of high catalyst carrying capacity, strong photocatalytic activity and there are many researches on their characterization and structural properties.

4.2.1.2 Properties of Mesoporous Nanoparticles

MNs have different structural characteristics, such as diverse morphology, controllable pore size, large pore volume, high specific surface area, increased solid–liquid interface, or adjustable frame composition for some specific applications (Yu et al. 2023; Eilers et al. 2020). These structural characteristics determine the performance diversity of mesoporous materials. For example, MNs have a large number of pores and high specific surface area, which can be used as catalyst carriers to improve the activity and stability of catalysts (Xu et al. 2023b). The pore size of MNs can be regulated and used to separate molecules, ions, and other substances for the preparation of efficient column chromatography materials (Ali 2023). Mesoporous materials can be used as the carrier of the photocatalyst to fix the photocatalyst in the pore to improve the utilization rate and stability of the photocatalyst (Li et al. 2020b). Mesoporous materials have a large number of pores and a high specific surface area and can be used to adsorb gas to prepare efficient gas adsorbents (Plavniece et al. 2023). The pore size of mesoporous materials can be continuously adjusted in the range of 2–50 nm and the characteristics of no physiological toxicity make them very suitable for the fixation and separation of enzymes, proteins, etc. (Lalzawmliana et al. 2020). Mesoporous materials can form continuous, tightly bonded membrane materials on different substrates, and these membranes can be directly separated from cells/DNA for the construction of microchip laboratories (Duan et al. 2021). Mesoporous materials can load porphyrin, pyridine, or fixed embedded protein and other biological drugs in the pore channels, and improve the durability of the drug effect by modifying the functional group of controlled release drugs (Li et al. 2023). Through the use of biological guidance, it can effectively and accurately hit the target such as cancer cells and lesions, and give full play to the efficacy of drugs (Fig. 4.6).

4.2.2 Application of Mesoporous Silica Nanomaterials

Mesoporous silica nanomaterials (MSNs) are inorganic silica nanomaterials with regular pore structure, which have the dual characteristics of both mesoporous materials and nanomaterials (Zhang et al. 2024). In addition, MSNs are biosafe and can spontaneously degrade in human tissues. Since the pore size directly affects the loading and release kinetics of drugs during delivery, the ability to precisely control pore diameter in silica will provide great advantages in biomedical applications (Xu et al. 2021b; Huang et al. 2021). As shown in Fig. 4.7, MSNs have been widely used in nanomedical field, especially in the research and development of antitumor drugs, because of their adjustable particle size and mesoporous diameter, good biocompatibility, and rich surface chemistry (Zhou et al. 2022).

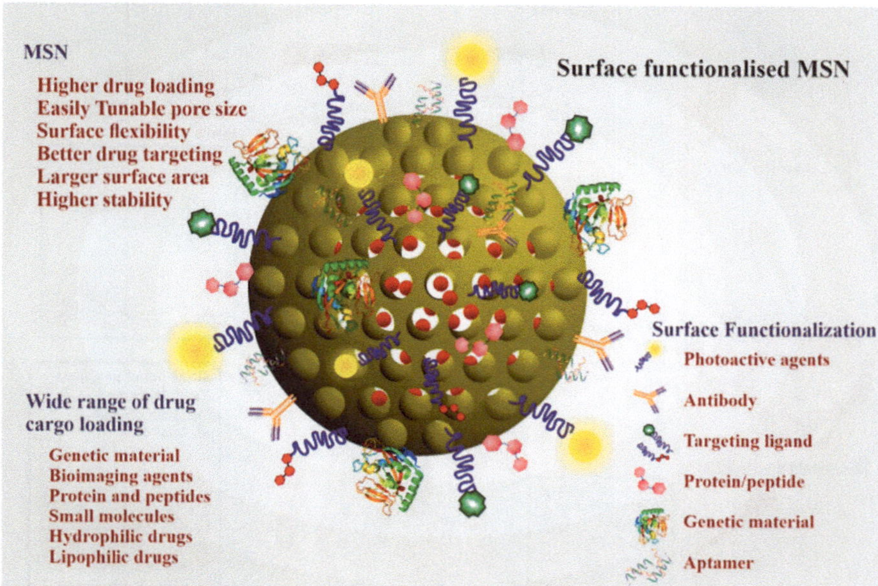

Fig. 4.6 Schematic of the functions of mesoporous nanoparticles (Reproduced from Ahmed et al. 2022). Reprinted from Ahmed et al. (2022). Copyright 2022 Elsevier

4.2.2.1 Mesoporous Silica Nanomaterials in Diagnosis of Tumor

Through reasonable design and modification of nanomaterials, two independent functions of cancer diagnosis and treatment can be integrated into the same nanocarrier, forming an integrated diagnosis and treatment nanoplatform (Xu et al. 2022a, b). The integrated platform can accurately diagnose the disease in real time and carry out synchronous treatment. Besides, real-time tracking of the treatment effect can be achieved during the treatment process to timely adjustment of the treatment plan and reduce side effects. Due to the unique physical and chemical properties of MSNs, the research based on MSNs diagnosis and treatment integration platform is a hotspot in nanobiomedical research (Zhou et al. 2022; Wu et al. 2019). MSNs can carry different therapeutic molecules such as chemotherapy drugs, gene drugs, and photosensitizer molecules. Moreover, MSNs can integrate probes or contrast agents such as fluorescence imaging, nuclear magnetic resonance imaging, ultrasound imaging, photoacoustic imaging, and PET imaging, which can be applied to gene therapy, image-guided chemotherapy, photodynamic therapy, photothermal therapy, immunotherapy, or a combination of multiple modalities (Richter et al. 2023; Escriche-Navarro et al. 2022). As shown in Fig. 4.8, Zhang et al. used mesoporous silica nanoparticles (MSNs) as drug carriers to load doxorubicin (DOX) and phorbol 12-myristate 13-acetate (PMA) for cancer diagnosis and synergetic therapy of tumors. The built multimode apoptosis pathway based on PMA and DOX could realize highly specific apoptosis of the HeLa cell. Moreover,

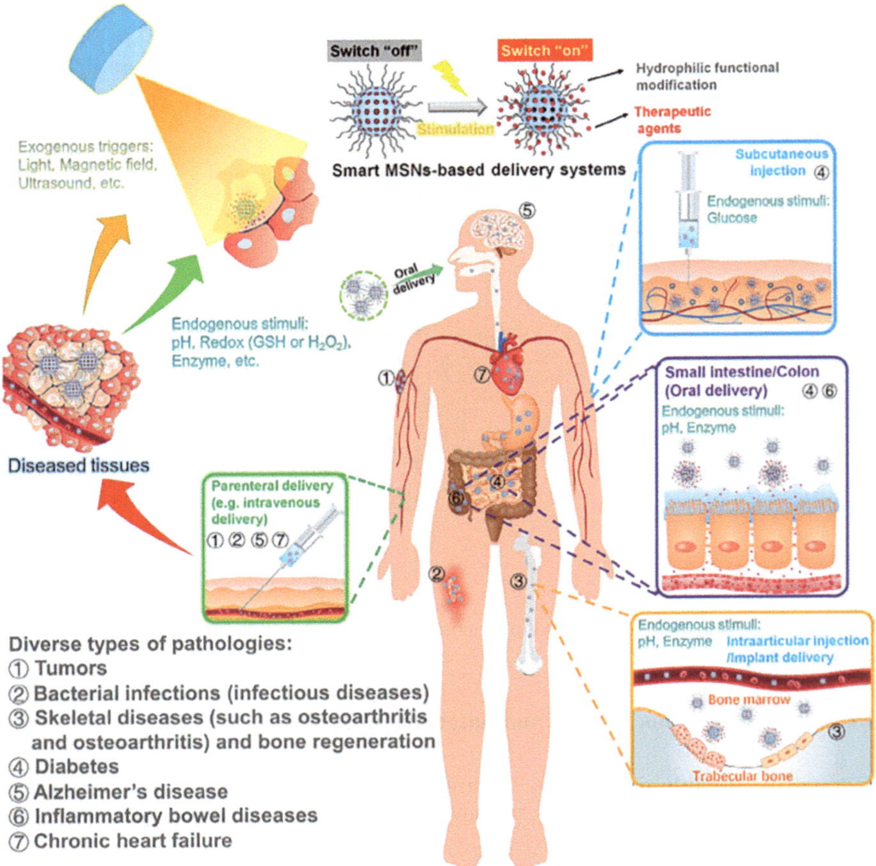

Fig. 4.7 Schematic of the application of mesoporous silica nanoparticles (Reproduced from Zhou et al. 2022). Reprinted from Zhou et al. (2022). Copyright 2022 Elsevier

the integration of diagnosis and therapy based on MSNs supports a theoretical foundation for the development of drug-nanoparticle in clinical applications (Gao et al. 2020).

4.2.2.2 Mesoporous Silica Nanomaterials in Medical Imaging of Tumor

With the development of imaging, molecular imaging provides accurate and reliable information in the diagnosis of tumors, and is widely used in preliminary studies and clinical trials. The common imaging techniques are computerized X-ray tomography (CT), magnetic resonance imaging (MRI), ultrasonic imaging (UI), positron emission tomography (PET), photoacoustic imaging (PA), and single photon diffraction imaging technology (SPECT), etc. (Verma and Gupta 2023; Duan et al. 2021; Nyayapathi et al. 2024). Various imaging techniques play an important role in the diagnosis of diseases. Nanomaterials as molecular imaging probes have

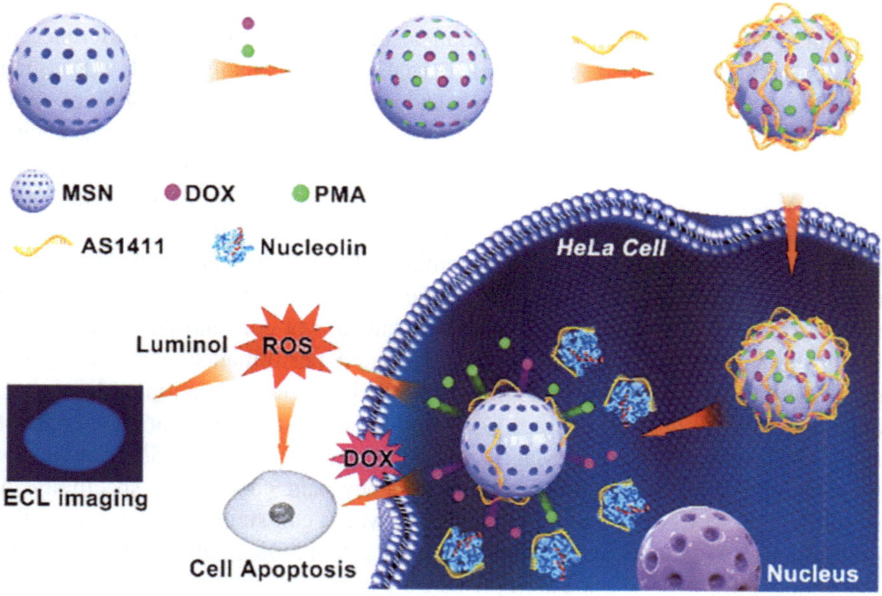

Fig. 4.8 Schematic of nucleolin imaging in a single tumor cell combined with synergetic therapy of tumor (Reproduced from Gao et al. 2020). Reprinted with the permission from Gao et al. (2020). Copyright 2020 American Chemical Society

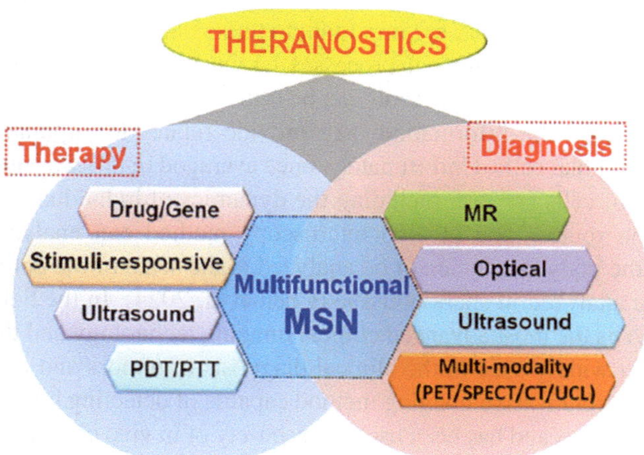

Fig. 4.9 Schematic of MSN in tumor diagnosis and therapy (Reproduced from Gao et al. 2020). Reprinted with the permission from Gao et al. (2020). Copyright 2020 American Chemical Society

important applications in the field of nanobiomedical research. As shown in Fig. 4.9, MSNs is widely used in MRI, UI, PET, CT, and other biomedical imaging because of its unique physical and chemical properties (Hosseinpour et al. 2020; Barani et al. 2021).

MRI is indispensable in the clinical diagnosis of cancer due to its high spatial resolution and three-dimensional anatomical information. Since the difference in relaxation time between diseased and normal tissue is usually small, MRI often requires the assistance of contrast agents, and it is estimated that about 40% of clinical MRI scans are performed with the assistance of gadolinium contrast agents (Iyad et al. 2023; Do et al. 2020). However, conventional gadolinium contrast agents such as Magenweil lack rapid renal clearance, tissue specificity, limited relaxation, which therefore require higher injection doses during tumor MRI, which can potentially cause harm to patients. The shortcomings of poor tumor targeting and moderate signal-to-noise ratio can be overcome by loading Magen dimension into nanocarriers, but the short blood circulation half-life of conventional nanocarriers limits the further improvement of contrast agent performance. Inspired by the deformability of red blood cells (improving their blood circulation), Chen et al. constructed novel deformable mesoporous silicone nanocallers. The new contrast agent constructed by loading Magen dimension on deformable MSNs carrier has long blood circulation, high relaxation efficiency, and good tissue specificity. A novel flexible MSNs carrier was constructed, which can significantly improve the performance of clinical contrast agents. The excellent biological properties of flexible nanocoliters make it good at improving the performance of contrast agents, and it is believed that in future studies, it can improve the efficacy of different clinical drugs and bring a wider range of application scenarios for existing drugs.

Clinical molecular imaging combines traditional medical imaging with molecular biology, so that molecular probes can monitor some physiological characteristics of the body. Established noninvasive imaging techniques, including MRI, PET, SPECT, and FL imaging, etc., are important to understand the role of cell transport in physiological and pathological processes (Azizi et al. 2020; Vermeulen et al. 2022). While molecular imaging tools can be utilized to estimate the distribution of cells noninvasively, the information they provide relates to population averages rather than individual cells. Unfortunately, once averaged in this way, many biological properties of cells are lost, including the dynamics of individual cells and specific migration routes. The in vivo microscopy method can analyze single-cell processes in the body, but it cannot be analyzed at the scale of the entire body, and it can only be analyzed in shallow tissue (Chen et al. 2024). In the field of nuclear medicine, PET is the most advanced clinical imaging technology and the high resolution makes it widely used in the clinical diagnosis of tumors and other diseases. PET is the only noninvasive imaging method capable of detecting human picomolar probe concentrations and has been used for a variety of in vivo cell tracking applications. However, despite the superior sensitivity, the potential of PET for single-cell imaging has yet to be proven (Ryuichi et al. 2021). The concept of using PET data to track discrete particles originated in the field of chemical engineering, where the method is used to estimate the dynamics of fluid and powder flow in complex opaque systems. Single-cell tracking can be used to determine the dynamics of cell transport and retardation in the early stages of a metastasis cascade, the transport of immune cells during cancer immunotherapy, and the distribution of cells after transplantation. Qiu et al. reported a tumor microenvironment responsive "head-to-foot" self-assembly nanoplatform for positron emission tomography imaging in living subjects (Fig. 4.10). They present an innovative 18F-labeled smart tracer ([^{18}F]1) engineered to target cancer-associated biotin receptors (BR) while exhibiting tumor microenvironment-responsive behavior through glutathione-triggered intracellular self-assembly into nanoparticles. Demonstrating remarkable specificity [^{18}F]1 selectively accumulates in BR-positive A549 and HeLa cancer cells, where it undergoes structural reorganization to form uniform nanoparticles under the reductive intracellular environment. This unique "head-to-foot" self-assembly mechanism significantly amplifies tracer uptake and prolongs tumor retention by converting small-molecule probes into nanoscale aggregates, thereby enhancing PET signal intensity and image contrast while minimizing nonspecific background signals. The resultant improvement in imaging fidelity positions [^{18}F]1 as a promising clinical tool for precise detection of BR-positive malignancies. Beyond its immediate diagnostic utility, this stimuli-responsive nanoplatform establishes a versatile paradigm for developing next-generation smart probes that synergistically integrate molecular targeting with programmable nanoscale transformations, offering broad potential for advancing precision oncology and theranostic applications.

4.2.2.3 Mesoporous Silica Nanomaterials in Chemotherapy of Tumor

MSNs can be modified with functional groups on the inner surface due to the silicon hydroxyl groups, and a large number of organic groups such as sulfhydryl, amino, phenyl, and vinyl are embedded in the mesoporous material. The dispersion

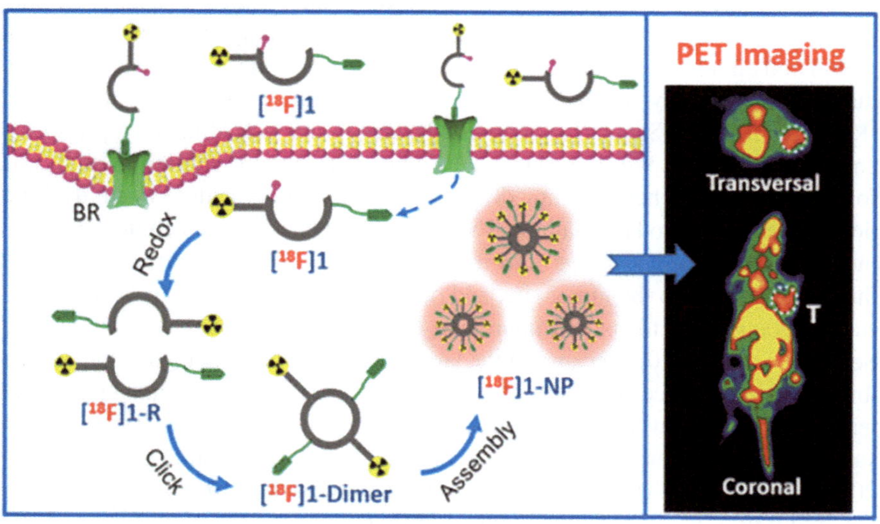

Fig. 4.10 Overview of tumor microenvironment responsive nanoplatform (Reproduced from Qiu et al. 2021). Reprinted by permission from Qiu et al. (2021). Copyright 2021 American Chemical Society

properties and processing properties of MSNs can be improved by surface functionality. At the same time, due to its excellent pore space or nanocapsule structure, small organic drugs, inorganic quantum dots, magnetic nanoparticles, metallic alkyl compounds, etc., can be introduced into the pore by chemical modification hand segment, which greatly improves the drug loading capacity (Dong et al. 2021; Zou et al. 2021).

The rationale for using MSNs as a platform for therapeutic purposes includes mesoporous properties, modifiable surface, biocompatibility, and body chemistry. (Vallet-Regí et al. 2022). Sol-gel treated mesopoles are generated in MSN structures with fine-tuned apertures, shapes, and arrangements that allow the incorporation of a large number of therapeutic molecules. In addition, the channel pore is an effective diffusion pathway for the slow release of drug molecules. Thus, the inherent mesoporous nature of MSNs provides an excellent platform for therapeutic delivery. Besides, the internal chemistry of MSNs is also customizable, allowing defects to form, which is an intrinsic source of imaging. Not only the volume, but also the surface chemistry of MSN is easily tuned to be versatile, such as imaging the conjugation of particles and dye molecules to adopt diagnostic properties. MSN has good cellular and histocompatibility and can attenuate/shield the toxic effects of metal imaging NPs. All the physicochemical and biological properties of MSNs make it a fascinating nanoplatform for chemotherapy (CT), photothermal therapy (PTT), photodynamics therapy (PDT), immunological therapy, and gene therapy (GT) of tumor (Fig. 4.11).

Chemotherapy (CT) is one of the most effective means of clinical treatment of cancer, using chemotherapy drugs to kill cancer cells to achieve the purpose of cancer treatment (Debela et al. 2021; Strasser and Vaux 2020). CT is a kind of systemic treatment, which is generally administered orally, intravenously, or through body cavity. Drugs are distributed throughout most tissues and organs of the body along with blood circulation, mainly treating some tumors that have a tendency to spread throughout the body and middle and advanced tumors that have metastasized (Rajput et al. 2023; Seynhaeve et al. 2020). Due to the lack of targeted properties, chemotherapy drugs will kill cancer cells while accidentally injuring normal cells, for example, some chemotherapy drugs will cause cardiovascular system dysfunction, and even lead to heart failure. Therefore, through effective technical means to integrate chemotherapy drugs into the nanoparticles, the nanoparticles can flow through the blood to the tumor site, and deliver the drugs directly to the tumor tissue, reducing side effects and improving the treatment effect (Hersh et al. 2022; Seynhaeve et al. 2020). Because of the large specific surface area, good biocompatibility and rich surface chemistry, etc., MSNs have emerged as a promising nanomaterials for drug delivery in brain tumors, where specific particle sizes and pore surface functions can be achieved by modulating synthesis strategies (Hu et al. 2024; Liu et al. 2024a). PEG is combined with an active targeting ligand to enhance the ability of MSN to specifically target brain tumors. As shown in Fig. 4.12, Wu et al. demonstrated a simple strategy using PEG molecules modified mesoporous silica nanoparticles to target brain tumors via the enhanced permeability and retention (EPR) effect and blood–brain barrier (BBB) penetration. MSN assembly strategies mainly include: (a) precise control of size and surface properties to optimize

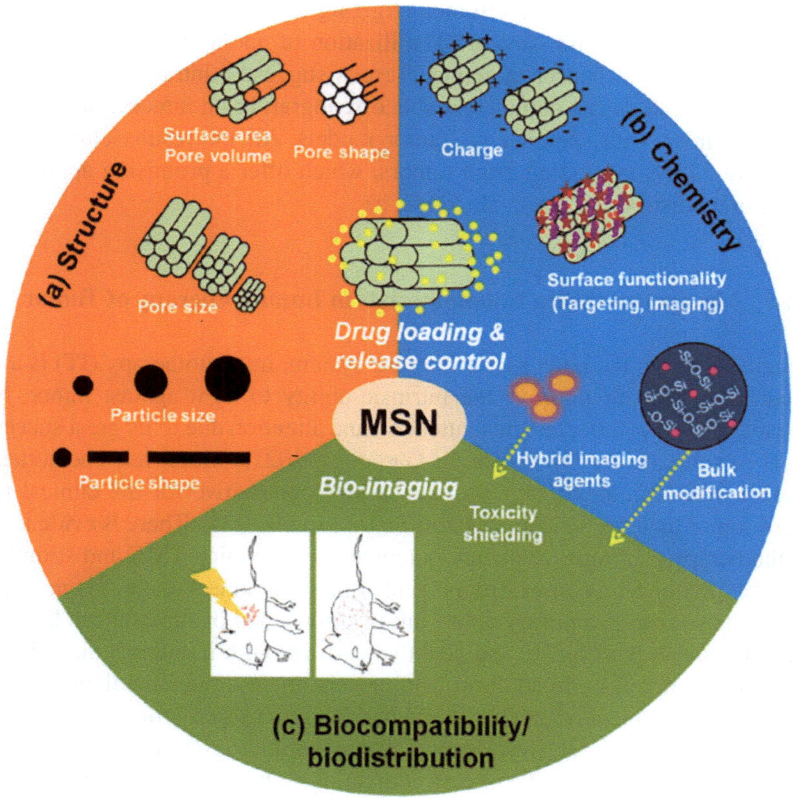

Fig. 4.11 Schematic of the merits of MSNs as a potential theranostic platform (Reproduced from Rajendra et al. 2017). Reprinted with the permission from Rajendra et al. (2017). Copyright 2017 American Chemical Society

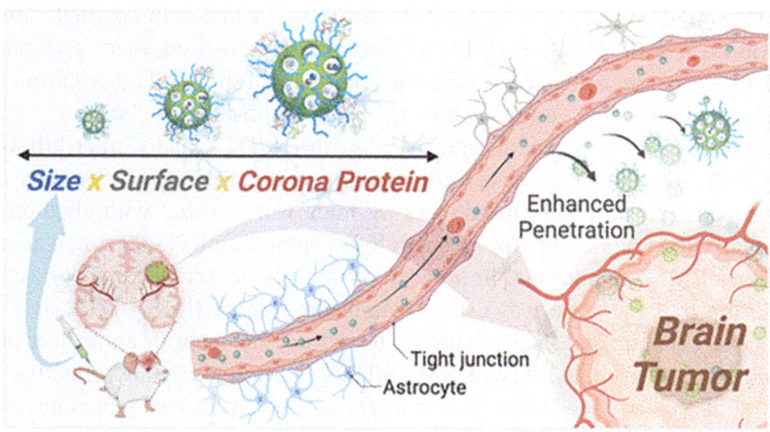

Fig. 4.12 Schematic of receptor ligand-free mesoporous silica nanoparticles (Reproduced from Chen et al. 2024a, b). Reprinted with the permission from Chen et al. (2024a, b). Copyright 2024 American Chemical Society

delivery, (b) design of effective tumor targeting using EPR effects, (c) ability to overcome the blood–brain barrier, (d) utilization of apolipoproteins in plasma to improve blood–brain barrier targeting. Results suggested intravenous administration of DOX@MSNs can completely ablate xenograft brain tumors and extend the lifespan of mice with spontaneous tumor models. All the results emphasize the capacity of MSNs in treating brain tumors, which offer a promising approach for treating brain diseases (Chen et al. 2024a, b).

4.2.2.4 Mesoporous Silica Nanomaterials in Immunotherapy of Tumor

Through activating the body's own immune system, immunotherapy (IT) is a kind of treatment mode to improve the intrinsic ability to fight against tumor. Now, immunotherapy has attracted great attention and attention due to its great success in clinical practice (Liu et al. 2022a, b; Gao et al. 2021). Moreover, chemotherapy combined with immunotherapy provides a new perspective for the clinical treatment of tumor patients (Zhu et al. 2021; Sharma et al. 2024). There is evidence that immunotherapy can improve patient outcomes by reshaping TME and stimulating antitumor immune responses. To prevent tumor immune escape, chemotherapy drugs such as taxanes, anthracyclines, mitoxantrone, and oxaliplatin can kill tumor cells and activate both innate and adaptive cells through immunogenic cell death (ICD) (Sun et al. 2021). In addition, with development of nanomedicine in cancer diagnosis and treatment research, therapeutic drugs can successfully enter the tumor area through passive or active targeting. Therapeutic drugs can successfully enter the tumor area through passive or active targeting. When stimulus–response strategies such as enzymes, pH, magnetism, light, and X-rays are used, the drug delivery system prepared obviously improves therapy effectiveness and reduces side effects (Zheng et al. 2023; Shi et al. 2022). Pei et al. designed a novel nanotherapeutic drug combined with immune checkpoint organic small molecule inhibitors. The hydrophobic anticancer drug docetaxel (DTX) was loaded into the pores of amylated MSN (MSN-NH_2/DTX) as a model drug. Bovine α-lactalbumin coupled with oleic acid was coated on MSN-NH_2/DTX by electrostatic attraction, then oleic acid was added to form the anticancer inclusion compound Bamlet. The resulting drug-carrying lipoprotein complex/silica nanocomposite is called BMSN/DTX. Cytotoxicity experiments showed that BMSN/DTX had similar killing efficiency to BAMLET on cancer cells, but had lower cytotoxicity on normally differentiated cells. Next, TNBC tumor-bearing mice were treated with chemotherapy combined with immunotherapy using self-assembled anti-PD-L1 small molecule inhibitors (BMS). In addition, in order to prevent the recurrence and metastasis of TNBC, the transcriptome was used to study the role of BMSN/DTX as a novel therapeutic adjuvant and its mechanism. Therefore, the superior anticancer properties and good biocompatibility of BMSN/DTX provide a safe and effective therapeutic strategy for the design and manufacture of anticancer nanoadjuvants by triggering a powerful immune response against TNBC (Yin et al. 2025).

As a kind of nonspecific immunoproliferative agent, immune adjuvant has the function of antigen transfer and immune enhancement (Kelleci et al. 2023; Darvishi et al. 2023). Therefore, it is of great significance to design new multifunctional immune adjuvants for the immunotherapy of cancer (Hu and Li 2020). However, the existing immune adjuvants still have some problems, such as poor cytocompatibility, low cell uptake rate, excessive particle size, poor therapeutic efficiency, and single function. Based on this, Ding et al. used a typical silica sol–gel reaction with trimethyltoluene as pore expansion agent to prepare macroporous silica coated up-conversion nanoparticles (UCMSs) with a size less than 100 nm, and the UCMSS was used as a new immune adjuvant (Ding et al. 2018). The obtained UCMSs not only had a high load of model protein (chicken ovalbumin (OVA)), photosensitizer merocyanine 540 (MC540), and tumor antigen (tumor cell fragment), but was also successfully used for efficient in vivo vaccine delivery. Under 980 nm near-infrared irradiation, the built UCMSs-MC540-OVA showed the best synergistic immune enhancement with the strongest immune responses of Th1 and Th2, and the highest frequencies of effector memory T cells and CD4+, CD8+. Moreover, the nanovaccine UCMSS-MC540-TF was more effective in inhibiting tumor growth and improving survival in BALB/c mice carrying colon cancer (CT26) than photodynamic therapy or immunotherapy alone, suggesting that UCMSs has higher immunotherapeutic efficacy and clinical potential as an immune adjuvant for cancer immunotherapy.

4.2.2.5 Mesoporous Silica Nanomaterials in Phototherapy of Tumor

Photodynamic therapy (PDT) is the use of specific wavelengths of light to stimulate photosensitizers, transfer energy to oxygen, and produce reactive oxygen species (ROS), so as to kill cancer cells and1 achieve the effect of tumor treatment (Correia et al. 2021; Niculescu and Grumezescu 2021). Specific wavelengths of light, photosensitizers, and oxygen are important components of PDT. The photosensitizer acts as a catalyst to transfer the energy of the light source and produce singlet oxygen ($^1O_2$) with active effect, which is the key to the success of PDT (Younis et al. 2021). Most of the current photosensitizers are suitable for ultraviolet or visible light excitation, the penetration ability is low, only suitable for the treatment of epidermal tissue, while the colored bodies of some tissues (hemoglobin, etc.) can strongly absorb visible light, which greatly interferes with the conversion efficiency of photosensitizers to light energy. In recent years, MSNs have been widely used as PDT carriers due to their unique properties (Prieto-Montero et al. 2023; Xu et al. 2022a, b).

Photothermal therapy (PTT) is a treatment method that uses near-infrared light (NIR) as an energy source to treat tumors (Lv et al. 2021; Li et al. 2021a, b). Materials with high photothermal conversion efficiency are selectively gathered near tumor tissues, and light energy is converted into heat energy under external near-infrared light irradiation to kill cancer cells (Xu et al. 2021a, b; Ge et al. 2023). In recent years, photothermal therapy based on nanomaterials has become a new means of cancer treatment. Photothermal therapy theoretically realizes the effective treatment of all solid tumors, including the effective treatment of chemoradiotherapy failure and drug-resistant tumors, while not causing side effects accompanying chemoradiotherapy and affecting the quality of life of patients (Yang et al. 2023a, b). In addition, photothermal nanomaterials can also play an imaging role through surface modification and other methods to achieve the integration of tumor diagnosis and treatment, or combined with photodynamic therapy, chemotherapy, gene therapy and immunotherapy, and other applications to achieve collaborative treatment of tumors (Qi et al. 2023; Li et al. 2023).

Because of the high biocompatibility and rich surface chemistry, MSNs are often used as coating materials to wrap on the surface of inorganic photothermal materials to improve the stability of inorganic photothermal materials, and can also carry other therapeutic molecules to achieve combined therapy. For instance, pH-responsive upconversion MSNs combined with PTT and PDT for multimodal diagnostic imaging and targeted cancer therapy have obtained considerable attention as potential alternatives to traditional cancer therapy. However, these approaches are still limited by poor stability, low solubility, and low targeting efficiency of photosensitizers (PS) and photothermic agents (PTA). To overcome these limitations, Palanikumar et al. designed biodegradable, biocompatible, and tumor-targeted upconversion nanospheres with imaging capabilities (Palanikumar et al. 2023). As shown in Fig. 4.13, this multifunctional nanospheres consist of a sodium yttrium fluoride nucleus doped with lanthanides and PTA bismuth selenide encased in mesoporous silica shells that encapsulate PS and dihydroporphyrin e6 (Ce6) in their pores. $NaYF_4$:Yb/Er converts deeply penetrating near-infrared (NIR) light into visible light, which stimulates Ce6 to produce cytotoxic reactive oxygen species (ROS), while Bi_2Se_3 efficiently converts absorbed near-infrared light into heat. In addition, Gd can perform magnetic resonance imaging of the nanospheres. The mesoporous silica shell is modified with DPPC/cholesterol/DSPE-PEG, thereby retaining encapsulated Ce6 and preventing serum protein adsorption and macrophage recognition from impeding tumor targeting. After that, the coating is conjugated with an acid-triggered rational membrane (ATRAM) peptide to promote specific, efficient internalization of malignant cells in the weakly acidic microenvironment of tumors. The nanospheres facilitate tumor magnetic resonance as well as thermal and fluorescent imaging, and exhibit effective near infrared laser-induced anticancer effects in vitro and in vivo through a combination of ROS generation and local hyperthermia, with negligible toxicity to healthy tissues, resulting in significantly prolonged survival. The results show that ALUMSN provide multimodal diagnostic imaging and targeted combination cancer therapy.

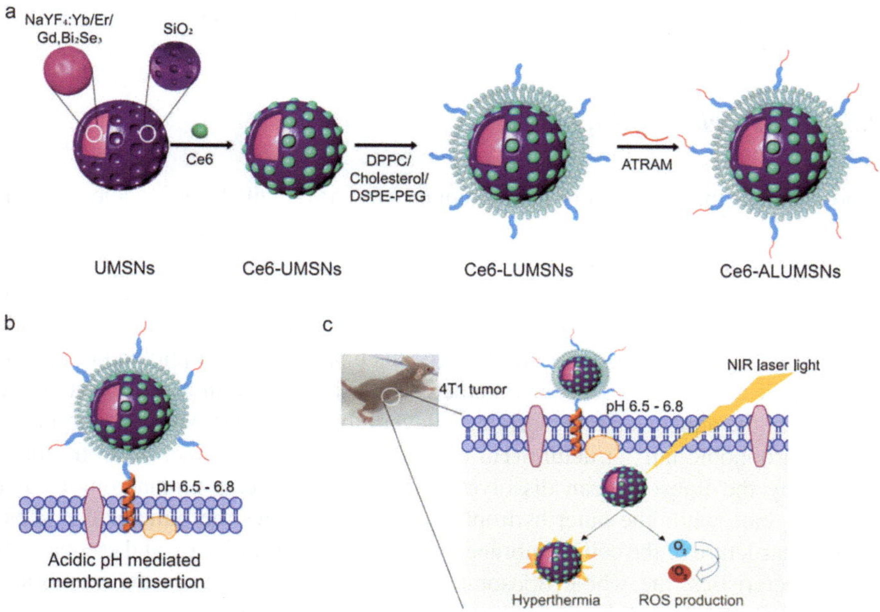

Fig. 4.13 Schematic of (**a**) representation of preparation of nanospheres, (**b**) acidic pH-mediated membrane insertion of Ce6-ALUMSNs, and (**c**) mode of action of the tumor-targeted nanospheres (Reproduced from Palanikumar et al. 2023). Reprinted with the permission from Palanikumar et al. (2023). Copyright 2023 American Chemical Society

4.2.3 Conclusions and Perspectives

Mesoporous nanomaterials (MNs) are inorganic biological carrier materials with large specific surface area, large pores, and adjustable pore size, and their surfaces can be modified with various chemical modifications to realize good biocompatibility and stability. As a drug carrier, MNs can control the release of drugs well, so as to achieve the ideal therapeutic effect. In addition, they are also widely used in biological delivery and biological imaging. In recent years, diagnostic reagents based on various functional modifications of MNs have been widely used in the early detection, diagnosis, and treatment of cancer. However, most studies remain at the preclinical stage, and whether their findings translate to practical clinical solutions requires further investigation. Critical safety concern including cytotoxicity, genotoxicity, and in vivo metabolic pathways also warrant additional study. In addition, there are many unresolved issues. For example, when diagnostic reagents are used in vivo, their distribution and biosafety need to be evaluated comprehensively and carefully. When the diagnostic reagents are introduced into the target cells, tissues, or organs, the targeting efficiency is relatively low. Efficiency and sensitivity are not high enough for in vivo diagnostic applications. A lot of efforts of researchers are still needed to promote the application of MNs in cancer diagnosis and treatment.

4.3 Application of Liposome in Tumor Diagnosis and Treatment

4.3.1 Overview of Liposome

Liposomes, as the precursor of lipid nanoparticles, are small closed vesicles spontaneously formed by the phospholipid bilayer (Thi et al. 2021). Discovered by Bangham et al. in the 1960s, liposomes were initially used as a simple model to study the structure and function of cell membranes (Trucillo et al. 2020; Gupta et al. 2020). Due to the amphiphilic structure, after amphotic molecules (phospholipids and sphingolipids) are dispersed in the aqueous phase, the hydrophobic tail groups of the molecules tend to cluster together, while the hydrophilic head groups are exposed in the aqueous phase, forming bilayer closed vesicles with a "hydrophilic head-hydrophobic tail" structure (Hu et al. 2021). Thus, the inner aqueous space formed by the liposome can dissolve hydrophilic chemicals or macromolecular enzymes, etc., while the outer hydrophobic layer simulates the bimolecular phospholipid structure of the cell membrane. According to the similarity and compatibility characteristics, the whole liposome structure can easily cross the membrane barrier and enter the cell interior. Since phospholipid bilayer is naturally occurring, liposomes have good biocompatibility and immunogenicity. At the same time, the loaded reagent or drug is transported into the cell to prevent it from being metabolized or attacked by the immune system (Dymek and Sikora 2022; Antimisiaris et al. 2021). At present, the United States Food and Drug Administration (FDA) approved the listing of liposome nanomaterials such as doxorubicin hydrochloride liposome, doxorubicin liposome, and paclitaxel liposome.

4.3.1.1 Properties of Liposome

According to the preparation methods and physicochemical properties of the formulation, lipid-based nanoparticles are divided into five categories. They are liposomes, lactosomes, metastases, nanostructured lipid carriers (nlc), and solid lipid nanoparticles (sln). In short, liposomes composed of phospholipids and cholesterol have strong drug protection and targeting capabilities. These properties result in liposomes with reduced toxicity and improved efficacy compared to free drugs in the systemic circulation (Shah et al. 2017). The main components of liposomes are usually cholesterol and amphiphilic phospholipids. Briefly, the bilayer structure is formed by amphiphilic phospholipids and supported by cholesterol (Gong et al. 2022; Zhang et al. 2023a, b). Commonly used phospholipids are sphingomyelin and glycerin, both of which have a hydrophilic head and a hydrophobic tail region. In the aqueous environment, phospholipids spontaneously arrange into liposomes driven by hydrophobic interactions and other intermolecular interactions. The function of cholesterol is to promote the accumulation of lipid chain and the formation of bilayer, reduce the fluidity of bilayer, and reduce the transmembrane transport of water-soluble drugs. Not only that, cholesterol can also reduce the interaction of liposomes with proteins in the body and reduce the loss of phospholipids, thereby improving the stability of liposomes (Chiu et al. 2022; Liu et al. 2017).

Due to its special structure, liposomes have the following advantages as tumor drug carriers. (1) The structure of the liposome is similar to the cell membrane, which makes it highly biodegradable and biocompatible. Therefore, the liposome can protect the drugs from enzymatic degradation before reaching the lesion site. Meanwhile, the drug is "hidden" inside the liposome through physical inclusion, which can reduce the toxicity of the drug, increase the dose of the drug, and improve the stability of the drug (Ding et al. 2022; Garg et al. 2022). (2) The bilayer surface produced by amphiphilic phospholipids can be structurally modified with ligands or other functional groups with physical or chemical methods to prolong the effective retention time of the liposomes at the lesion site, make the liposomes tissue-targeting, and even realize efficient targeted drug delivery (Wang et al. 2022; Spitzenberg et al. 2025). (3) By changing the charge of the bilayer surface, cationic liposomes can be used for the inclusion and delivery of DNA and RNA drugs (Liang et al. 2016; Tang et al. 2021).

Due to the special structure of liposomes, stability is an important factor that limit the drug delivery. The stability of liposomes can be divided into chemical stability, physical stability and biological stability (Almeida et al. 2020; Wang et al. 2023). In order to keep the integrity of the vesicle structure of the liposomes, it is important to maintain the balance of various interaction forces within and between the liposomes. As a thermodynamically unstable colloidal dispersion system, the vesicles of liposomes tend to condense and flocculate, and the surface charge can affect the aggregation behavior of liposomes. In general, neutral liposomes tend to accumulate or merge vesicles, while liposomes with strong potential can maintain stable dispersion due to electrostatic mutual repulsion. The type and dosage of membrane material, drug loading, particle size, and particle size distribution all affect the physical stability of liposome (Wang et al. 2023; Erickson et al. 2023).

The bilayer vesicles of liposomes are mainly composed of phospholipids, but phospholipids contain a large number of functional groups that are easy to oxidize and hydrolyze (such as unsaturated double bonds and ester groups), so liposomes have chemical stability problems that are easy to oxidize and hydrolyze (Bar et al. 2023; Dymek et al. 2022). In order to avoid oxidation and hydrolysis of phospholipids, saturated phospholipids can be used, which not only has good chemical stability, but also has a high glass transition temperature. If it is inevitable to use unsaturated phospholipids in the prescription of liposomes, try to choose phospholipids with low unsaturation, such as containing fatty acids as monounsaturated fatty acids (Dong et al. 2021). Since the oxidation of fatty acids is a free radical chain reaction without the involvement of specific oxidants, the presence of metal ions or oxygen to form a peroxide will trigger an oxidation reaction. Therefore, it is necessary to avoid the use of materials containing metal ions and peroxides as much as possible, and metal ion chelating agents can also be added, and the preservation conditions are as mild as possible and not easy to oxidize (Nakashima et al. 2016;

Liu et al. 2024a, b). Although the addition of antioxidants can slow the oxidation of phospholipids, antioxidants are rarely added in the development of liposomes. The hydrolysis of liposomes is due to the hydrolysis of phospholipid ester bonds, and although the degree of hydrolysis can be reduced by freeze-drying the liposome suspension to avoid contact with water, the freeze-drying process is not suitable for all liposome products (Marín et al. 2018; Popova and Hincha 2016). Cholesterol has the function of stabilizing lipid bilayer membrane structure and can produce steric hindrance effect to slow down phospholipid hydrolysis. In addition, the ionic strength of the external aqueous phase, the type of buffer salt, and the pH also affect the hydrolysis of liposome (Guo et al. 2023; Wang et al. 2019a, b).

The biological stability of liposomes is more due to the prescription composition and preparation process; for example, the phospholipid polyethylene glycol in the liposome can form a hydration film on the surface of the liposome to avoid the rapid identification and clearance of the liposome by MPS. The addition of sphingomyelin can also improve the stability of liposomes (Lin et al. 2024; Onguka et al. 2021).

4.3.1.2 Construction of Liposome Nanocarriers

There are four types of liposomal nanomedicine including traditional liposomes, spatially stable liposomes, ligand targeting liposomes, and the synthesis of the above three liposomes (Tran et al. 2017). As the first generation of liposome nanomaterials, traditional liposome nanomaterials carriers have been used for in vivo delivery of clinical anticancer drugs (adriamycin and amphotericin) since 1980 (Koning and Storm 2003). Compared with direct intravenous injection of drugs, although the application of traditional liposome nanomaterials reduces the toxicity of drugs to the body, its easy removal from blood vessels also limits its efficacy. Especially in the liver and spleen, macrophages of the reticuloendothelial system (RES) play a major role in clearance. In order to improve the stability of liposomes and prolong the retention time of liposomes in blood circulation, the polymeric polymer polyethylene glycol (PEG) was modified on the surface of liposomes, so as to construct PEG-modified liposomes with more stable structure than the first-generation liposomes (De et al. 2021; Fulton and Najahi-Missaoui 2023). PEG modification can produce steric hindrance on the surface of liposomes, which prevents the rapid recognition and clearance of liposome nanomaterials by RES, thus prolonging the retention time of liposome nanomaterials in blood circulation and reducing their toxic side effects (Kowalska et al. 2021). The spatial stability generated by PEG modification also affects the pharmacokinetics of liposome nanomaterials and increases their half-life (Shi et al. 2021; Inglut et al. 2020). Ligand-targeting liposomes are based on different tumor cell types to select specific targets, so as to achieve the purpose of targeted tumor recognition. For example, Jiang et al. exploited pegylated doxorubicin liposomes to elucidate the intrahepatic fate of liposomes and payloads in male mice (Fig. 4.14). Kupfer cells initiate and dominate the intrahepatic capture of doxorubicin liposomes and subsequently transport the released

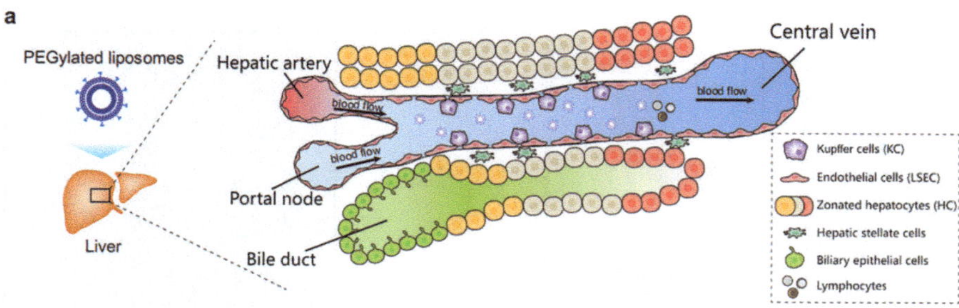

Fig. 4.14 Schematic of atomic engineering of PEGylated liposomal doxorubicin (Reproduced from Jiang et al. 2024). Reprinted from Jiang et al. (2024). Copyright 2024 Springer Science+Business Media

doxorubicin to hepatocytes in a zonal distribution along the lobule-portal-central axis. Increasing Kupfer cell capture promotes the accumulation of doxorubicin in hepatocytes, which reveals the process of Kupfer cell capture—payload release—hepatocyte accumulation. In contrast, Kupfer cells ignore free adriamycin, which is rapidly distributed into hepatocytes by passing directly through the endothelium of the perforated hepatic sinuses. Compared with free DOX, liposomal DOX demonstrates sustained metabolism/excretion due to the additional capture-release process. This study reveals the critical role of Kupfer cells in the intrahepatic transport of pegylated liposome therapy and quantitatively describes the intrahepatic transport/distribution/elimination process, providing key information to guide the further development of nanomaterials (Jiang et al. 2024).

Liposomes are nanoparticles that can be used as drug delivery carriers, biosensors, chemical reactors, synthetic organelles, and biofilm models (Vallet-Regí et al. 2022; Zong et al. 2023). However, due to the limitations of existing generation methods, this type of soft matter nanoparticles shows limited morphological changes, ultimately narrowing their functional range. Therefore, it would be beneficial to control parameters such as the composition, size, and number of lipid bilayer (thin layer) to build lipid systems with enhanced structural complexity. Microfluidic hydrodynamic focusing (MHF) is carried out through a nucleation growth mechanism where lipids reach a solubility threshold along a well-defined diffusion gradient, nucleate and self-assemble into the most favorable energy form. Based on this, Elani et al. developed a novel microfluidic technique for the precise manufacture of nanoscale liposomes that can be released in multiple layers for drug and vaccine delivery (Fig. 4.15). This technique can achieve fine regulation of the structure and release properties of liposomes, which improve the efficacy and safety of treatment. It also has the potential to simulate cell structure in synthetic biology, and supports high-throughput production and integration with other laboratory equipment. The researchers expect this technology to advance the development of drug delivery vehicles to meet the needs of the pharmaceutical industry in the future (Pilkington et al. 2024).

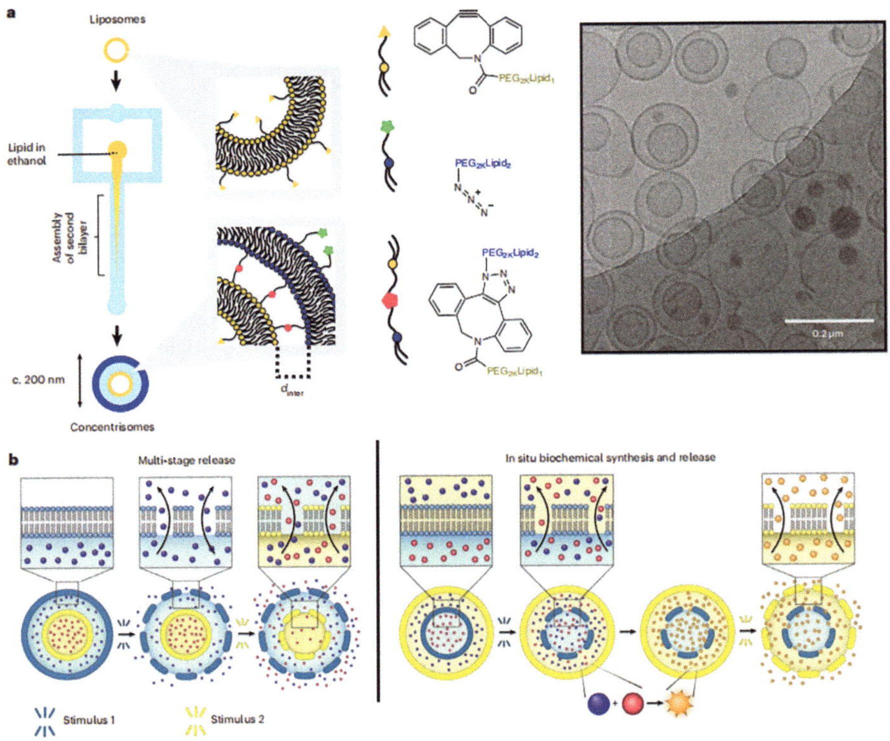

Fig. 4.15 Schematic of (**a**) synthesis and characterization of engineering nanoscale liposomes, (**b**) the release process of engineering nanoscale liposomes (Reproduced from Pilkington et al. 2024). Reprinted from Pilkington et al. (2024). Copyright 2024 Springer Science+Business Media

4.3.2 Application of Liposome Nanocarriers

Due to the biocompatibility, stability, and low toxicity of liposomes, their clinical applications have received considerable attention. The construction of multifunctional liposome nanocarriers to maximize the service of liposome for cancer diagnosis and drug delivery in vivo has been a focus of clinical research in recent years (Ghazanfari et al. 2017; Kong et al. 2024). Liposome has enough space in its structure to enable it to load both diagnostic and therapeutic drugs at the same time, which makes it the best nanomaterials choice for integrated diagnosis and treatment platforms. Hydrophilic therapeutic drugs can be loaded in the hollow inner cavity of the liposome, while hydrophobic therapeutic drugs can be embedded in the hydrophobic bicomeric sublayer (Koehler et al. 2023; Seleci et al. 2017). The purpose of loading diagnostic and therapeutic drugs can be achieved through intracavitary encapsulation, intercalation of phospholipid bilayer, or connection/adsorption on the surface of liposomes, so that two or more drugs with a single therapeutic effect can be combined on a carrier to realize the integration of diagnosis and treatment of liposome nanocardiac drug carriers (Cheraghi et al. 2017; Hong et al. 2023).

4.3.2.1 Liposome Nanocarriers in Tumor Imaging

Liposome nanocarriers have been widely used in clinical drug delivery and slow release. At the same time, the imaging functional groups can also be modified to the surface of liposome by simple means, and the research of liposome nanocarriers in vivo imaging is also relatively mature. For optical imaging, hydrophobic fluorescent dyes can be embedded in phospholipid bilayers. For example, fluorescent dye has been used in in vivo imaging of edema models in mice (Meng et al. 2024). Quantum dots are also often modified in the hydrophilic parts of liposomes as fluorophores (Woźniak et al. 2022). When the hydrophobic parts of liposomes are connected with chelators and incubated with radioisotopes, radioisotopically labeled liposome nanocarriers can be obtained, which can be used as in vivo positron emission computed tomography (PET) and single photon emission computed tomography (SPECT) imaging methods (Das et al. 2022).

Although small-molecule fluorescent dyes have been widely used in the aggregation-induced quenching properties, the photochemical efficiency of small-molecule fluorescent dyes decreases sharply when the loading concentration is high or when they aggregate in the lipid bilayer, due to the formation of π-π accumulation between molecules. Therefore, stable and effective tracers could not be achieved, and the discovery of aggregation-induced luminescence (AIE) molecules was born (Ye et al. 2023; Zhao and Chen 2021). AIE molecules basically do not emit light in the dispersed state, but their photochemical effect is significantly enhanced in the aggregated state, so they have better stability and fluorescence intensity at high concentrations or loaded nanocarriers (Chen et al. 2023a, b). For example, Wang et al. developed a novel tetraphenylvinyl-coated liposomes (TPE-LPs), which were constructed by embedding TPE-trimethylanthocyanin into liposomes for the first time, and used as probes for in vivo tumor bioimaging (Fig. 4.16). TPE

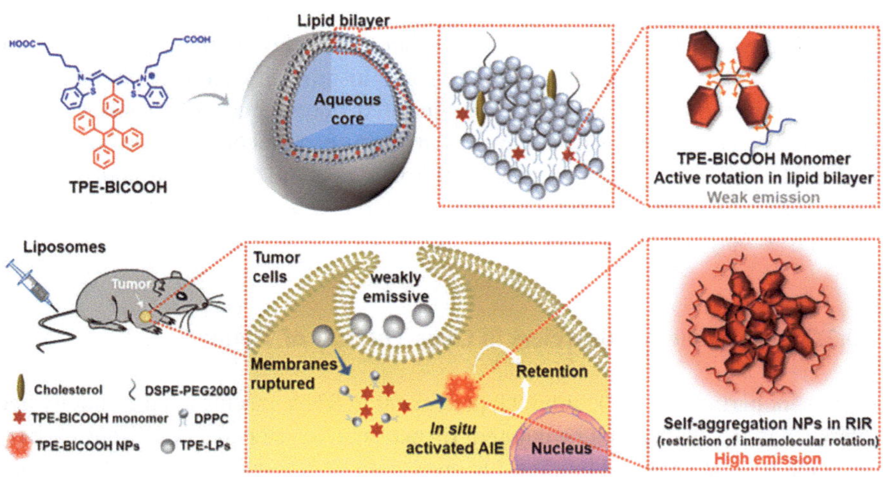

Fig. 4.16 Illustration of the mechanism of TPE-LPs in imaging tumor in vivo (Reproduced from Zhang et al. 2018a, b). Reprinted with the permission from Zhang et al. (2018a, b). Copyright 2018 American Chemical Society

functionalized trimethylcyanine was synthesized by a novel and efficient one-pot reaction. In TPE-LP, it was found that the TPE-functionalized bicarboxylate benzoindole-trimethocyanine (TPE-BICOOH) fluorophones were well dispersed in the lipid bilayer during blood circulation (with unrestricted rotation) and then aggregated in tumor tissue when the liposomes broke up (with the restriction of intramolecular rotation). Therefore, high target signals and low background for tumor imaging are obtained. In situ activated AIE probes not only accumulated a large amount at the tumor site after intravenous injection in 4T1 tumor-bearing mice, but also showed excellent signal-to-noise ratio, excellent biocompatibility, and low cytotoxicity. This designed strategy is considered a convenient and powerful method for sensitive detection of tumors (Zhang et al. 2018a, b).

Liposome nanoparticles are ideal for drug and developer delivery because of their robust manufacturing capabilities, high biocompatibility, ease of surface modification, and the ability to carry multiple payloads. These properties have facilitated further development of liposomes, in which phospholipids are the key building blocks for many drug delivery (Yi et al. 2023; Zhong and Zhang 2019). There has always been an "all-in-one" approach to transforming liposomes from a single drug delivery vehicle into a multimodal therapeutic nanoplatform, that is, combining different functional parts together for multifunctional use (Katari and Jain 2021; Akanda et al. 2023). This inherent complexity presents significant challenges for their clinical application. Zhang et al. reported the synthesis and self-assembly of a novel azaze-fluoroboropyrrole lipid building block into liposome nanostructures (BODIPYome), which exhibit near-infrared J-aggregation redshift. They explored various formulations of BODIPYome and investigated their structural morphology, fluorescence quenching efficiency, and J-aggregation properties (Fig. 4.17). In addition, they also examined the inherent optical stability of J-aggregated BODIPYome and found that there may be a J-dimer effect in the structure of the liposome caused by aza-BODIPY restriction. Cholesterol is added to improve colloidal stability for in vivo applications. The cholesterol-containing BODIPYome provides a stable nanoparticle with J-dimer-induced near-infrared optical properties. This allows the complete cholesterol-containing BODIPYome to provide excellent photoacoustic properties in the mold and restore near-infrared fluorescence when taken up and destroyed by cells. Finally, in vivo results in a mouse model of in situ prostate tumor highlight the biocompatibility of BODIPYome as a biocompatible and excellent optical imaging probe for photoacoustic and fluorescence imaging, as shown in the literature supplement (Zhang et al. 2023a, b).

4.3.2.2 Liposome Nanocarriers in Gene Therapy of Tumor

Lipid nanoparticles (LNP), although prominent in RNA vaccines, have toxicity and tolerance issues that cannot be ignored. Brown et al. developed a new "gene carrier" of fusion protein and lipid binding-fusion protein lipid carrier (FAST-PLVs), which not only showed excellent gene delivery efficiency in vivo and in vitro, but also had

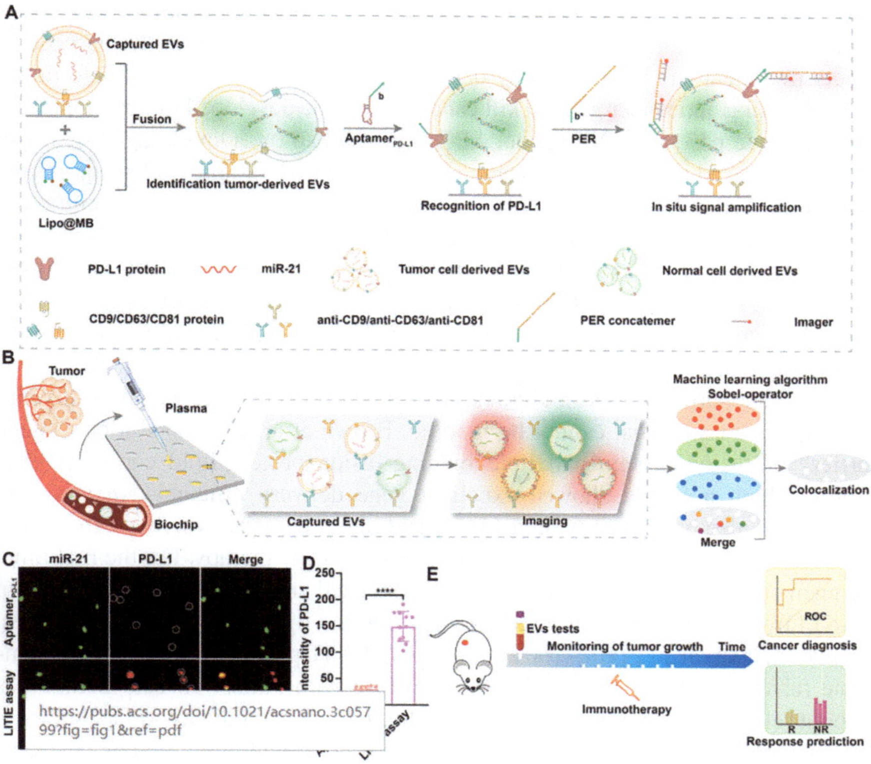

Fig. 4.17 Illustration of (**a**) the design of the in situ signal amplification, (**b**) the extraction of plasma from a tumor, and (**c**) imaging results for miR-21 and PD-L1. LITIE assay (Reproduced from Zhang et al. 2023a, b). Reprinted with the permission from Zhang et al. (2023a, b). Copyright 2023 American Chemical Society

low immunogenicity and high tolerance, which is expected to pave the way for future repeatable drug delivery of gene therapy. For the first time, the team successfully introduced a fusion-associated small transmembrane protein (FAST) from a non-envelope fused reovirus into the liposomal system. Despite their small size (100–200 amino acids), FAST proteins have a unique membrane fusion capability that allows them to fuse with almost all cell membranes without relying on specific cell receptors. Through microfluidic hybrid technology, this technology can achieve large-scale production, providing a low-cost and efficient solution for gene drug delivery.

Gene therapy is the introduction of normal genes or therapeutic genes into diseased cells or tissues of the human body to correct gene defects or play a therapeutic role. Gene therapy for cancer can be achieved through several strategies: reducing the expression of cancerous genes, introducing tumor suppressor genes, enhancing the sensitivity of tumors to radiotherapy and chemotherapy, and enhancing the host's antitumor immunity (Paskeh et al. 2022). The mutation of tumor suppressor gene is an important factor in the incidence of cancer. Therefore, the purpose of cancer treatment can be realized by delivering normal anticancer gene into tumor cells and using anticancer gene to kill cancer cells. The developed tumor suppressor genes include p53 gene, NF1 gene, APC gene, and so on (Ooi et al. 2024; Bagci 2024). The introduction of foreign anticancer genes into biological cells requires the help of corresponding carriers. Viral vector has strong gene transport ability, but there are many security risks. Liposomes are the first nonviral vectors studied and have been commercialized. In order to protect nucleotides, liposomes or lipid nanoparticles (LNPs, including SLN and or NLCs) are used as drug delivery systems that exhibit adequate encapsulating and tissue-cell targeting capabilities (Ojansivu et al. 2025; Liu et al. 2023).

Lipid nanoparticles (LNP), although prominent in RNA vaccines, have toxicity and tolerance issues that cannot be ignored. Brown et al. developed a new "gene carrier" of fusion protein and lipid binding-fusion protein lipid carrier (FAST-PLVs), which not only showed excellent gene delivery efficiency in vivo and in vitro, but also had low immunogenicity and high tolerance, which is expected to pave the way for future repeatable drug delivery of gene therapy. For the first time, the team successfully introduced a fusion-associated small transmembrane protein (FAST) from a non-envelope fused reovirus into the liposomal system (Fig. 4.18). Despite their small size (100–200 amino acids), FAST proteins have a unique membrane fusion capability that allows them to fuse with almost all cell membranes without relying on specific cell receptors. Through microfluidic hybrid technology, this technology can achieve large-scale production, providing a low-cost and efficient solution for gene drug delivery.

4.3.2.3 Liposome Nanocarriers in Synergistic Therapy of Tumor

Due to the biocompatibility, stability and low toxicity capable, the clinical application of liposomes has been paid much attention. The construction of multifunctional liposome nanocarriers to maximize the service of liposome for cancer diagnosis and drug delivery in vivo has been a focus of clinical research in recent years (Zhang et al. 2015; Lombardo et al. 2016). The liposome has enough space in its structure to enable it to load both diagnostic and therapeutic drugs at the same time, which makes it the best nanomaterials choice for integrated diagnosis and treatment platforms (Bilal et al. 2021; Domenico et al. 2016). Hydrophilic therapeutic drugs can be loaded in the hollow inner cavity of the liposome, while hydrophobic therapeutic drugs can be embedded in the hydrophobic bicomeric sublayer. The purpose of loading diagnostic and therapeutic drugs can be achieved through intracavitary encapsulation, intercalation of phospholipid bilayer, or connection/adsorption on the surface of liposomes, so that two or more drugs with a single therapeutic effect can be combined on a carrier to realize the integration of diagnosis and treatment of liposome nanocardiac drug carriers (Yang et al. 2018; Kulkarni et al. 2020).

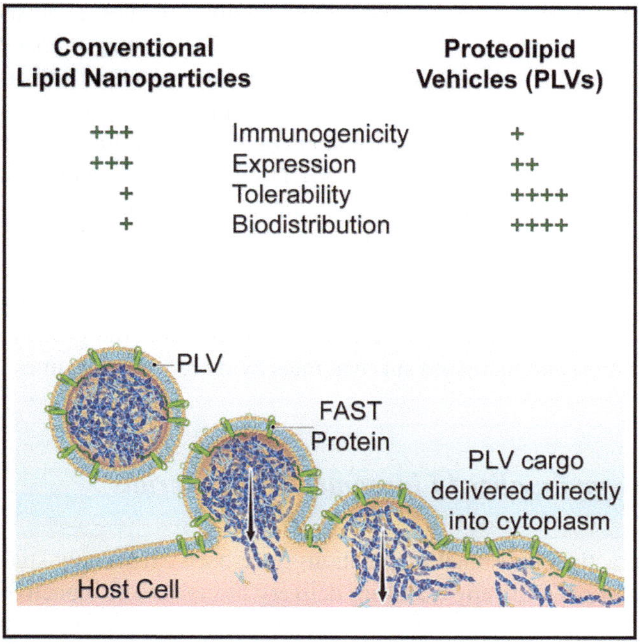

Fig. 4.18 Illustration of in vivo delivery of DNA and RNA using proteolipid vehicles (Reproduced from Brown et al. 2024). Reprinted from Brown et al. (2024). Copyright 2024 Elsevier

Gene therapy and acoustic dynamic therapy (SDT), as an emerging therapeutic method, has great potential in cancer treatment (Tan et al. 2009; Zhang et al. 2023a, b). However, there are significant challenges in the delivery of gene and sound sensitizers in vivo during therapy, which ultimately affects treatment outcomes. In this study, an ultrasound-sensitive targeted liposome nanoparticle system (MLipsiBcl-2) was developed by Wang et al. to deliver acoustic sensitizer and small interfering RNA (siRNA) for synergistic treatment of hepatocellular carcinoma. MLipsiBcl-2 can trigger the production of reactive oxygen species (ROS) through ultrasonic stimulation, leading to the rupture of liposomes and the release of sound sensitivities and siRNA (Wang et al. 2024c). In addition, ROS can disrupt lysosomal membranes and promote gene release, thereby downregulating levels of overexpressed anti-apoptotic proteins in cancer cells. The results of in vitro and in vivo studies demonstrate the synergistic therapeutic efficacy and high biocompatibility of MLipsiBcl-2 on liver cancer cells under ultrasonic stimulation. Advances in this ultrasound-sensitive targeted gene delivery system show potential as an easy-to-operate, multifunctional therapeutic platform, offering the prospect of synergistic therapeutic approaches across multiple cancer types.

As a drug carrier, liposome has the advantages of adjustable carrying capacity and good biological distribution. However, due to the dysfunction of the targeted part and the loss of payload during preparation, immune liposomes have not been well used in commercial production. Immune liposomes have strong payload capacity, improved pharmacokinetics, enhanced drug tolerance, and specific distribution (Mayer et al. 2019; Rahman et al. 2024). The existing preparation process involves chemical modifications that compromise the stability of the targeted part. Based on this, Wan team reports a chemical-free method to generate immune liposomes by self-assembly of chimeric nanomeric (cNB) into liposomal bilayer, one-step process (Rahman et al. 2024). cNB is composed of antihuman epidermal growth factor receptor 2 (HER2) nanobodies (NB), flexible peptide linkers, and hydrophobic single transmembrane domains (STMD). The results showed that 64% of the therapeutic compounds could be encapsulated into the immune liposomes, demonstrating that drug-loaded immune liposomes increased cytotoxicity 10–20 times to HER2-overexpressing cancer cell lines, inhibited the growth of xenograft tumors 3.4 times, and increased survival rates by more than two times.

4.3.3 Commercialized Liposome Nanomaterials

In clinical practice, liposomal preparations have been extensively studied for their applications in gene therapy and drug delivery as carriers of bioactive substances. Due to its significant stability and favorable toxicity compared to traditional therapies, liposomes can encapsulate hydrophilic biomolecules or drugs, enhancing their internalization and solubility within the cellular lipid bilayer (Li et al. 2020a, b; Maritim et al. 2021). Over the past decade, research interest in liposome formulations has increased significantly, and it has been demonstrated that liposome formulations have a higher safety profile compared to viral vectors. This is due to their lower immunogenicity, lower toxicity, and ability to transport larger cargo to target locations (Ghosh et al. 2019; Alrbyawi et al. 2022). At present, many liposome products are undergoing clinical trials, and in-depth research on lipid carriers and liposome formulations has promoted the development of novel liposome-mediated drug delivery. In 1995, Doxorubicin hydrochloride liposome injection (Doxil) was approved by the FDA as the first pegylated long-circulating liposome drug for advanced ovarian cancer, multiple bone marrow cancer, and Kaposi's sarcoma with HIV (Kabilova et al. 2018; Lotfabadi et al. 2018). Doxil was developed by Sequus Pharmaceutical in the United States. Doxil was prepared by using ammonium sulfate gradient method to achieve active doxorubicin loading. The drug molecules combine with sulfate ions in the internal water phase to form crystalline sulfate precipitation, which makes the doxorubicin encapsulation rate high and stable, and drug leakage is not easy. Classic liposome nanomaterials are amphoteric B liposome, daunorubicin liposome, doxorubicin liposome, paclitaxel liposome, and so on (Van Eerden et al. 2019; Atrafi et al. 2020). Common tumors in approved indications include: breast cancer, ovarian cancer, myeloma, hepatocellular carcinoma, leukemia, non-Hodgkin lymphoma, non-small cell lung cancer, and pancreatic cancer (Sharma et al. 2024).

4.3.4 *Conclusions and Perspectives*

In recent decades, great achievements have been made in the application of liposome nanocarriers in the integrated research of tumor diagnosis and treatment. Dozens of liposome drugs have been marketed worldwide, and many more drugs are in the preclinical or clinical research stage. Through the ingenious construction and surface modification of nanomaterials, diagnosis and treatment, two clinically separate processes, are integrated on a single nanoplatform, forming the "integrated nano platform for diagnosis and treatment." This nanoplatform can diagnose the condition in real time with high accuracy and carry out treatment simultaneously. Besides, it can monitor the efficacy and adjust the drug administration plan at any time during the treatment process, which is conducive to reduce toxic and side effects and achieve the best treatment effect. However, the research and development and clinical transformation of liposome nanomaterials are both challenges and opportunities, and the clinical needs of liposome nanoplatform are still unsatisfied. The resources and input have been mostly directed toward liposomal technology, Nevertheless, it could hardly bring a revolutionary breakthrough in the short term. In order to promote the clinical transformation of liposome nanomedicine, researchers should deeply explore the construction and mechanism of liposome nanocarriers. Increasing the system stability of liposome nanocarriers, improving their tissue distribution, overcoming the biological barrier and immunogenicity of liposome nanocarriers in vivo, and improving the tumor tissue targeting ability of liposome nanocarriers are all key problems that need to be solved urgently.

4.4 Application of Polymeric Nanoparticles in Tumor Diagnosis and Treatment

Polymer nanomaterials, synthesized from natural or synthetic materials as monomers or polymers, can be designed into smart nanoparticles with a variety of structures, biocompatible and biodegradable, which is supposed to be good nanomedical carriers (Liu et al. 2024c, d; Vinicius et al. 2021). In addition, polymeric nanomaterials can be modified to achieve precise control of nanomaterial characteristics. As shown in Fig. 4.19, polymeric nanomaterials can deliver drugs, proteins, and genetic materials to target tissues through surface target modification (Dorel 2019).

4.4.1 *Overview of Polymeric Nanoparticles*

At present, polymer nanomaterials can be synthesized by a variety of technical means, such as emulsification (solvent displacement or diffusion), nanoprecipitation, and ion gelation and microfluidics. Drugs can be encapsulated inside the

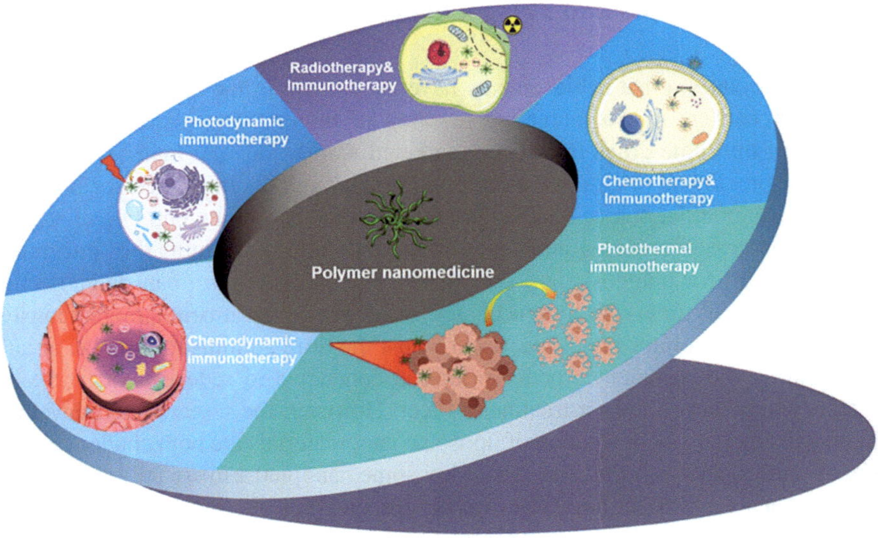

Fig. 4.19 Illustration of the development of polymer nanomedicines for synergistic immunotherapy and radiotherapy, chemotherapy, photothermal therapy, photodynamic therapy, and chemodynamic therapy (Reproduced from Wang et al. 2024a). Reproduced from Wang et al. (2024a). Copyright 2024 John Wiley & Sons Ltd

nanomaterial, encapsulated in polymer molecules, chemically bound to the polymer and bound to the surface of the nanomaterial, so polymer nanomaterials can deliver hydrophobic and hydrophilic compounds such as biomacromolecules, small molecules, proteins, and vaccines (Wang et al. 2024b). By adjusting the stability, composition, responsiveness, and surface charge of polymer nanomaterials, the drug loading rate and release kinetics of delivered drugs can be precisely controlled (Marlon et al. 2020). Polymer nanomedical preparations have attracted more and more attention due to the advantages such as long cycle, reduced immunogenicity, and small adverse reactions, which have become the focus of nanomedical preparations research.

Among different nanotherapeutics, polymer nanoparticles are versatile and outstanding nanocarriers that can control drug delivery to targeted tissues. The first-generation polymer nanoparticles are used to encapsulate drug molecules and have a slow release effect (Liu et al. 2021a, b). In order to efficiently deliver drugs to targets, numerous strategies have been developed in second-generation polymer nanoparticles by incorporating stimulus responsiveness, such as temperature, pH, or light activation (Tiffany et al. 2023). The third-generation polymer nanocarriers exhibit multifunctionality, such as targeting and multidrug delivery properties (Shang et al. 2024). The most common forms of polymer nanoparticles are nanospheres (solid matrix systems) and nanocapsules (chambers surrounded by polymer membranes or shells). In these two categories, nanomaterials are further divided into polymeric objects, micelles, and dendritic polymers (Zhang et al. 2021a, b;

Wang et al. 2020). The polymeric object is an artificial vesicle whose membrane is made of amphiphilic block copolymer. They are comparable to liposomes, but they have drug retention efficiency and higher stability, so they can be used as effective carriers to deliver therapeutic drugs to the cytoplasm (Michael et al. 2020).

In 1969, Speiser and his team first proposed the concept of polymer nanoparticles for drug delivery. The synthetic polymers such as polylactic acid (PLA), polylactic acid glycolic acid (PLGA), polyglycolic acid (PGA), polymethyl methacrylate (PMMA), and the natural polymers such as collagen, dextran, chitosan, gelatin, alginate, and heparin are preferred for designing polymer-based nanoparticle (Zhang et al. 2021a, b; Han et al. 2023). Among the synthetic polymers, PLGA-based nanoparticles have attracted much attention due to their drug release tunability and good biocompatibility. The properties of PLGA, such as drug release rate, molecular weight, biodegradability, and hydrophobicity, can be regulated by regulating the molecular weight (Mw) of PLGA and lactate-glylactone (L:G) ratio (Silva et al. 2022; Sheffey et al. 2022). Among the natural polymers, chitosan-based nanoparticles have become one of the most popular biopolymers in nanomedicine due to the low toxicity, good biocompatibility and biodegradability, hemostasis, adhesion, and simple preparation methods and antimicrobial properties (Zaiki et al. 2023; Elkomy et al. 2022). Chitosan nanoparticles have a strong affinity for in vivo site-specific drug delivery and negatively charged surface of cell. The drug released from chitosan nanoparticles is dependent on polymer–drug interactions, polymer erosion, and their swelling degree, which is affected by the pH of the medium (Xie and Liu 2020; Yuan et al. 2010).

Bader et al. first proposed polymer micelles as delivery systems in 1984, which have spherical core-shell structures and are formed by self-assembly of amphiphilic block copolymers in aqueous solutions (Su et al. 2021; Zheng et al. 2024). The hydrophobic kernel serves as a storage space for hydrophobic drugs (Park et al. 2020). However, micelles mostly exhibit sudden drug release and lack of targeting, for which stimulus-responsive polymer micelles were developed in the late 1980s (Ren et al. 2021; Ramesh et al. 2020). Nanogels are hydrophilic or amphiphilic polymer particles, which can absorb water and expand, are chemically or physically crosslinked polymer hydrogel networks, with a size of about 100–200 nm, and have swelling behavior with high water retention capacity. Nano-gels have several advantages over other nano-carriers, such as greater stability, higher drug loading capacity, surface area, control over particle size, and greater responsiveness to temperature, pH, enzyme stimulation, and REDOX conduction (Li et al. 2022; Fabian et al. 2020). The researchers attempted to develop multi-stimulus-responsive nanogels for co-encapsulation of hydrophobic or hydrophilic drugs for the therapy of cancer in order to reduce toxic side effects and improve efficacy (Yang et al. 2023a, b; Liu et al. 2022a, b). Dendritic macromolecules are highly branched spherical synthetic polymer macromolecules with a size of 1–100 nm. Controlling its size and chemical properties makes it an ideal nanocarrier for drug delivery (Najafi et al. 2020; Song et al. 2024a, b). Drugs can be physically encapsulated into the covalently or void bonded to functional groups on the surface. A large number of terminal units in tree-shaped macromolecular networks enable them to achieve multiple functions (Rebecca et al. 2021; Silvia et al. 2022).

4.4.2 Application of Polymeric Nanoparticles

Polymeric nanoparticles have variable drug delivery capabilities. The therapeutic agent may be encapsulated in the NP core, encased in a polymer matrix, chemically coupled with the polymer or bound to the NP surface. This makes a variety of payloads, including hydrophilic and hydrophobic compounds, as well as cargo with different molecular weights, such as biomacromolecules, small molecules, vaccines, and proteins that make polymer NPs ideal for co-delivery applications (Zhang et al. 2020; Knight et al. 2019). By adjusting properties such as stability, composition, surface charge, and responsiveness, the release kinetics and loading effect of these therapeutic agents can be precisely controlled. To sum up, polymer NPs are ideal candidates for drug delivery because they are biodegradable, water-soluble, biocompatible, biomimetic, and stable during storage (Wu et al. 2022a, b; Zhang et al. 2018a, b). Polymeric nanoparticles have surfaces that can be easily modified to allow them to deliver drugs, genetic material, and proteins to target tissues, making them useful in gene therapy, cancer medicine, and diagnostics (Michael et al. 2020). However, the disadvantages of polymer NPs include an increased risk of toxicity and particle aggregation. Only a small number of polymer nanomedicines are currently approved by the FDA for clinical use, but polymer nanocarriers are currently being tested in a large number of clinical trials.

4.4.2.1 Polymeric Nanoparticles in Tumor Imaging

The field of optical nanoprobes is booming, and conjugated polymer nanoparticles for multifunctional diagnosis and treatment are widely favored by researchers because of their low preparation difficulty, strong modifiability, and low price and are widely used in tumor imaging, tumor therapy, drug release, intelligent monitoring, and other fields (Feng et al. 2020; Chen et al. 2020). However, because of their large molecular weight and particle size, polymer nanoparticles are not easy to be metabolized rapidly from organisms and have potential long-term toxicity, which is a problem that cannot be ignored in clinical transformation. Polystyrene (PPV) is a class of commercial conjugated polymers widely used in the manufacture of organic photovoltaic devices. In recent years, it has also made great achievements in biological fields such as biosensing, fluorescence imaging, and afterglow imaging and has been proved to have good biodegradation ability (Song et al. 2024a, b; Amal et al. 2015; Chen et al. 2023a, b). However, due to the wide band gap of PPV, the fluorescence emission can only reach the first NIR window of 600 nm–700 nm, which limits its ability to be further used for deep tissue imaging. By introducing electron-withdrawing groups into the conjugated main chain to construct push-pull electron polymer, the band gap can be effectively reduced and the emission of conjugated polymer can be regulated to the second near-infrared window above 1000 nm (Lei et al. 2021; Shou et al. 2017). The fluorescence in the second region of near-infrared has the advantages of smaller light scattering, higher signal-to-noise ratio, and deeper penetration depth (generally greater than 5 mm). It can visualize deep biomolecular organization and structural organization and has better application and development prospect than NIR zone 1 fluorescence. Therefore, the development of a biodegradable NIR two-zone diagnosis and treatment probe can effectively promote the development of optical diagnosis and treatment to clinical application.

Gu proposes a rational design strategy for the development of biodegradable conjugated polymers (BSPN) as nanotherapeutic agents for IR-II fluorescence image-guided photodynamic therapy (Gu et al. 2024). According to our previous studies, PPV-type SPN can be degraded in the presence of myeloperoxidase (MPO), hydrogen peroxide (H_2O_2), and chloride. This property allows for the effectiveness of degradation in vivo, as both MPO and H_2O_2 are enriched in macrophages. On the basis of PPV structure, a biodegradable NIR-II fluorescent polymer (BSP) was constructed by introducing pyrrole and pyrrole dione (DPP) fragment and extending its absorption and emission wavelengths. In order to make BSP water-soluble, under the irradiation of 808 nm laser, BSPN can not only emit the IR-II fluorescence signal, but also convert molecular oxygen into singlet oxygen for PDT. After intravenous administration of BSPN, it may partially accumulate to the tumor site through enhanced osmotic retention. Herein, BSPN effectively illuminates tumors and kills tumor cells through its IR-II fluorescence signal and PDT effect, respectively.

4.4.2.2 Polymeric Nanoparticles in Antitumor Treatment

The polymer drug transport system has the advantages of long cycle time and low toxicity and side effects, overcoming the problems of poor biocompatibility of traditional chemotherapy drugs, easy elimination in blood circulation, and greater toxicity to normal tissues and organs (Guo and Jiang 2020; Luo et al. 2020). However, how to construct a polymer drug transport system with high drug loading rate, high kinetic stability, and high anticancer activity is still a challenge. Recently, Xu et al. used anticancer polypeptide (FKLAK) directly as an aqueous monomer to participate in emulsion interface polymerization, and constructed polymer polypeptide nanomedics with a drug loading rate of 77%, good kinetic stability, and high anticancer activity. The researchers designed and synthesized FKLAK polypeptides with six side chain amino groups as aqueous monomers, and acid anhydride monomers with anhydride groups at both ends and long alkyl chains in the middle as oil-phase monomers. With the aid of ultrasound, the amino group of the aqueous monomer can react with the anhydride group of the oily monomer at the emulsion interface to form a maleic acid amide bond with acid-response properties. The bond can remain stable under normal physiological environment, so that the toxic and side effects of nanomedicine on normal cells are low. However, in the acidic environment of tumor interstitium, it can be rapidly hydrolyzed and efficiently release anticancer polypeptides to exert its anticancer effect. Nano-drugs with high drug loading rate, high kinetic stability, and high anticancer activity were synthesized by directly participating in emulsion interface polymerization of anticancer peptides as monomers. This method is expected to be applied to other types of chemotherapy drugs and provides a new idea for the development of efficient and low-toxicity drug transportation system (Yang et al. 2022).

The rational design of polymers with specific and reversible supramolecular structures also gives them unique advantages: a broad and modular platform with adaptable biological and chemical properties (De et al. 2021; Dorca et al. 2019). In addition, the dynamic and reversible nature of non-covalent interactions gives biomaterials the ability to sense and respond to physiological signals from organisms, thus facilitating the development of smart supramolecular therapeutic stents and drug delivery libraries (Simonson et al. 2020). Combining luminescence properties with supramolecular nanomedicine can achieve accurate imaging of tumors and real-time monitoring of treatment processes. Aggregation-induced luminescence (AIE) fluorophores are ideal for fluorescence sensing applications. Tetraphenylethylene (TPE), a widely studied AIE fluorophore, has been successfully used for fluorescence imaging and biosensing due to its low cytotoxicity, biocompatibility and significant emission properties in solid state (Isla et al. 2019; Yang et al. 2022). Athina Anastasaki group reported on a hydrogen bond interaction based polymer nanomaterial with AIE properties for multi-pathway intelligent anti-tumor therapy. To this end, they first designed a novel tetraphenylvinyl (TPE)-based hydrogen bonding prodrug TPE-(FUA)4, which is characterized by four chemotherapeutic fluorouracil-1-acetic acid (FUA) groups connected to the core of TPE (Fig. 4.20). Then, a six-arm star-shaped amphiphilic polymer carrier P(DAP-co-OEGEA)6 was prepared by ARGET-ATRP, which contained a hydrophilic and biocompatible polyoligoethylene glycol ethyl acrylate (POEGEA) fragment, and a hydrophobic and hydrogen-bonding polydiaminopyridine acrylamide (PDAP) fragment. Through the polymerization process of ATRP, the obtained multi-arm polymer is a hyperbranched single molecule micelle, which has the characteristics of compact structure, spherical structure, and large surface area, and its drug carrying property is better than that of the linear polymer similar micelle (Li et al. 2023).

4.4.3 Conclusions and Perspectives

In general, polymeric nanomaterials are water-soluble, biodegradable, imitative, biocompatible, and storage stable, which make them a good carrier for drug delivery. They can deliver genetic material, drugs and proteins to target tissues through surface modification and have a wide range of research and applications in gene

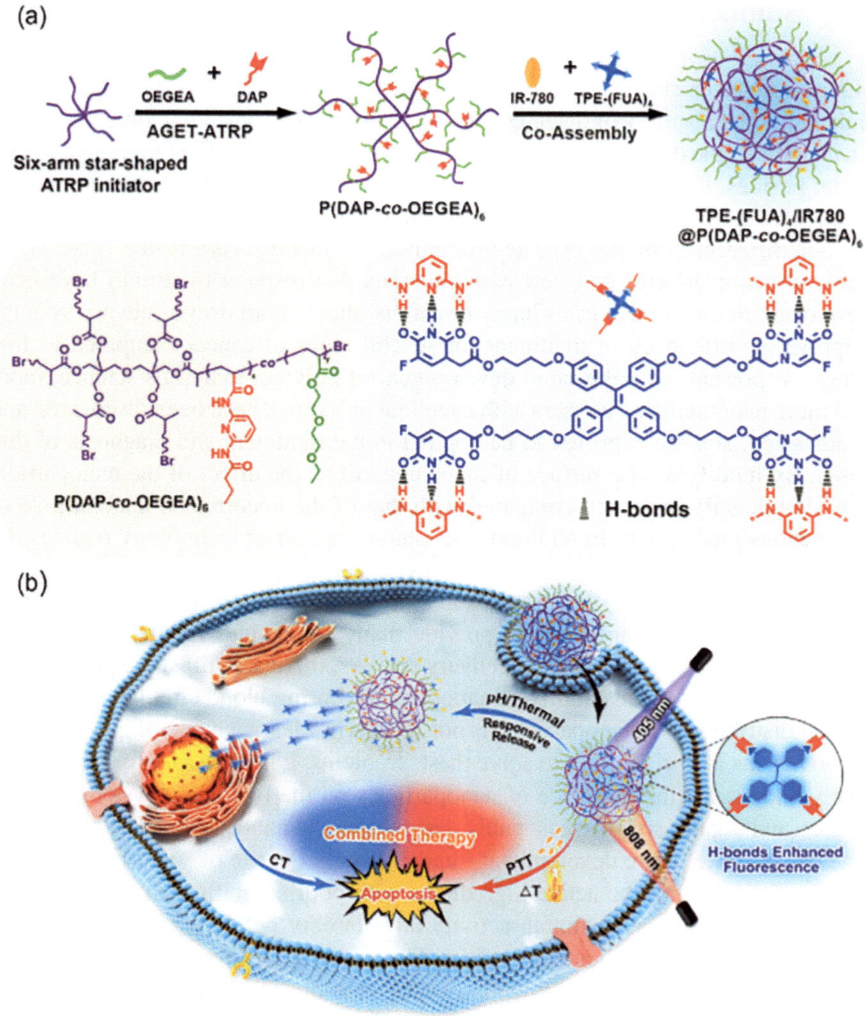

Fig. 4.20 Illustration of the (**a**) formation of H-bonded supramolecular polymeric micelles, (**b**) the application of the polymeric micelles in a cellular environment (Reproduced from Li et al. 2023). Reprinted with the permission from Li et al. (2023). Copyright 2023 American Chemical Society

therapy, cancer medicine, and diagnostics. However, polymer nanomaterials also have their shortcomings, such as particle aggregation, toxicity and pharmacokinetic behavior is not ideal, and only a few polymer nanomaterials are currently approved by the FDA and used in the clinic, but there are many polymer nanomaterials currently being tested in clinical trials.

4.5 Summary

Nanomedicine carriers have gained a great deal of attention, and technologies in this field help treat a variety of diseases, including tumor. The development of therapeutic nanomedicine delivery systems is a challenge for the research community worldwide; besides, there is a high demand for systems that combine active ingredient drugs with efficient delivery vectors. The clinical response of classical nanocarriers is often different from that expected for anticancer therapy. As a result, smart drugs delivery nanoplatforms and new nanomaterials that respond to stimuli have been developed. Some clinical trials have shown that these smart drugs delivery systems improve the efficiency of treatment for several types of cancer compared to free drugs. At present, with the rapid development of polymer materials science, more and more nanomedicine carriers with excellent properties have been discovered and synthesized, and are expected to be applied in the treatment and diagnosis of diseases. By modifying the surface of the nanoparticle, the effect of the nanoparticle was significantly improved compared with that of the uncorrected nanoparticle of intravenous medication. In addition, the nano-drug carrier technology realizes the effective control of the drug dose for the targeted organ, prolongs the action time for the targeted organ to a certain extent, and the actual use effect has been significantly improved. Although nanoparticles provide many advantages for the treatment of cancer and other diseases as drug delivery carriers, there are still some problems to be solved, such as the instability of nanoparticles during blood circulation; the biological distribution of nanoparticles is not ideal. The clearance rate of nanoparticles by organisms is low. In order to solve these problems, it is necessary to comprehensively understand the influence of nanoparticles on the biological system in cells, tissues, and organs, so as to reasonably design the ideal nanocarriers. For nanomedicine carriers, the future development trend is mainly the use of suitable nanomedicine carriers to wrap the active ingredients with poor solubility but good curative effect, so that they can maintain activity and stability before reaching the site of action, so as to play the curative effect. The preparation of nano-drug carriers is combined with specific drugs to obtain smart nano-drugs with targeted and quantitatively timed drug release, so as to solve the diagnosis and treatment of major diseases such as cancer.

References

Ahmed H, Gomte SS, Prathyusha E et al (2022) Biomedical applications of mesoporous silica nanoparticles as a drug delivery carrier. J Drug Deliv Sci Technol 76:103729

Akanda M, Mithu MSH, Douroumis D (2023) Solid lipid nanoparticles: an effective lipid-based technology for cancer treatment. J Drug Deliv Sci Technol 86:104709

Ali M (2023) What function of nanoparticles is the primary factor for their hyper-toxicity? Adv Colloid Interface 314:102881

Almeida B, Nag OK, Rogers KE et al (2020) Recent progress in bioconjugation strategies for liposome-mediated drug delivery. Molecules 25(23):5672

Alrbyawi H, Poudel I, Annaji M et al (2022) Recent advancements of stimuli-responsive targeted liposomal formulations for cancer drug delivery. Pharm Nanotechnol 10(1):3–23

Amal KM, Sivaramapanicker S, He TC et al (2015) Three-photon-excited luminescence from unsymmetrical Cyanostilbene aggregates: morphology tuning and targeted bioimaging. ACS Nano 9(5):4796–4805

Antimisiaris SG, Marazioti A, Kannavou M et al (2021) Overcoming barriers by local drug delivery with liposomes. Adv Drug Deliver Rev 174:53–86

Anwar A, Muhammad I, Hafiz MNI (2023) Smart chemistry and applied perceptions of enzyme-coupled nano-engineered assemblies to meet future biocatalytic challenges. Coord Chem Rev 493:215329

Atrafi F, van Eerden RA, van Hylckama Vlieg MA et al (2020) Intratumoral comparison of nanoparticle entrapped docetaxel (CPC634) with conventional docetaxel in patients with solid tumors. Clin Cancer Res 26(14):3537–3545

Azizi M, Dianat-Moghadam H, Salehi R et al (2020) Interactions between tumor biology and targeted nanoplatforms for imaging applications. Adv Funct 30(19):1910402

Baek DS, Joo SH (2022) Non-siliceous ordered mesoporous materials via nanocasting for small molecule conversion electrocatalysis. Bull Korean Chem Soc 43(10):1156–1168

Bagci O (2024) Association of KRAS gene and microRNA-124-3p in sporadic colorectal Tumours. JBM 12(01):150–161

Bakandritsos A, Kadam RG, Kumar P et al (2019) Mixed-valence single-atom catalyst derived from functionalized graphene. Adv Mater 31:1900323

Bar L, Villanueva ME, Martín C et al (2023) Stability of supported hybrid lipid bilayers on chemically and topographically-modified surfaces. Colloid Surface A 664:131125

Barani M, Mukhtar M, Rahdar A et al (2021) Recent advances in nanotechnology-based diagnosis and treatments of human osteosarcoma. Biosensors 11(2):55

Bilal M, Qindeel M, Raza A et al (2021) Stimuli-responsive nanoliposomes as prospective nanocarriers for targeted drug delivery. J Drug Deliv Sci Technol 66:102916

Brown DW, Wee P, Bhandari P et al (2024) Safe and effective in vivo delivery of DNA and RNA using proteolipid vehicles. Cell 187(19):5357–5375

Cauda V, Canavese G (2020) Mesoporous materials for drug delivery and theranostics. Pharmaceutics 12(11):1108

Chang BS, Zhang LQ, Wu SL et al (2022) Engineering single-atom catalysts toward biomedical applications. Chem Soc Rev 51:3688–3734

Chellasamy G, Varathan E, Sekar Y et al (2023) Single-atom catalysts for biosensing: progress in theoretical and mechanistic understanding. Coord Chem Rev 502:215606

Chen C, Ou HL, Liu RH et al (2020) Regulating the photophysical property of organic/polymer optical agents for promoted cancer phototheranostics. Adv Mater 32:1806331

Chen H, Luo X, Huang Q et al (2023a) Platelet membrane fusion liposome loaded with type I AIE photosensitizer to induce chemoresistance cancer pyroptosis and immunogenic cell death for enhancing cancer immunotherapy. Chem Eng J 476:146276

Chen X, Teng S, Li J et al (2023b) Gadolinium (III)-chelated deformable mesoporous organosilica nanoparticles as magnetic resonance imaging contrast agent. Adv Mater 35(20):2211578

Chen Y, Yang YW, Zhang F et al (2024) Noninvasive in vivo microscopy of single neutrophils in the mouse brain via NIR-II fluorescent nanomaterials. Nat Protoc 19:2386–2407

Chen C, Zhang X, Gao Z et al (2024a) Preparation of AIEgen-based near-infrared afterglow luminescence nanoprobes for tumor imaging and image-guided tumor resection. Nat Protoc 19:2408–2434

Chen ZA, Wu CH, Wu SH et al (2024b) Receptor ligand-free mesoporous silica nanoparticles: a streamlined strategy for targeted drug delivery across the blood–brain barrier. ACS Nano 18:12716–12736

Cheraghi M, Negahdari B, Daraee H et al (2017) Heart targeted nanoliposomal/nanoparticles drug delivery: an updated review. Biomed Pharmacother 86:316–323

Chiu D-C, Jeremy M, Baskin (2022) Maging and editing the phospholipidome. Acc Chem Res 55(21):3088–3098

Correia JH, Rodrigues JA, Pimenta S et al (2021) Photodynamic therapy review: principles, photosensitizers, applications, and future directions. Pharmaceutics 13(9):1332

Darvishi M, Tosan F, Nakhaei P et al (2023) Recent progress in cancer immunotherapy: overview of current status and challenges. Pathol Res Pract 241:154241

Das S, Ranjan OP, Rao V et al (2022) Multifunctional liposome-quantum dot hybrid nanocarriers for drug targeting to brain tumors. In: Nanocarriers for drug-targeting brain tumors. Elsevier, Amsterdam, pp 649–677

De LV, Milano F, Agostiano et al (2021) Recent advancements in polymer/liposome assembly for drug delivery: from surface modifications to hybrid vesicles. Polymers 13(7):1027

Debela DT, Muzazu SGY, Heraro KD et al (2021) New approaches and procedures for cancer treatment: current perspectives. SAGE Open Med 9:20503121211034366

Ding B, Shao S, Yu C et al (2018) Large-pore mesoporous-silica-coated upconversion nanoparticles as multifunctional immunoadjuvants with ultrahigh photosensitizer and antigen loading efficiency for improved cancer photodynamic immunotherapy. Adv Mater 30(52):1802479

Ding Y, Wang L, Li H et al (2022) Application of lipid nanovesicle drug delivery system in cancer immunotherapy. J Nanobiotechnol 20(1):214

Do C, DeAguero J, Brearley A et al (2020) Gadolinium-based contrast agent use, their safety, and practice evolution. Kidney 360(6):561–568

Dong D, Zhang Y, Xiao Y et al (2020) High performance aqueous supercapacitor based on nitrogen-doped coal-based activated carbon electrode materials. J Colloid Interface Sci 15(580):77–87

Dong L, Li Y, Wu H (2021) Platelet activating-factor acetylhydrolase II: a member of phospholipase A2 family that hydrolyzes oxidized phospholipids. Chem Phys Lipids 239:105103

Domenico, Lombardo Pietro, Calandra Davide, Barreca Salvatore, Magazù Mikhail, Kiselev (2016) Soft Interaction in Liposome Nanocarriers for Therapeutic Drug Delivery Nanomaterials 6(7) 125 https://doi.org/10.3390/nano6070125

Douglas W., Brown Ping, Wee Prakash, Bhandari Amirali, Bukhari Liliya, Grin Hector, Vega Maryam, Hejazi Deborah, Sosnowski Jailal, Ablack Eileen K., Clancy Desmond, Pink Jitendra, Kumar Maria Paola, Solis Ares Suellen, Lamb Rodrigo, Quevedo Bijal, Rawal Fahed, Elian Natasha, Rana Luis, Morales Natasha, Govindasamy Brendan, Todd Angela, Delmage Somnath, Gupta Nichole, McMullen Duncan, MacKenzie Perrin H., Beatty Henry, Garcia Manoj, Parmar Jennifer, Gyoba Chandra, McAllister Matthew, Scholz Roy, Duncan Arun, Raturi John D., Lewis (2024) Safe and effective in vivo delivery of DNA and RNA using proteolipid vehicles Cell 187(19) 5357–5375.e24 https://doi.org/10.1016/j.cell.2024.07.023

Dorca YE, Greciano ES, Valera J (2019) Hierarchy of asymmetry in chiral supramolecular polymers: toward functional, helical supramolecular structures. Chem Eur J 25:5848–5864

Dorel F (2019) Polymers and polymer nanocomposites for cancer therapy. Appl Sci 9(18):3899–3923

Du FX, Liu LC, Wu ZH et al (2021) Pd-single-atom coordinated biocatalysts for chem-/sono-/photo-trimodal tumor therapies. Adv Mater 33:2101095

Duan X, Li N, Chen X et al (2021) Characterization of tissue scaffolds using synchrotron radiation microcomputed tomography imaging. Tissue Eng Part C 27(11):573–588

Duan LL, Wang CY, Zhang W et al (2021) Interfacial assembly and applications of functional mesoporous materials. Chem Rev 121(23):14349–14429

Dymek M, Sikora E (2022) Liposomes as biocompatible and smart delivery systems–the current state. Adv Colloid Interf Sci 309:102757

Eilers A, Witt S, Walter J (2020) Aptamer-modified nanoparticles in medical applications. Adv Biochem Eng Biotechnol 174:161–193

Elibol K, Mangler C, O'Regan DD et al (2021) Single indium atoms and few-atom indium clusters anchored onto graphene via silicon heteroatoms. ACS Nano 15(9):14373–14383

Elkomy MH, Ali AA, Eid HM et al (2022) Chitosan on the surface of nanoparticles for enhanced drug delivery: a comprehensive review. J Control Release 351:923–940
Erickson S, Vizmeg J, Shuckra J et al (2023) Effect of active pharmaceutical ingredient, lipid composition, and thermal processing on the physical stability of local anesthetic-loaded lipid microparticles. J Pharm Innov 18(3):1466–1475
Escriche-Navarro B, Escudero A et al (2022) Mesoporous silica materials as an emerging tool for cancer immunotherapy. Adv Sci 9(26):2200756
Fabian M, Ana LL, Markus N et al (2020) From colloidal dispersions to aerogels: how to master nanoparticle gelation. Nano Today 30:100827
Fei H, Dong JC, Feng YX et al (2018) General synthesis and definitive structural identification of MN4C4 single-atom catalysts with tunable electrocatalytic activities. Nat Catal 1:63–72
Feng GX, Zhang GQ, Ding D et al (2020) Design of superior phototheranostic agents guided by Jablonski diagrams. Chem Soc Rev 49:8179–8234
Feng L, Wang YQ, Bi ZR et al (2023) Single-atom nanoenzyme-based autoluminescence system for cancer cell imaging and mitochondrial-targeted therapy. ACS Appl Bio Mater 6:5086–5096
Fulton MD, Najahi-Missaoui W (2023) Liposomes in cancer therapy: how did we start and where are we now. Int J Mol Sci 24(7):6615
Gao W, Liu Y, Zhang H et al (2020) Electrochemiluminescence biosensor for nucleolin imaging in a single tumor cell combined with synergetic therapy of tumor. ACS Sens 5(4):1216–1222
Gao S, Yang X, Xu J et al (2021) Nanotechnology for boosting cancer immunotherapy and remodeling tumor microenvironment: the horizons in cancer treatment. ACS Nano 15(8):12567–12603
Garg J, Pathania K, Sah SP et al (2022) Nanostructured lipid carriers: a promising drug carrier for targeting brain tumours. Future J Pharm Sci 8(1):25
Ge RL, Yan PN, Liu Y et al (2023) Recent advances and clinical potential of near infrared photothermal conversion materials for photothermal hepatocellular carcinoma therapy. Adv Funct Mater 33(29):2301138
Geng B, Hu J, He X et al (2024) Single atom catalysts remodel tumor microenvironment for augmented sonodynamic immunotherapy. Adv Mater 36:2313670
Ghazanfari MR, Jaafari MR, Shams SF et al (2017) Design and fabrication of multifunctional temperature-sensitive magnetoliposomal nanostructures. Mater Today Commun 13:102–111
Ghosh S, Carter KA, Lovell JF (2019) Liposomal formulations of photosensitizers. Biomaterials 218:119341
Gloag L, Somerville SV, Gooding JJ et al (2024) Co-catalytic metal–support interactions in single-atom electrocatalysts. Nat Rev Mater 9:173–189
Gong FL, Wang ZB, Mo R et al (2022) Nano-sponge-like liposomes remove cholesterol crystals for antiatherosclerosis. J Control Release 349:940–953
Gu XX, Shen JL, Xu ZW et al (2024) Rational design of biodegradable semiconducting polymernanoparticles for NIR-II fluorescence imaging-guided photodynamic therapy. J Nano Res 17:5399–5408
Guo Q, Jiang C (2020) Delivery strategies for macromolecular drugs in cancer therapy. Acta Pharm Sin B 10:976–986
Guo S, Ma C, Hu Y et al (2023) A notable impact of lipid matrices on cholesterol bioaccessibility from phytosterols-loaded nanostructured lipid carriers during in vitro intestinal digestion. Food Biosci 52:102500
Gupta P, Mazumder R, Padhi S (2020) Challenges of new generation liposomes–a review. IJPSN 13(2):4815–4825
Han WQ, Liu FY, Li YY et al (2023) Advances in natural polymer-based transdermal drug delivery systems for tumor therapy. Small 35(19):2301670
He WY, Wu JC, Liu JL et al (2024) Single-atom nanozymes for catalytic therapy: recent advances and challenges. Adv Funct Mater 34(16):2312116
Hersh AM, Alomari S, Tyler BM (2022) Crossing the blood-brain barrier: advances in nanoparticle technology for drug delivery in neuro-oncology. Int J Mol Sci 23(8):4153

Hong J, Yoon S, Choi Y et al (2023) Rational design of nanoliposomes by tuning their bilayer rigidity for the controlled release of oxygen. J Mol Liq 370:121003
Hosseinpour S, Walsh LJ, Xu C (2020) Biomedical application of mesoporous silica nanoparticles as delivery systems: a biological safety perspective. J Mater Chem B 8(43):9863–9876
Hu HG, Li YM (2020) Emerging adjuvants for cancer immunotherapy. Front Chem 8:601
Hu ZM, Liu SY, Yang HY et al (2021) Research progress of liposome drug delivery system in stomatology. Zhonghua Kou Qiang Yi Xue Za Zhi 56(3):294–300
Hu L, Ying Y, Wang A et al (2024) Tumor microenvironment stimuli activatable dendritic mesoporous cerium-based nanoplatform boosting synergistic antitumor efficacy. ACS Mater Lett 6:2846–2857
Huang T, Li S, Fang J et al (2021) Antibody-activated trans-endothelial delivery of mesoporous organosilica nanomedicine augments tumor extravasation and anti-cancer immunotherapy. Bioact Mater 6(7):2158–2172
Huo MF, Wang LY, Wang Y et al (2019) Nanocatalytic tumor therapy by single-atom catalysts. ACS Nano 13(2):2643–2653
Inglut CT, Sorrin AJ, Kuruppu T et al (2020) Immunological and toxicological considerations for the design of liposomes. Nano 10(2):190
Isla M, Hu Z, Wang QS et al (2019) Pyrene-based aggregation-induced emission luminogens and their applications. Mater Chem Front 3:762–781
Iyad N, Ahmad MS, Alkhatib SG et al (2023) Gadolinium contrast agents-challenges and opportunities of a multidisciplinary approach: literature review. Eur J Radiol Open 11:100503
Ji SF, Chen YY, Wang XL et al (2020) Chemical synthesis of single atomic site catalysts. Chem Rev 120(21):11900–11955
Jiang K, Tian K, Yu Y et al (2024) Kupffer cells determine intrahepatic traffic of PEGylated liposomal doxorubicin. Nat Commun 15(1):6136
Jiao L, Xu WQ, Wu Y et al (2020) Single-atom catalysts boost signal amplification for biosensing. Chem Soc Rev 50:750–765
Jiang-Hui, Wang Dominic J., Gessler Wei, Zhan Thomas L., Gallagher Guangping, Gao (2024) Adeno-associated virus as a delivery vector for gene therapy of human diseases Abstract Signal Transduction and Targeted Therapy 9(1) https://doi.org/10.1038/s41392-024-01780-w
Kabilova TO, Shmendel EV, Gladkikh DV et al (2018) Targeted delivery of nucleic acids into xenograft tumors mediated by novel folate-equipped liposomes. Eur J Pharm Biopharm 123:59–70
Katari O, Jain S (2021) Solid lipid nanoparticles and nanostructured lipid carrier-based nanotherapeutics for the treatment of psoriasis. Expert Opin Drug Deliv 18(12):1857–1872
Kelleci K, Allahverdiyev A, Bağirova M et al (2023) Particulate and non-particle adjuvants in Leishmaniasis vaccine designs: a review. J Vector Borne Dis 60(2):125–141
Knight FC et al (2019) Mucosal immunization with a pH-responsive nanoparticle vaccine induces protective CD8+ lung-resident memory T cells. ACS Nano 13:10939–10960
Koehler JK, Gedda L, Wurster L et al (2023) Tailoring the lamellarity of liposomes prepared by dual centrifugation. Pharmaceutics 15(2):706
Kong Y, Xu L, Cao J (2024) Preparation of a novel multifunctional cationic liposome drug-carrying system and its functional study on lung cancer. Anti-Cancer Agents Med, Chem
Koning GA, Storm G (2003) Targeted drug delivery systems for the intracellular delivery of macromolecular drugs. Drug Discov Today 11(8):482–483
Kowalska M, Broniatowski M, Mach M et al (2021) Effect of lipopolymer (DSPE-PEG750) on phospholipid monolayers and bilayers differing in the structure of the polar head group. J Mol Liq 344:117715
Kulkarni P, Rawtani D, Kumar M et al (2020) Cardiovascular drug delivery: a review on the recent advancements in nanocarrier based drug delivery with a brief emphasis on the novel use of magnetoliposomes and extracellular vesicles and ongoing clinical trial research. J Drug Deliv Sci Technol 60:102029
Kumar P, Kannimuthu K, Zeraati AS et al (2023) High-density cobalt single-atom catalysts for enhanced oxygen evolution reaction. J Am Chem Soc 145:8052–8063

Lalzawmliana V, Anand A, Roy M et al (2020) Mesoporous bioactive glasses for bone healing and biomolecules delivery. Mater Sci Eng C-Mater 106:110180

Lei ZH, Zhang F et al (2021) Molecular engineering of NIR-II fluorophores for improved biomedical detection. Angew Chem Int Ed 60:16294–16308

Liang XF, Shi BZ, Wang K et al (2016) Development of self-assembling peptide nanovesicle with bilayers for enhanced EGFRtargeted drug and gene delivery. Biomaterials 82:194–207

Li Q, Chen C, Li C et al (2020a) Bicontinuous mesoporous polymeric semiconductor photocatalyst. ACS Nano 14(10):13652–13662

Li YY, Ou CZ, Zhu JL et al (2020b) Ultrahigh and durable volumetric lithium/sodium storage enabled by a highly dense graphene-encapsulated nitrogen-doped carbon@Sn compact monolith. Nano Lett 20(3):2034–2046

Li J, Zhang W, Ji W et al (2021a) Near infrared photothermal conversion materials: mechanism, preparation, and photothermal cancer therapy applications. J Mater Chem B 9(38):7909–7926

Li Z, Xiong X, Peng S et al (2021b) Effect of pluronic block composition on the structure, stability, and cytotoxicity of liposomes. J Am Chem Soc 42(11):1651–1659

Li Z, Zhu Y, Zeng H et al (2022) Mechano-boosting nanomedicine antitumour efficacy by blocking the reticuloendothelial system with stiff nanogels. Nat Commun 14:1437

Li ZK, Chen SB, Zhu JT et al (2023) Hydrogen-bonded polymer nanomedicine with AIE characteristic for intelligent cancer therapy. ACS Macro Lett 12:1384–1388

Liang X, Fu NH, Yao SC et al (2022) The progress and outlook of metal single-atom-site catalysis. J Am Chem Soc 144(40):18155–18174

Lim T, Seo BH, Kim SJ et al (2024) Nitrogen-doped activated hollow carbon nanofibers with controlled hierarchical pore structures for high-performance, binder-free, flexible supercapacitor electrodes. ACS Omega 9(7):8247–8254

Lin SY, Pan M, Chiou Y et al (2024) Stability enhancement of proanthocyanidin-loaded liposomes via surface decoration with oxidized konjac glucomannan. Int J Biol Macromol 275:133230

Liu D, Ai HQ, Lou WT et al (2020) Substrate strain engineering: an efficient strategy to enhance the catalytic activity of SACs on waved graphene for e-NRR. Sustain Energy Fuels 4:3773–3779

Liu QL, Wang YF, Hu ZZ et al (2021a) Iron-based single-atom electrocatalysts: synthetic strategies and applications. RSC Adv 11:3079–3095

Liu Q, Zhang TX, Zheng YD et al (2021b) Calixarene-embedded nanoparticles for interference-free gene–drug combination cancer therapy. Small 17(8):2006223

Liu C, Yang M, Zhang D et al (2022a) Clinical cancer immunotherapy: current progress and prospects. Front Immunol 13:961805

Liu MQ, Yang M, Wan XZ et al (2022b) From nanoscopic to macroscopic materials by stimuli-responsive nanoparticle aggregation. Adv Mater 20(35):2208995

Liu J, Gong L, Chen HY et al (2023a) Fe-single-atom nanozyme catalysts for sensitive and selective detection of nitrite via colorimetry and test strips. ACS Appl Nano Mater 7:5879–5888

Liu Y, Wang B, Zhu J et al (2023b) Single-atom nanozyme with asymmetric electron distribution for tumor catalytic therapy by disrupting tumor redox and energy metabolism homeostasis. Adv Mater 35:2208512

Liu K, Wu C, Zhu M et al (2024a) Structural characteristics of alpha-fetoprotein, including N-glycosylation, metal ion and fatty acid binding sites. Commun Biol 7(1):505

Liu PJ, Xu LG, Li JL et al (2024b) Advanced science and technology of polymer matrix nanomaterials. Materials 17(2):461–465

Liu W, Chen Q, Wu J et al (2024c) Asymmetric coordination of iron single-atom nanozymes with efficient self-cascade catalysis for ferroptosis therapy. Adv Funct Mater 34:2312308

Liu Y, Yu S, Jiang X et al (2024d) Functional paclitaxel-manganese-doped mesoporous silica nanoparticles for orthotopic brain glioma targeted therapy. Mater Des 238:112715

Liu WL, Wei FQ, Ye AQ et al (2017) Kinetic stability and membrane structure of liposomes during in vitro infant intestinal digestion: effect of cholesterol and lactoferrin. Food Chem 230:6–13

Lombardo D, Calandra P, Barreca D et al (2016) Soft interaction in liposome nanocarriers for therapeutic drug delivery. Nano 6(7):125

Lotfabadi NN, Kouchesfehani HM, Sheikhha MH et al (2018) Development of a novel cationic liposome: evaluation of liposome mediated transfection and anti-proliferative effects of miR-101 in acute myeloid leukemia. J Drug Deliv Sci Technol 45:196–202
Lu X, Gao S, Lin H et al (2021) Single-atom catalysts for nanocatalytic tumor therapy. Small 17:2004467
Luo FX, Yu Y, Li MQ et al (2020) Polymeric nanomedicines for the treatment of hepatic diseases. J Nanobiotechnol 20:488–516
Lv Z, He S, Wang Y et al (2021) Noble metal nanomaterials for NIR-triggered photothermal therapy in cancer. Adv Healthc Mater 10(6):2001806
Marín D, Alemán A, Sánchez-Faure A et al (2018) Freeze-dried phosphatidylcholine liposomes encapsulating various antioxidant extracts from natural waste as functional ingredients in surimi gels. Food Chem 245:525–535
Maritim S, Boulas P, Lin Y (2021) Comprehensive analysis of liposome formulation parameters and their influence on encapsulation, stability and drug release in glibenclamide liposomes. Int J Nanomedicine 592:120051
Marlon O, Estefanía M, Tonny N et al (2020) Recent advances in polymer nanomaterials for drug delivery of adjuvants in colorectal cancer treatment: a scientific-technological analysis and review. Molecules 25(10):2270–2305
Mayer LD, Tardi P, Louie AC (2019) CPX-351: a nanoscale liposomal co-formulation of daunorubicin and cytarabine with unique biodistribution and tumor cell uptake properties. Int J Nanomedicine 14:3819–3830
Meng Y, Gao J, Zhou P et al (2024) NIR-II conjugated electrolytes as biomimetics of lipid bilayers for in vivo liposome tracking. Angew Chem Int Ed 63(13):e202318632
Miina, Ojansivu Hanna M. G., Barriga Margaret N., Holme Stefanie, Morf James J., Doutch Samir EL, Andaloussi Tomas, Kjellman Markus, Johnsson Justas, Barauskas Molly M., Stevens (2025) Formulation and characterization of novel Ionizable and cationic lipid nanoparticles for the delivery of splice-switching oligonucleotides abstract Adv Mater 37(17) 10.1002/adma.v37.17 https://doi.org/10.1002/adma.202419538
Michael JM, Margaret MB, Rebecca MH et al (2020) Engineering precision nanoparticles for drug delivery. Nat Rev Drug Discov 20:101–124
Mikšík F, Miyazaki T, Inada M (2019) Detailed investigation on properties of novel commercial mesoporous silica materials. Microporous Mesoporous Mater 289:109644
Najafi F, Salami-Kalajahi M, Roghani-Mamaqani H (2020) A review on synthesis and applications of dendrimers. J Iran Chem Soc 18:503–517
Nakashima S, Tatewaki Y, Okada S et al (2016) Polymerization of a divalent metal salt of an unsaturated carboxylic acid stimulated by ion exchange with a metal salt of fatty acids. Polym J 48(8):855–858
Nawaz S, Khan Y, Abdelmohsen SAM et al (2022) Polyaniline inside the pores of high surface area mesoporous silicon as composite electrode material for supercapacitors. RSC Adv 12(27):17228–17236
Niculescu AG, Grumezescu AM (2021) Photodynamic therapy—an up-to-date review. Appl Sci 11(8):3626
Nyayapathi N, Zheng E, Zhou Q et al (2024) Dual-modal photoacoustic and ultrasound imaging: from preclinical to clinical applications. Front Photonics 5:1359784
Onguka O, Babin BM, Lakemeyer M et al (2021) Toxoplasma gondii serine hydrolases regulate parasite lipid mobilization during growth and replication within the host. Cell Chem Biol 28(10):1501–1513
Ooi YJ, Wen Y, Zhu J et al (2024) Codelivery of doxorubicin and p53 gene by β-cyclodextrin-based supramolecular nanoparticles formed via host–guest complexation and electrostatic interaction. Biomacromolecules 25(5):2980–2989
Palanikumar L, Kalmouni M, Houhou T et al (2023) pH-responsive upconversion mesoporous silica nanospheres for combined multimodal diagnostic imaging and targeted photodynamic and photothermal cancer therapy. ACS Nano 17(19):18979–18999

Park J, Smit S, Wood CD et al (2020) Core hyper-cross-linked star polymers from block polymer micelle precursors. Polym Chem 11:7178–7184

Paskeh MDA, Saebfar H, Mahabady MK et al (2022) Overcoming doxorubicin resistance in cancer: siRNA-loaded nanoarchitectures for cancer gene therapy. Life Sci 298:120463

Pilkington CP, Gispert I, Chui SY et al (2024) Engineering a nanoscale liposome-in-liposome for in situ biochemical synthesis and multi-stage release. Nat Chem 16(10):1612–1620

Plavniece A, Dobele G, Djachkovs D et al (2023) "Sweetwoods" lignin as promising raw material to obtain micro-mesoporous carbon materials. Materials 16(17):6024

Popova AV, Hincha DK (2016) Effects of flavonol glycosides on liposome stability during freezing and drying. BBA-Biomembranes 1858(12):3050–3060

Prieto-Montero R, Arbeloa T, Martínez-Martínez V (2023) Photosensitizer-mesoporous silica nanoparticles combination for enhanced photodynamic therapy. Photochem Photobiol 99(3):882–900

Qi K, Sun B, Liu S et al (2023) Research progress on carbon materials in tumor photothermal therapy. Biomed Pharmacother 165:115070

Qian, Liu Chenglu, Zhao Kai, Sun Yinlu, Deng Zhenghe, Li (2023) Engineered biocontainable RNA virus vectors for nontransgenic genome editing across crop species and genotypes Mol Plant 16(3) 616–631 https://doi.org/10.1016/j.molp.2023.02.003

Qiu L, Li K, Dong WY et al (2021) Tumor microenvironment responsive "head-to-foot" self-assembly Nanoplatform for positron emission tomography imaging in living subjects. ACS Nano 15(11):18250–18259

Rahman MM, Ara MG, Alim MA et al (2021) Mesoporous carbon: a versatile material for scientific applications. Int J Mol Sci 22(9):4498

Rahman MM, Wang J, Wang G et al (2024) Chimeric nanobody-decorated liposomes by self-assembly. Nat Nanotechnol 19:818–824

Rajendra KS, Kapil DP, Kam WL et al (2017) Progress in Nanotheranostics based on mesoporous silica nanomaterial platform. ACS Appl Mater Interfaces 9:10309–10337

Rajput S, Sharma PK, Malviya R (2023) Fluid mechanics in circulating tumour cells: role in metastasis and treatment strategies. Med Drug Discov 18:100158

Ramesh G, Pavan W, Veluru RN et al (2020) Acetal containing polymers as pH-responsive nano-drug delivery systems. J Control Release 328:736–761

Rebecca K, Jan BH, Aldrik HV et al (2021) Dendroids, discrete covalently cross-linked dendrimer superstructures. ACS Nano 15(1):1666–1674

Ren J, Cao YM, Li L et al (2021) Self-assembled polymeric micelle as a novel mRNA delivery carrier. J Control Release 338:537–547

Richter RF, Vater C et al (2023) Treatment of critical bone defects using calcium phosphate cement and mesoporous bioactive glass providing spatiotemporal drug delivery. Bioact Mater 28:402–419

Ryuichi H, Yoshimi H, Michinori E et al (2021) 18F-SMBT-1: a selective and reversible PET tracer for monoamine oxidase- B imaging. J Nucl Med 62(2):253–258

Seleci M, Ag Seleci D, Scheper T et al (2017) Theranostic liposome–nanoparticle hybrids for drug delivery and bioimaging. Int J Mol Sci 18(7):1415

Seynhaeve ALB, Amin M, Haemmerich D et al (2020) Hyperthermia and smart drug delivery systems for solid tumor therapy. Adv Drug Deliv Rev 163:125–144

Shang SL, Li XM, Wang HR et al (2024) Targeted therapy of kidney disease with nanoparticle drug delivery materials. Bioact Mater 37:206–221

Sharma A, Jasrotia S, Kumar A (2024) Effects of chemotherapy on the immune system: implications for cancer treatment and patient outcomes. Naunyn Schmiedeberg's Arch Pharmacol 397(5):2551–2566

Shah VM, Nguyen DX, Alfatease A et al (2017) Characterization of pegylated and non-pegylated liposomal formulation for the delivery of hypoxia activated vinblastine-N-oxide for the treatment of solid tumors. J Control Release 253:37–45

Sheffey VV, Siew EB, Tanner EEL et al (2022) PLGA's plight and the role of stealth surface modification strategies in its use for intravenous particulate drug deliver. Adv Healthc Mater 8(11):2192–2640

Shen J, Chen J, Qian YP et al (2024) Atomic engineering of single-atom nanozymes for biomedical applications. Adv Mater 36(21):2313406

Shi L, Zhang J, Zhao M et al (2021) Effects of polyethylene glycol on the surface of nanoparticles for targeted drug delivery. Nanoscale 13(24):10748–10764

Shi Z, Liu J, Tian L et al (2022) Insights into stimuli-responsive diselenide bonds utilized in drug delivery systems for cancer therapy. Biomed Pharmacother 155:113707

Shou KQ, Qu CR, Sun Y et al (2017) Multifunctional biomedical imaging in physiological and pathological conditions using a NIR-II probe. Adv Funct Mater 27:1700995

Silva F, Agner T, Sayer C, Lona LMF (2022) Curcumin encapsulation in functional PLGA nanoparticles: A promising strategy for cancer therapies. Advances in Colloid and Interface Science 300:102582. https://doi.org/10.1016/j.cis.2021.102582

Simonson A, Aronson M, Medina S et al (2020) Supramolecular peptide assemblies as antimicrobial scaffolds. Molecules 25(12):2751–2784

Song C, Ouyang ZJ, Liu JJ et al (2024a) Core-shell tecto dendrimers co-deliver drug/gene for immunomodulation of macrophages and dendritic cells and checkpoint blockade-promoted breast cancer chemoimmunotherapy. Nano Today 58:102437

Song YC, Mao CQ, Zhang WJ et al (2024b) Catalytic hairpin assembly-based AIEgen/graphene oxide nanocomposite for fluorescence-enhanced and high-precision spatiotemporal imaging of microRNA in living cells. Biosens Bioelectron 259:116416

Spitzenberg C, Bruckschlegel C, Holzhausen F et al (2025) Encapsulants affect liposome surface interactions with biological systems. Small 2505312

Strasser A, Vaux DL (2020) Cell death in the origin and treatment of cancer. Mol Cell 78(6):1045–1054

Su T, Cheng FR, Pu YJ et al (2021) Polymeric micelles amplify tumor oxidative stresses through combining PDT and glutathione depletion for synergistic cancer chemotherapy. Chem Eng J 411:128561

Sun ZH, Deng GJ, Peng XH et al (2021) Intelligent photothermal dendritic cells restart the cancer immunity cycle through enhanced immunogenic cell death. Biomaterials 279:121228

Tan MH, Smith AJ, Pawlyk B et al (2009) Gene therapy for retinitis pigmentosa and Leber congenital amaurosis caused by defects in AIPL1: effective rescue of mouse models of partial and complete Aipl1 deficiency using AAV2/2 and AAV2/8 vectors. Hum Mol Genet 18(12):2099–2114

Tang L, Mei Y, Shen Y et al (2021) Nanoparticle-mediated targeted drug delivery to remodel tumor microenvironment for cancer therapy. Int J Nanomedicine 2021:5811–5829

Tang Y, Liu YW, Xia YD et al (2023) Simultaneous detection of ovarian cancer-concerned HE4 and CA125 markers based on cu single-atom-triggered CdS QDs and Eu MOF@Isoluminol ECL. Anal Chem 95(10):4795–4802

Thi TTH, Suys EJ, Lee JS et al (2021) Lipid-based nanoparticles in the clinic and clinical trials: from cancer nanomedicine to COVID-19 vaccines. Vaccine 9(4):359

Tiffany H, Keegan G, Ding LL et al (2023) Novel strategy to drive the intracellular uptake of lipid nanoparticles for photodynamic therapy. Angew Chem Int Ed 16(62):e202218218

Tran S, DeGiovanni PJ, Piel B et al (2017) Cancer nanomedicine: a review of recent success in drug delivery. Clin Transl Med 6:1–21

Trucillo P, Campardelli R, Reverchon E (2020) Liposomes: from bangham to supercritical fluids. PRO 8(9):1022

Vallet-Regí M, Schüth F, Lozano D et al (2022) Engineering mesoporous silica nanoparticles for drug delivery: where are we after two decades? Chem Soc Rev 51(13):5365–5451

Van Eerden RA, Atrafi F, vanHylckama VMA et al (2019) Comparison of intratumoral docetaxel exposure in cancer patients between nanoparticle entrapped docetaxel (CPC634) and conventional docetaxel (cd): the CriTax study. Ann Oncol 30:184

Verma A, Gupta SK (2023) History and techniques of bioimaging. In: Magnetic quantum dots for bioimaging. CRC Press, Boca Raton, pp 111–132
Vermeulen I, Isin EM, Barton P et al (2022) Multimodal molecular imaging in drug discovery and development. Drug Discov Today 27(8):2086–2099
Vinicius DOA, Victor JRRP, Maria DFV et al (2021) Ultra-high molecular weight polyethylene nanocomposites reinforced with novel surface chemically modified sonic-exfoliated graphene. J Mater Res Technol 11:1932–1941
Wang G, Huang W, Xia Y et al (2019a) Cholesterol-lowering potentials of lactobacillus strain overexpression of bile salt hydrolase on high cholesterol diet-induced hypercholesterolemic mice. Food Funct 10(3):1684–1695
Wang Y, Guo H, Luo X et al (2019b) Nonsiliceous mesoporous materials: design and applications in energy conversion and storage. Small 15(32):e1805277
Wang LQ, Ma J, Hong W et al (2020) Nanoscale diffusion of polymer-grafted nanoparticles in entangled polymer melts. Macromolecules 53(19):8393–8399
Wang S, Hu ZF, Wei QL et al (2022) Diatomic active sites nanozymes: enhanced peroxidase-like activity for dopamine and intracellular H_2O_2 detection. Nano Res 15:4266–4273
Wang YP, Attam A, Fan HG et al (2023) Engineering of siloxanes for stabilizing silicon anode materials. Small 19(50):2303804
Wang G, Lu H, Pan Y et al (2024a) Ultrasound-sensitive targeted liposomes as a gene delivery system for the synergistic treatment of hepatocellular carcinoma. Small 20:2406182
Wang Q, Mohamed SO, Dongpyo K et al (2024b) Advanced strategies for combinational immunotherapy of cancer based on polymeric. Nanomedicine 2:398–414
Wang Z, Li W, Jiang Y et al (2024c) Cholesterol-modified sphingomyelin chimeric lipid bilayer for improved therapeutic delivery. Nat Commun 15(1):2073
Woźniak M, Płoska A, Siekierzycka A et al (2022) Molecular imaging and nanotechnology—emerging tools in diagnostics and therapy. Int J Mol Sci 23(5):2658
Wu J, Williams GR, Niu S et al (2019) A multifunctional biodegradable nanocomposite for cancer theranostics. Adv Sci 6(14):1802001
Wu HA, Peng B, Mohammed FS et al (2022a) Brain targeting, antioxidant polymeric nanoparticles for stroke drug delivery and therapy. Small 22(18):2107126
Wu J, Wei Y, Lan J et al (2022b) Screening of protein-based ultrasmall nanozymes for building cell-mimicking catalytic vesicles. Small 18:2202145
Wu L, Lin H, Cao X et al (2024) Bioorthogonal cu single-atom nanozyme for synergistic nanocatalytic therapy, photothermal therapy, cuproptosis and immunotherapy. Angew Chem Int Ed 63:e202405937
Xia J, Wang B, Di J et al (2022) Construction of single-atom catalysts for electro-, photo- and photoelectro-catalytic applications: state-of-the-art, opportunities, and challenges. Mater Today 53:217–237
Xie PW, Liu P (2020) pH-responsive surface charge reversal carboxymethyl chitosan-based drug delivery system for pH and reduction dual-responsive triggered DOX release. Carbohyd Polym 236:11609
Xie W, Ren Y, Jiang F et al (2023) Solvent-pair surfactants enabled assembly of clusters and copolymers towards programmed mesoporous metal oxides. Nat Commun 4(1):8493
Xu G, Song S, Nawaz MAH et al (2021a) Fluorescent mesoporous silica nanomaterials for targeted tumor chemotherapy. Mater Lett 304:130520
Xu WQ, Song WY, Kang YK et al (2021b) Axial ligand-engineered single-atom catalysts with boosted enzyme-like activity for sensitive immunoassay. Anal Chem 93:12758–12766
Xu Q, Yang Y, Lu J et al (2022a) Recent trends of mesoporous silica-based nanoplatforms for nanodynamic therapies. Coord Chem Rev 469:214687
Xu W, Qing X, Liu S et al (2022b) Hollow mesoporous manganese oxides: application in cancer diagnosis and therapy. Small 18(15):2106511

Xu J, Xu J, Shi T et al (2023a) Probiotic-inspired nanomedicine restores intestinal homeostasis in colitis by regulating redox balance, immune responses, and the gut microbiome. Adv Mater 35(3):e2207890

Xu WQ, Wu Y, Wang XS et al (2023b) Bioinspired single-atom sites enable efficient oxygen activation for switching anodic/cathodic electrochemiluminescence. Angew Chem Int Ed 62(29):e202304625

Yang JC, Shang Y, Li YH et al (2018) An "all-in-one" antitumor and anti-recurrence/metastasis nanomedicine with multi-drug co-loading and burst drug release for multi-modality therapy. Chem Sci 9(36):7210–7217

Yang J, Li W, Wang D, Li Y (2020) Electronic metal–support interaction of single-atom catalysts and applications in electrocatalysis. Adv Mater 32:2003300

Yang JP, Wang H, Yin ZS et al (2022) Emulsion interfacial polymerization of anticancer peptides: fabricating polypeptide nanospheres with high drug-loading efficiency and enhanced anticancer activity. Sci China Chem 65:2252–2259

Yang J, Dai D, Zhang X et al (2023a) Multifunctional metal-organic framework (MOF)-based nanoplatforms for cancer therapy: from single to combination therapy. Theranostics 13(1):295

Yang YT, Li M, Pan GY et al (2023b) Multiple stimuli-responsive nanozyme-based cryogels with controlled no release as self-adaptive wound dressing for infected wound healing. Adv Funct Mater 31(33):2214089

Ye FY, Hu M, Zheng YS (2023) Advances and challenges of metal ions sensors based on AIE effect. Coord Chem Rev 493:215328

Yi XZ, Gao SX, Gao X et al (2023) Glycolipids improve the stability of liposomes: the perspective of bilayer membrane structure. Food Chem 412:135517

Yin LX, Fu ZR, Wang MM et al (2025) A prodrug nanodevice co-delivering docetaxel and ROR1 siRNA for enhanced triple negative breast cancer therapy. Acta Biomater 193:498–513

Younis MR, He G, Qu J et al (2021) Inorganic nanomaterials with intrinsic singlet oxygen generation for photodynamic therapy. Adv Sci 8(21):2102587

Yu S, Guo Z, Zhou Y et al (2023) Research progress of MOFs/carbon nanocomposites on promoting ORR in microbial fuel cell cathodes. Environ Sci Pollut Res Int 41:93422–93434

Yuan Q, Shah J, Hein S et al (2010) Controlled and extended drug release behavior of chitosan-based nanoparticle carrier. Acta Biomater 6(3):1140–1148

Zaiki Y, Iskandar A, Wong TW et al (2023) Functionalized chitosan for cancer nano drug delivery. Biotechnol Adv 67:108200

Zhang Y, Zhang K, Wu Z et al (2015) Evaluation of transdermal salidroside delivery using niosomes via in vitro cellular uptake. Int J Pharm 478(1):138–146

Zhang FW, Ni QQ, Jacobson O et al (2018a) Polymeric nanoparticles with a glutathione-sensitive heterodimeric multifunctional prodrug for in vivo drug monitoring and synergistic cancer therapy. Angew Chem Int Ed 24(57):7066–7070

Zhang X, Wang B, Xia Y et al (2018b) In vivo and in situ activated aggregation-induced emission probes for sensitive tumor imaging using tetraphenylethene-functionalized trimethincyanines-encapsulated liposomes. ACS Appl Mater 10(30):25146–25153

Zhang L et al (2020) Microfluidic-assisted polymer-protein assembly to fabricate homogeneous functional nanoparticles. Mater Sci Eng C 111:110768

Zhang BK, Li J, Hu JM et al (2021a) Theory of polymer diffusion in polymer–nanoparticle mixtures: effect of nanoparticle concentration and polymer length. Soft Matter 17:4632–4642

Zhang X, Shao XT, Cai ZZ et al (2021b) The fabrication of phospholipid vesicle-based artificial cells and their functions. New J Chem 45:3364–3376

Zhang R, Zang P, Yang D et al (2023a) A phase engineering strategy of perovskite-type $ZnSnO_3$: Nd for boosting the sonodynamic therapy performance. Adv Funct Mater 33(25):2300522

Zhang Z, Feng Z, Zhao X et al (2023b) Functionalization and higher-order organization of liposomes with DNA nanostructures. Nat Commun 14:5256

Zhang H, Li Z, Yang X et al (2024) Modulator directed synthesis of size-tunable mesoporous MOFs and their derived nanocarbon-based electrocatalysts for oxygen reduction. Chem Eng J 486:150088

Zhao E, Chen S (2021) Materials with aggregation-induced emission characteristics for applications in diagnosis, theragnosis, disease mechanism study and personalized medicine. Mater Chem Front 5(8):3322–3343

Zheng K, Zhu X, Guo S et al (2023) Gamma-ray-responsive drug delivery systems for radiation protection. Chem Eng J 463:142522

Zheng YT, Yavuz O, Gu YM et al (2024) Rational design of polymeric micelles for targeted therapeutic delivery. Nano Today 55:102147

Zhong Q, Zhang L (2019) Nanoparticles fabricated from bulk solid lipids: preparation, properties, and potential food applications. Adv Colloid Interf Sci 273:102033

Zhou S, Zhong Q, Wang Y et al (2022) Chemically engineered mesoporous silica nanoparticles-based intelligent delivery systems for theranostic applications in multiple cancerous/non-cancerous diseases. Coord Chem Rev 452:214309

Zhu S, Zhang T, Zheng L et al (2021) Combination strategies to maximize the benefits of cancer immunotherapy. J Hematol Oncol 14(1):156

Zong W, Shao X, Li J et al (2023) Synthetic intracellular environments: from basic science to applications. Anal Chem 95(1):535–549

Zou Y, Huang B, Cao L et al (2021) Tailored mesoporous inorganic biomaterials: assembly, functionalization, and drug delivery engineering. Adv Mater 33(2):2005215

Chapter 5
Toxicological Analysis of Nanomaterials for Tumor Diagnosis and Treatment

Yun Luo

Abstract To date, various cancer nano-biomaterials have demonstrated success in preclinical studies, improving therapeutic outcomes, prolonging survival, and/or decreasing side effects. However, the translation from bench to bedside remains challenging. While a number of nano-biomaterials have entered clinical trials, only a few have been approved for clinical applications. Before the application of every nano-biomaterial, it is necessary to evaluate its biological toxicity and enable its safe use in patients. Further clinical trials are carried out under the premise of ensuring that the material exhibits low toxicity in vivo and in vitro. Here, we mainly focused on six types of nanomaterials (hollow mesoporous manganese oxides nanomaterials, titanium-based nanomaterials, graphene-based nanomaterials, two-dimensional theranostic nanomaterials, liposomal nanocarriers, metal-organic framework) and introduced their toxicity. Ultimately, we discussed the present challenges and future perspectives for the next stage of toxic research in nanomedicine.

Keywords Toxicity · Nanomaterials · Hollow mesoporous manganese oxides nanomaterials · Titanium-based nanomaterials · Graphene-based nanomaterials · Two-dimensional theranostic nanomaterials · Liposomal nanocarriers · Metal-organic framework

Y. Luo (✉)
Shandong Provincial Key Laboratory of Tumor Imaging Equipment Development and Tumor Diagnosis and Treatment Integration Technology, School of Medicine, Linyi University, Linyi, Shandong, People's Republic of China
e-mail: luoyun@lyu.edu.cn

© The Author(s), under exclusive license to Springer Nature Singapore Pte Ltd. 2026

P. Shi et al. (eds.), *Functional Organic Materials for Tumor Diagnostics and Therapy*, https://doi.org/10.1007/978-981-96-9478-5_5

5.1 Toxicity of Hollow Mesoporous Manganese Oxides Nanomaterials (HM-MONs)

Hollow mesoporous manganese oxides nanomaterials (HM-MONs), mainly including manganese oxide (MnO), manganese dioxide (MnO_2), manganese sesquioxide (Mn_2O_3), and manganese tetroxide (Mn_3O_4), play a great role in applications of tumor diagnosis and therapy with their unique properties such as tumor microenvironment stimuli-responsiveness, prominent catalytic activity, excellent biodegradation, and outstanding magnetic resonance imaging ability.

The characteristics of tumor microenvironment (TME) are hypoxia, mild acidity, excess hydrogen peroxide (H_2O_2), and high glutathione (GSH) concentration. MONs are ideal TME responsive nanomaterials with the ability to consume GSH and catalyze the generation of oxygen from hydrogen ions (H^+)/H_2O_2, further reshaping TME. The generated oxygen can improve the efficiency of tumor treatment, including photodynamic therapy (PDT), sonodynamic therapy (SDT), radiation therapy (RT), starvation therapy, chemodynamic therapy (CDT), and immunotherapy. At the same time, clearing GSH effectively weakens the clearance of reactive oxygen species (ROS), further enhancing PDT, SDT, and CDT. In addition, the divalent manganese ions (Mn^{2+}) generated by the reaction of MONs with H^+/H_2O_2 exhibit significant potential in magnetic resonance imaging (MRI) and CDT. Due to its TME responsiveness and degradability, MONs as a nanocarrier have superior abilities to achieve targeted and controlled release of chemotherapy drugs, genes, and drugs at tumor sites. Meanwhile, MONs have been widely used as contrast agents in fields such as photoacoustic imaging (PAI), photothermal imaging (PTI), ultrasound imaging (USI), photothermal therapy (PTT), fluorescence imaging, etc. (Xu et al. 2022).

The biodegradability and nontoxicity of MONs have been confirmed, further indicating that MONs are potential biomaterials with good biocompatibility. HM-MONs exhibited high cell viability under relatively high concentrations in vitro experiments, indicating their low cytotoxicity and good biocompatibility (Table 5.1) (Ding et al. 2020). On the contrary, Greene et al. believe that hollow MnO_2 exhibits certain toxicity to both healthy cells and cancer cells (Greene et al. 2020). And this toxicity may originate from the accumulation of hollow MnO_2, which is more toxic to cancer cells than healthy cells, regardless of whether MnO_2 is modified by dopamine. In addition, in the early stages of treatment, the degradation of MnO_2 by antioxidants has a positive impact on healthy cells and a negative impact on cancer cells. Hollow MnO_2 nanoparticles have a fast dissolution rate and good biodegradability. Therefore, they still insist that hollow MnO_2 is a potential preferred candidate for targeted drug delivery in cancer therapy.

In in vivo experiments, the manganese ions produced by the decomposition of HM-MONs can be rapidly cleared by the kidneys. The hematoxylin and eosin (H&E) staining images of major organs (heart, liver, spleen, lungs, and kidneys) further demonstrate their low tissue toxicity and high biocompatibility. In addition, in Zhu and colleagues' study, HM-MONs were intravenously injected into healthy

Table 5.1 The synthesis and applications of MON-based composite nanomaterials

Materials	Surface modification	Cell lines	Cell viability	Method	References
HM-MnO_2	@m-PEG	HeLa	>88% (24 h, 100×10^{-6} m)	MTT	Zhuang et al. (2020)
HM-MnO_2	@PLTM	L929	>80% (24 h, 40 μg mL^{-1})	MTT	Wang et al. (2020a)
HM-MnO_2	@ PDA-NH_2-PEG-FA	MCF-7, HeLa, NP69	>85% (24 h, 200 μg mL^{-1})	MTT	Zeng et al. (2020)
HM-Mn_2O_3	@mRBC-TKD	MCF-7, RAW264.7	>90% (24 h, 20 μg mL^{-1})	SRB	Zhang et al. (2021a)
HM-MnO_2	@PAH/ PAH-PEG-NH_2	OSCC	>90% (24 h, 200 μg mL^{-1})	MTT	Zhou et al. (2021a)
HM-MnO	@ZDS	SMCC-7721	>90% (24 h, 200 μg mL^{-1})	MTT	Wei et al. (2019)

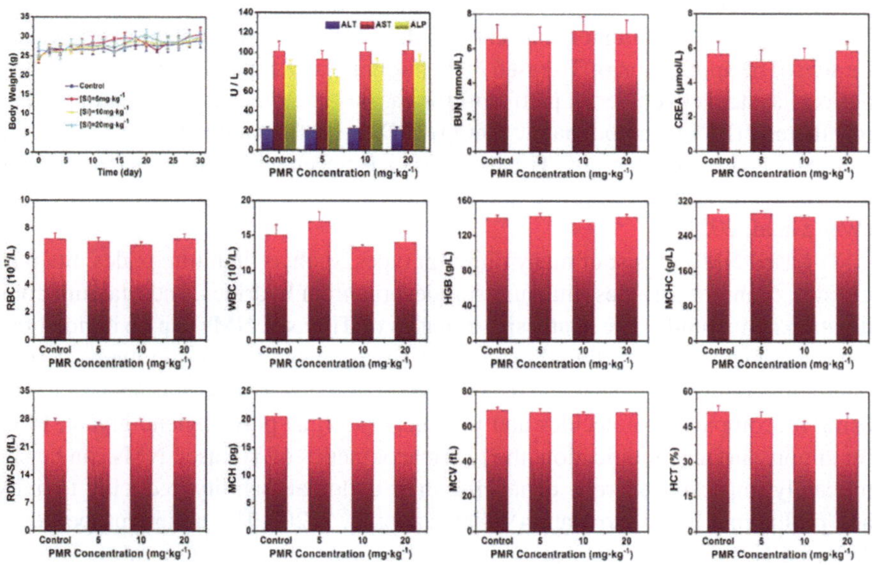

Fig. 5.1 In vivo body-weight curve and hematological index. The changes of body weight of Kunming mice with time during 1-month feeding and the hematological assays of mice 30 days posttreatment with different doses of PMR (5, 10, and 20 mg.kg^{-1}). (Reproduced from Zhu et al. 2018). Reprinted with permission from Zhu et al. (2018). Copyright 2018 American Chemical Society

female Kunming mice (Fig. 5.1). All blood indicators, including alanine aminotransferase (ALT), aspartate aminotransferase (Zanoni et al. 2015), alkaline phosphatase (ALP), blood urea nitrogen (BUN), creatinine (CREA), red blood cell count (RBC), white blood cell count (WBC), hemoglobin (HGB), mean corpuscular hemoglobin concentration (MCHC), red blood cell distribution width (RDW-SD), mean corpuscular hemoglobin (MCH), mean corpuscular volume (MCV), and hematocrit level (HCT), showed no significant differences compared to the control group (Zhu et al. 2018). The lack of significant weight changes further demonstrates the superior biocompatibility of HM MONs. Although HM MONs have low toxicity and good biocompatibility, careful evaluation of their biological toxicity is still necessary before clinical application.

5.2 Toxicity of Titanium-Based NMs

In recent years, titanium-based nanomaterials (Ti-based NMs) have been used as nontoxic transition metals to construct various nanomaterials. Ti-based NMs have unparalleled structure, composition, physicochemical characteristics, and various aluminum-inducing properties. These excellent and unique properties endow Ti-based NMs with extensive potential for biomedical applications. Ti-based NMs mainly include titanium oxide, titanium carbide, titanium nitride, titanium sulfide, titanium hydride, titanium organic coordination compounds, and their nanocomposite sites. The application of Ti-based NMs has gradually expanded from simple drug delivery systems to universal platforms for various treatments, including photothermal therapy (PTT), photodynamic therapy (PDT), sonodynamic therapy (SDT), etc. Compared with monotherapy, Ti-based NMs have been widely used in combination therapy to significantly improve tumor treatment efficiency and reduce side effects (Wang et al. 2021).

Ti-based NMs consist of many different types, such as titanium oxides, titanium nitrides, titanium carbides, titanium sulfides, titanium hydrides, and titanium coordination compounds. The synthesis strategies of Ti-based NMs can be divided into top-down methods (e.g., liquid exfoliation strategy) and bottom-up methods (e.g., high-temperature organic method, and hydrothermal/solvothermal approach). By using biocompatible molecules and polymers (such as polyethylene glycol, PEG) for simple surface modification, the pharmacokinetics of Ti-based NMs can be significantly improved, thereby enhancing their colloidal stability, reducing toxicity, and prolonging blood circulation (Shahbazi et al. 2020). The unique properties of Ti-based NMs make them widely used in tumor imaging and treatment. Ti-based NMs are mainly divided into the following aspects based on their properties. (1) Firstly, the large specific surface area of Ti-based NMs can serve as drug-loading platforms to load different types of chemotherapy drugs (such as paclitaxel PTX and doxorubicin DOX), achieving enhanced chemotherapy and effectively reducing drug side effects. (2) Secondly, Ti-based NMs with good absorption in the near-infrared region (including near-infrared I and near-infrared II windows) can effectively convert light energy into thermal energy, making them great potential for application in photoacoustic (PA) imaging and photothermal therapy (PTT). (3) Thirdly, the excellent catalytic performance of Ti-based NMs (mainly titanium oxides such as TiO_2) make them widely used for photocatalytic nanomaterials and photosensitizers. Under light irradiation, it can effectively convert oxygen (O_2) into toxic ROS (such as $^1O_2$) to damage cancer cells, making it widely used in photodynamic therapy (PDT) (Lan et al. 2019). (4) Fourthly, Ti-based NMs (mainly titanium oxide) can also generate a large amount of ROS under US irradiation. The generation of ROS can not only destroy superficial tumors but also have good killing efficiency against deep tumors. In recent years, they have received widespread attention and recognition in tumor sonodynamic therapy (SDT) (Cao et al. 2019).

After systemic administration, nanomaterials will lead to nonspecific accumulation in healthy organs and tissues and thus induce toxicity, which will greatly limit the preclinical translation of nanomaterials (Peng et al. 2019). The difference between Ti-based nanomaterials and other materials in terms of biosafety is that they use highly safe Ti element and their compounds are composed of Ti and other nonmetallic elements such as C, N, O, S, P, H, etc., which have good safety. However, even so, they may have adverse effects on local tissues and organs, and even cause systemic reactions, leading to long-term toxicity issues. But the related mechanisms of reaction are currently unclear. Therefore, the effectiveness, pharmacokinetic characteristics, bioavailability, biocompatibility, metabolic pathways, and potential long-term adverse reactions of titanium-based NMs should be comprehensively evaluated in animal models before conducting preclinical trials in humans (Yan et al. 2019). For example, to evaluate the in vivo toxicity of H-PEG-TiO_2 NPs, researchers administered different doses (1, 5, 25 mg/kg) intravenously to mice and conducted blood routine and biochemical tests 1 month later. These indicators showed no significant differences compared to the control group. In addition, H-PEG-TiO_2 NPs have no significant acute toxicity or inflammatory response to the major organs of mice (heart, liver, spleen, kidney, lungs) and have significant in vivo biocompatibility. Similarly, TiN NPs, Ti_3C_2 NPs, tis2nps, TiP NPs, etc., have also been shown to have excellent biosafety (Yan et al. 2019). The synthesis of ultrasmall or biodegradable nanomaterials has received widespread attention because they can be rapidly metabolized from the body after achieving therapeutic functions, significantly reducing the long-term toxicity caused by the retention of nanomaterials in the body (Wang et al. 2020b; Miao et al. 2019).

5.3 Toxicology of Graphene-Based Nanomaterials

With the extending applications, graphene-family nanomaterials begin to enter people's life via various ways, largely increasing the exposure frequencies. In spite of the increasing toxicological studies, the biosafety of graphene-based nanomaterials still remains elusive. Graphene oxide (GO), an oxidation derivative of graphene, is considered as one of the recently emerging nanomaterials attractive for biomedical applications. Accompanied with the prospect of applications are the great concerns about its biosafety for human and environment. The currently reported toxicity of GO mainly includes inhalation toxicity, ingestion toxicity, dermal toxicity, and hemocompatibility depending on exposure routes (Chen et al. 2020).

5.3.1 Respiratory Toxicity of GO

In animal experiments, exposure to 0.05–500 mg/kg GO (0.1–10 μm) or nanographene oxide (NGO) by intratracheal instillation, tail intravenous injection, or intraperitoneal injection was reported to lead to acute and chronic pulmonary inflammatory damage with progressive dose-dependent aggravation in mice. In addition, obvious pathological changes were observed in the lungs, which are mainly characterized by neutrophil infiltration, increased pulmonary capillary permeability, microthrombosis, alveolar wall thickening, epithelioid granulomas, pulmonary edema, and peribronchiolar collagen deposition on day 7 after exposure. And elevated levels of LDH, alkaline phosphatase (ALP), TNF-α, IL-6, and IL-1β were also shown in a marked dose-dependent manner (Zhang et al. 2021b). Furthermore, the size of GO influences its toxicity (Taheriazam et al. 2023). The micro-sized GO particles were reported to exhibit a reduced clearance rate in the lungs and induce more severe inflammation and fibrosis compared to nano-sized GO particles (Rodrigues et al. 2020; Loret et al. 2022).

Furthermore, the impact of GO on the induction of lung damage in animals is also influenced by other factors, such as oxygen content of GO, concentration of the administered dose, duration of exposure, pathways into the body, as well as characteristics of animal species, exogenous toxicant, and underlying disease.

5.3.2 Dermal Toxicity of GO

With its electrical conductivity, flexibility, and optical transparency, graphene-family nanomaterials have been extensively applied to in vivo visual electrophysiology studies as wearable electronic devices and smart contact lens. Herein, the contact between eyes and graphene-based materials is unavoidable. The evaluation toward the potential ocular toxicity of GO is of great significance. Yet, the available information on this issue is limited. Yan et al. explored the biocompatibility of GO sheets using human retinal pigment epithelium cells and white rabbits. The experimental animals were intravitreally injected with GO. Their results showed no obvious toxicity in GO-exposed animals, with undetectable changes in ocular appearance, eyesight, intraocular pressure, and histological examination. Also, no obvious cytotoxicity was observed in GO-incubated human retinal pigment epithelium cells. In addition, Yin et al. reported that no corneal irritation or abnormal behavior was observed in animals applied with graphene contact lens electrodes for electroretinography measurements. However, Wu et al. drew a different conclusion that GO nanosheets induced oxidative stress-based cytotoxicity to human conjunctiva epithelial cells and corneal epithelial cells. Moreover, the repeated GO exposure led to the damage of cornea epithelium in Sprague Dawley rats (Chen et al. 2020).

5.3.3 *Cytotoxicity of GO*

As far as the toxicity of graphene-based nanomaterials is concerned, the sharp edges of graphene are the main contributor to its high cytotoxicity. The increase in active edge sites and surface area contributes to the increased cytotoxicity; however, surface modification may reduce the toxicity level (Martín et al. 2019). The cytotoxicity of various 2D graphene is more severe than that of 2D transition metal dichalcogenide (TMDC) (Fadeel et al. 2018).

5.4 Toxicity of Two-Dimensional Theranostic Nanomaterials in Cancer Treatment

Generally, inorganic nanomaterials are relatively challenging to degrade and excrete in biological applications. They have a long residence time in living organisms causing adverse effects, including inflammation and tissue cysts (Zhou et al. 2020). Hence, consideration of the biosafety profile of theranostic 2D nanomaterials is of great importance for their cancer treatment. The toxicity of nanomaterials depends on the complex interaction of several physicochemical properties, such as size, shape, surface modification, oxidative state, dispersion state, synthesis procedure, exposure time, dose, and route of administration (Li et al. 2019).

The degradation of transition metal dichalcogenides (TMDCs) depends on the external environment, including oxidation, routes of administration, and pH, while studying their toxicity profile. While the transition metals evolved in TMDCs raised concerns about the risk of toxicity, the oxidants could degrade them and subsequently eliminate rapidly from the body (Gazzi et al. 2019). Mei et al. investigated translocation and correlation between the degradation and toxicity levels of polyvinylpyrrolidone modified 2H-phase MoS_2 nanosheets (MoS_2-PVP NSs). Their results showed that MoS_2 NSs demonstrated a different level of biodegradation with H_2O_2, catalase and human myeloperoxidase (hMPO) in the following order: H_2O_2 < catalase < hMPO. The MoS_2 NSs accumulated in the gastrointestinal organs and were excreted with the feces within 48 h of post-intragastric administration. Remarkable clearance of NSs was observed due to decreased accumulation in the liver and spleen within 30 days of post-intravenous injection. When the NSs were administered through the intraperitoneal route, they accumulated slightly in the spleen (Mei et al. 2019). TMDCs have a different level of oxidation, transformation, and excretion, depending on the elemental composition of different transition metals and their various physicochemical properties. The tungsten disulfide nanosheets (WS_2 NSs) possess high stability and are hardly degraded in the physiological environment and hence retain in RES organs for a long time. TiS_2 NSs are unstable, and their gradual transformation into the water-insoluble TiO_2 aggregates slowly eliminates from the body (Zhou et al. 2020).

Generally, MnO_2 NSs can transform into degraded products within a few hours in the physiological environment. Though MnO_2 NSs are relatively stable under neutral and alkaline pH conditions, they easily decompose into $Mn2^+$ and O_2 at acidic pH. The free Mn^{2+} ions exhibit an outstanding MRI contrast effect due to its five unpaired 3d electrons and hence poses a high relaxivity than that of MnO_2 nanoparticles. The degradation process of MnO_2 NSs can be monitored by MRI, due to the much-improved MRI contrast enhancement after decomposition (Zhou et al. 2020). The literature report reveals that the toxicity of Mn^{2+} is much higher than that of MnO_2 NSs. Although many studies have demonstrated that the degradation products (Mn^{2+}) of MnO_2 NSs can be excreted by renal clearance, the elimination amount is not yet precise.

The black phosphorus nanosheets (BP NSs) strongly react with oxygen and water and degrade to non-harmful phosphate and phosphonate ions since the elemental P atoms are connected by weak forces in the structure. Hence, their accumulation in the body organs is better than other 2D nanomaterials (Wang et al. 2019). When The biodegradation rate of exfoliated BP NSs was investigated under distinct atmospheres, viz., Ar, air, and O_2, it was discovered that the NSs degrades faster in the O_2 atmosphere. There are several reports to prove that the toxicity of palladium nanosheets (Pd NSs) is majorly dependent on its size. Chen et al. synthesized 2D Pd NSs with diameters ranging from 5 nm to 80 nm and investigated their size-dependent biodistribution, elimination, toxicity, and genomic gene expression profiles. The results showed that 5-nm Pd NSs evaded from the reticuloendothelial system with a longer blood half-life and were cleared by renal excretion, while larger sized Pd NSs mainly accumulated in the liver and spleen. Although Pd NSs did not induce any significant toxicity at the cellular level, lipid accumulation in the liver and inflammation was observed slightly in the spleen. Genomic gene expression analysis revealed that 80-nm Pd NSs interacted with more cellular components and influenced more biological processes in the liver when compared to 5-nm Pd nanosheet (Raja et al. 2020). In another study, Li et al. reported that 5 nm Pd NSs effectively accumulated on the tumor sites of the animal models through the enhanced permeability and retention (EPR) effect and could be excreted out of rats through urine (Li et al. 2019).

The investigations on the biosafety of 2D theranostic nanomaterials are still at an infancy stage. It is envisaged that a systematic study on the biological effects of the nanomaterials, toxicity profiles, and the relevant clinical trials would accelerate the commercialization of 2D theranostic nanomaterials soon.

5.5 The Safety Profiles of Liposomal Nanocarriers

Liposomes designed for cancer drug delivery are typically vesicles ranging in size from 50 nm to 300 nm, composed of one or more bilayers made from natural or synthetic lipids. Despite extensive research in this area, liposomes remain a focus of active investigation and are among the most clinically successful nanocarriers. As of January 2020, approximately 2400 clinical studies have involved liposomes, with around 1800 specifically related to cancer (data from clinicaltrials.gov).

5.5.1 Liposomal Doxorubicin (Doxil)

In 1995, the US Food and Drug Administration (FDA) authorized Doxil as the first nanomedicine for use in cancer treatment. The doxorubicin (DOX) hydrochloride liposome formulation has a diameter of approximately 85 nm and is made up of lipids that include cholesterol (Chol), hydrogenated soy phosphatidylcholine (HSPC), and poly(ethylene glycol) (PEG, weight-average molecular weight ≈2000) conjugated phospholipid (DSPE–PEG2k) (Fig. 5.3a). Following its initial authorization for the treatment of Kaposi's sarcoma associated with AIDS, it was subsequently granted approval for the treatment of multiple myeloma and ovarian cancer.

Doxil has demonstrated a significantly improved toxicity profile compared to nonencapsulated anthracyclines. Key benefits include enhanced cardiac safety and reduced incidences of myelosuppression, alopecia, nausea, and vomiting. This advantage was quantitatively confirmed through a meta-analysis involving 2220 patients (1112 treated with liposomal formulations and 1108 with conventional anthracyclines), which revealed a lower incidence of congestive heart failure in the liposome-treated group (OR 0.34, 95% CI, 0.24–0.47), among other safety improvements. As one of the longest-circulating liposomal formulations, the superior safety profile of Doxil compared to free DOX primarily results from its altered tissue distribution and pharmacokinetics (PK) (Fig. 5.2b, left). Intravenously administered liposomes are unable to easily pass through the tight capillary junctions in normal tissues, such as those in the heart muscle and gastrointestinal tract. In contrast, free drug infusion results in deep tissue penetration, leading to significant cardiotoxicity through various mechanisms, as shown in multiple experimental models. These mechanisms include mitochondrial damage caused by reactive oxygen species

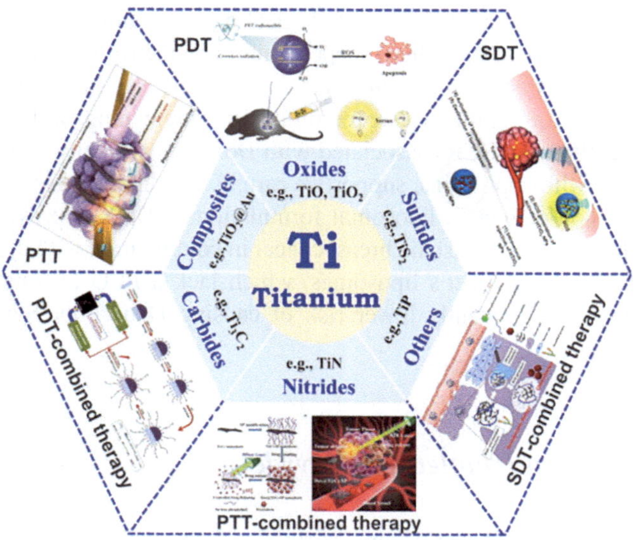

Fig. 5.2 Multifunctional Ti-based NMs for cancer theranostic applications. (Reproduced from Zhuang et al. 2020). Reprinted from Zhuang et al. (2020), Copyright 2020, with permission from Elsevier

(ROS), increased DNA damage, suppression of cardiomyocyte-specific genes, induction of cardiomyocyte apoptosis, and the degeneration of myofibers. In cancer patients, DOX treatment has frequently been linked to myocyte damage, often resulting in left ventricular dysfunction and, in some cases, heart failure, including fatal outcomes. The primary mechanism is now understood to involve the inhibition of topoisomerase 2β, which triggers cell death pathways and suppresses mitochondrial biogenesis (Fig. 5.2b, upper right). For patients at higher risk of cardiotoxicity, Doxil offers a valuable alternative therapy. However, using Doxil does not completely eliminate DOX-related toxicity. In fact, new side effects associated with Doxil have been reported (Fig. 5.3c). The primary dose-limiting side effect of liposomal DOX is palmar-plantar erythrodysesthesia (PPE), also known as hand-foot syndrome. This condition is characterized by redness, inflammation, and occasionally blisters on the palms and soles. PPE is thought to result from the accumulation of long-circulating liposomes in healthy, vulnerable tissues such as the skin. Figure 5.3b (lower right) outlines the potential mechanisms behind PPE caused by liposomal DOX. Molecular toxicology studies have identified the pivotal role of Cu^{2+} ions in the skin, which significantly contribute to the generation of reactive oxygen species (ROS). This disruption of skin tissue involves keratinocyte apoptosis, the release of chemokines (such as IL-8, GRO, and Fractalkine), and the induction of an inflammatory response through the release of cytokines like IL-1β, IL-1α, and IL-6 (Fig. 5.3b, lower right). Currently, the most effective strategy to mitigate this effect is to extend the interval between treatments. However, this approach may reduce the maximum tolerated dose (MTD), potentially compromising the drug's antitumor efficacy. Another side effect associated with Doxil is infusion-related hypersensitivity reactions (HSRs), which manifest as flushing and shortness of breath. These reactions are a type of complement activation-related pseudo-allergy, potentially triggered by the interaction of negatively charged PEG-PE with Doxil. Studies have also shown that the morphological transformation of liposomes from spherical to nonspherical (oval-like) shapes, caused by the formation of intraliposomal fiber-like drug crystals, may contribute to adverse events in patients, likely due to their reduced deformability. However, unlike life-threatening cardiotoxicity, side effects such as PPE and HSRs associated with Doxil are typically manageable and often reversible with appropriate supportive care. Additionally, it is worth mentioning Myocet, a non-pegylated liposomal formulation of DOX approved in Europe and Canada for treating metastatic breast cancer in combination with cyclophosphamide. Unlike Doxil, Myocet's liposomes, which lack a PEG coating, have been shown to carry a significantly lower risk of causing Hand-Foot Syndrome (Liu et al. 2020).

5.5.2 Liposomal Irinotecan (Onivyde)

Beyond Doxil, other liposomal formulations have also been approved for oncology applications. Onivyde, also referred to as MM-398 or PEP02, is a liposomal irinotecan injection that received FDA approval in 2015. These liposomes, approximately 100 nm in diameter, are constructed with bilayers composed of

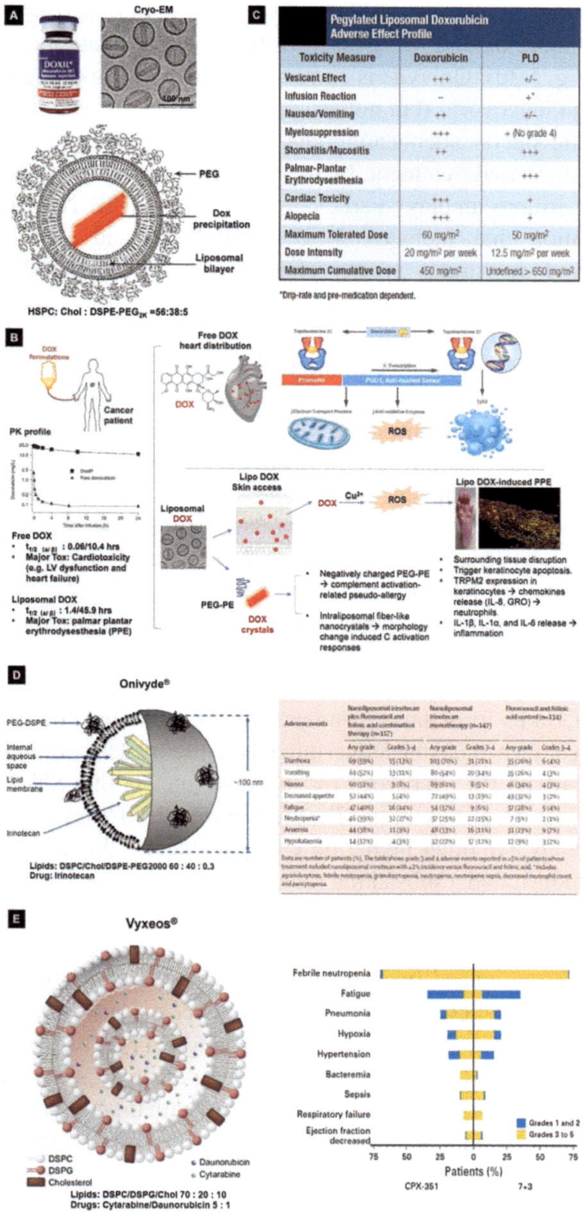

Fig. 5.3 Comparative demonstration of free DOX and Doxil's PK/biodistribution, targeted organs, and safety features. (**a**) Appearance, cryoEM image, and scheme of doxil formulation. *HSPC* hydrogenated soy phosphatidylcholine, *Chol* cholesterol, *DSPE–PEG2k* 1,2-distearoyl-sn-glycero-3-phosphoethanolamine-N-[methoxy(polyethylene glycol)-2000]. (**b**) Pharmacokinetics (PK) profiles and main cardiotoxicity and palmar plantar erythrodysesthesia effects/mechanism of doxorubicin and doxil formulation. (**c**) Clinical adverse effect profile of pegylated liposomal doxorubicin. (**d**) Scheme and clinical adverse effect profile of liposomal irinotecan onivyde. (**e**) Scheme and clinical adverse effect profile of cytarabine and daunorubicin codelivery liposome Vyxeos (Reproduced from Liu et al. 2020). Reproduced from Liu et al. (2020) by permission of John Wiley & Sons Ltd

distearoylphosphatidylcholine (DSPC), cholesterol (Chol), and DSPE–PEG2k in a molar ratio of 60:40:0.3, allowing for the efficient encapsulation of drug crystals or precipitates (Fig. 5.3d). Onivyde is currently approved for use in combination with fluorouracil and leucovorin (but not as a monotherapy) to treat patients with metastatic pancreatic ductal adenocarcinoma (PDAC) following disease progression after gemcitabine-based treatment. In the treatment of pancreatic ductal adenocarcinoma (PDAC), irinotecan plays a crucial role in the FOLFIRINOX regimen, a four-drug combination that includes 5-FU, oxaliplatin, and irinotecan. Although FOLFIRINOX demonstrates greater clinical efficacy compared to gemcitabine (GEM), its use is significantly limited due to its high toxicity, making it suitable only for advanced-stage PDAC patients with good functional performance. Irinotecan is a major contributor to the toxicity of FOLFIRINOX, causing severe side effects such as bone marrow suppression (e.g., neutropenia in approximately 70% of patients) and gastrointestinal issues (e.g., nausea, vomiting, diarrhea, and anorexia, also affecting roughly 70% of patients). To mitigate these toxicities while preserving the drug's therapeutic effectiveness, irinotecan has been encapsulated in liposomes. Examples include (1) the MM-398 formulation, which uses a polyanionic trapping agent (triethylammonium sucrose octasulfate, TEA8SOS) to stably incorporate irinotecan hydrochloride into liposomes, and (2) liposomal systems employing active loading of irinotecan via transmembrane proton gradients facilitated by ionophores (e.g., A23187) or ammonium sulfate. Like Doxil, Onivyde exhibits significant pharmacokinetic (PK) advantages, despite utilizing a lower percentage of PEG in its formulation. Patients treated with Onivyde showed markedly slower drug clearance, an extended terminal half-life ($t½$) for circulating total irinotecan, and improved PK properties for its active metabolite, SN-38, compared to those treated with free irinotecan. The dose-normalized AUC_0-∞ value of SN-38 was approximately five times higher for Onivyde than for the free drug. While studies in mice demonstrated a substantial reduction in toxicity and a more than fourfold increase in the maximum tolerated dose (MTD) with liposomal irinotecan compared to free irinotecan, clinical trials revealed no notable differences in the primary adverse events experienced by patients. In patients receiving Onivyde combined with 5-FU/LV, the most common grade 3 or 4 adverse events included neutropenia (27%), diarrhea (13%), vomiting (11%), and fatigue (14%) (Fig. 5.3d). Although irinotecan liposomes exhibit a high drug-loading capacity, their instability under shear and osmotic stress in the bloodstream, along with bilayer disruption caused by serum proteins, has been linked to toxicity stemming from premature drug release (Liu et al. 2020).

5.5.3 Ratiometric Cytarabine and Daunorubicin Codelivery Liposome (Vyxeos)

Most FDA-approved liposome carriers deliver a single active pharmaceutical ingredient (API). However, the recent approval of Vyxeos (also known as CPX-351), which encapsulates cytarabine and daunorubicin in a 5:1 molar ratio (Fig. 5.3e), highlights the potential of nano-enabled drug synergy in the treatment of acute myeloid leukemia (AML). In traditional chemotherapy regimens, the combination of these two drugs was often tested by progressively increasing the dose of one drug while maintaining the recommended dose of the other, until the cumulative toxicity became dose-limiting. A major benefit of ratiometric co-encapsulation is the ability to achieve drug synergy in vivo through a synchronized pharmacokinetic (PK) and biodistribution profile. This is crucial, as the degree of synergy (or antagonism) often depends on the ratio and concentrations of the active pharmaceutical ingredients (APIs) in the combination. Additionally, this approach helps avoid the maximum dose of each drug, providing an important safeguard for reducing side effects. In a bone marrow (BM) P388 leukemia tumor model, CPX-351 liposomes were shown to not only preserve the synergistic drug ratio in plasma but also in the BM, resulting in superior therapeutic activity compared to free-drug combinations. In a separate BM-engrafting CCRF-CEM leukemia model, CPX-351 liposomes demonstrated greater efficacy against BM leukemia cells than the free-drug cocktail. The drug distribution data further revealed that drug concentrations were significantly higher in the BM with the liposomal formulation compared to the free-drug combination. Interestingly, the CPX-351 formulation was shown to exhibit greater selective cytotoxicity against AML cells in vitro, as compared to normal progenitor cells from patients, when used in combination with other drugs. These pharmacological benefits—such as preferential uptake of liposomes by leukemia cells and reduced drug exposure in nontarget tissues—may help explain the strong efficacy and manageable safety profile of Vyxeos observed in clinical trials (Mayer et al. 2019). In clinical trials, Vyxeos significantly improved median overall survival compared to the standard-of-care chemotherapy regimen of cytarabine plus daunorubicin ("7 + 3"), with a survival of 9.56 months versus 5.95 months ($P = 0.003$). There is limited information on the toxicity mechanisms of Vyxeos, which was approved only in 2017. The side effect profile of Vyxeos was found to be similar to that of the free drug combination in patients (Fig. 5.3e). However, toxicity must be carefully considered as a potential outcome. While ratiometric codelivery aims to optimize drug synergy for enhanced in vivo potency (increased killing effect), this approach may lead to unforeseen adverse effects in healthy organs.

What can we learn from these approved liposomes? There is substantial evidence showing that the safety profiles of liposomal nanocarriers depend on their structure. As the technology has evolved, the quality, efficacy, and safety of liposomal carriers (and drug products containing nanomaterials) have proven to be highly sensitive to the manufacturing process. Key factors to consider include lipid composition, size, lamellarity, charge, drug-loading methods, loading conditions (such as feed ratio, temperature, and time), type and concentration of trapping agents, intraliposomal osmotic pressure and pH, drug localization (whether in the interior, surface-attached, or lipid-incorporated), drug release profiles, and whether or not targeting mechanisms are incorporated. In the case of Doxil, the primary component, HSPC, has a high melting temperature (Tm), which keeps the liposome membrane in a gel state below this Tm. This gel state prevents drug diffusion through the bilayer at body temperature. Another common approach involves using a trapping agent, such as ammonium sulfate for DOX or TEA8SOS for irinotecan. To achieve this, liposomes are typically prepared with the desired concentration of the trapping agent. In the case of Doxil, after removing free ammonium sulfate from the external medium, the intraliposomal NH^{4+} dissociates into NH_3. This NH3 then diffuses out of the liposome, resulting in a lower pH inside the liposome as the H^+ ions are retained within the bilayer. The nonionized, weak base form of DOX can pass through the lipid bilayer at temperatures above its Tm, where it becomes ionized upon entering the liposome. The resulting cationic form then interacts with $SO42^-$ anions to form long, fiber-like salt crystals. Doxil's success, including its safety features, is partly attributed to the effects of the trapping agent, which helps reduce premature DOX release, minimize osmotic pressure within the liposome, and enhance drug-loading capacity and efficiency. This underscores the importance of a well-integrated nanomanufacturing process and rigorous quality control (QC) standards. These principles form the foundation of "nano Chemistry Manufacturing and Controls" (nano-CMC), which could significantly impact nanosafety (Liu et al. 2020).

5.6 The Toxicity Assessment of Metal-Organic Framework (MOF)

Metal-organic frameworks (MOFs) are a new category of crystalline, porous materials made up of metal ions and organic linkers. Due to their high porosity and large specific surface area, these hybrid materials hold significant potential for various applications, including gas separation, energy storage, biomedical imaging, and drug delivery. As MOFs are increasingly considered for biomedical use, it is crucial to conduct thorough toxicity assessments to understand their safety.

Current knowledge raises valid concerns about the safety of MOFs in biological systems. Any material intended for use in contact with the human body must meet stringent requirements before being applied. In addition to fulfilling its intended function with the appropriate structure and properties, the material must also be evaluated for its stability, biodegradability, and biocompatibility. The initial step in assessing the toxicity of such materials involves conducting both in vitro and in vivo studies. Each type of test offers valuable, complementary insights, making it essential to perform both for a comprehensive evaluation. In vitro studies allow for a detailed analysis of cytotoxic effects on specific mammalian cells, such as those from target organs, without ethical concerns. These studies are particularly valuable for investigating potential toxicity mechanisms. In contrast, in vivo studies provide insights into the absorption, distribution, and overall behavior of substances within the entire organism, offering a broader perspective compared to focusing solely on individual cells. They also help identify previously unrecognized side effects, enabling a more thorough evaluation of potential risks. Therefore, despite the promising outlook for MOFs in biomedicine, their use cannot advance until comprehensive toxicity assessments are conducted (Wisniewska et al. 2023).

Several studies have reported that MOFs can interact with living cells and tissues, sometimes causing damage, while others suggest they are safe. Additionally, research has proposed that the toxicity of MOFs may be influenced by their physicochemical properties (Ettlinger et al. 2022). Building on these observations, this analysis examines and clarifies data on the toxicity of various MOFs, with a focus on carefully chosen critical factors. Both in vitro and in vivo studies are explored in detail.

Nanoparticle (NP) toxicity is likely driven by multiple distinct mechanisms. One common cause is oxidative stress, which results from the excessive production of reactive oxygen species (ROS), including $O_2^{\bullet-}$, $OH^{\bullet}$, $OOH^{\bullet}$. These ROS can be generated through the interaction of transition metal-based NPs with H_2O_2 in a process known as the Fenton reaction, as illustrated in Eqs. (5.1) and (5.2). Additionally, ROS can also be triggered by free radicals present on the reactive surface of nanoparticles, a phenomenon often observed following the functionalization of NPs.

$$Fe^{2+} + H_2O_2 = Fe^{3+} + OH^- + OH\bullet \tag{5.1}$$

$$Fe^{3+} + H_2O_2 = Fe^{2+} + H^+ + OOH\bullet \tag{5.2}$$

While a moderate level of reactive oxygen species (ROS) is essential for physiological processes like signal transduction and gene expression, an excessive amount can cause significant harm. It can damage mitochondrial membranes as well as the proteins, lipids, and mitochondrial DNA contained within them. This, in turn, can trigger inflammation signaling, apoptosis, or necrosis. Numerous studies have reported such effects in correlation with elevated ROS levels. For instance, Chen et al. (2021) observed mitochondrial membrane depolarization and the formation of apoptotic bodies in human embryonic kidney cells exposed to the copper-based MOF (HKUST-1). At elevated concentrations of HKUST-1, ROS levels were found to be over six times higher than those in the control sample. Similarly, Yan et al. reported that exposure of microglial cells to the cobalt-based MOF (ZIF-67) induced apoptosis by interfering with the protein signaling pathway. The study confirmed that ROS generation played a key role in the toxicity of ZIF-67, as the addition of an antioxidant effectively prevented apoptosis (Yan et al. 2021).

The toxic effects of nanoparticles (NPs) are often linked to the release of metal ions, which occurs when NPs dissolve in aqueous solutions, producing metal cations that are inherently toxic. The harmful impact of these released metal ions has been widely documented, particularly for ZnO and CuO nanoparticles, as Zn^{2+} and Cu^{2+} ions are especially harmful to living organisms.

Nanoparticle toxicity has also been linked to endocytosis, a cellular uptake process where particles are internalized and enclosed within vacuoles. The primary forms of endocytosis include phagocytosis, pinocytosis, and clathrin- or caveolae-mediated pathways. Toxic effects associated with this process often stem from the unrestricted movement of nanoparticles within the cell. For instance, nanoparticles taken up through pinocytosis may localize to various cellular components, such as the membrane, cytoplasm, lipid vesicles, mitochondria, or nucleus. Depending on their location, they can inflict damage on DNA or organelles, ultimately resulting in cell death. Eom and Choi demonstrated that silica nanoparticle (SiNP) uptake via clathrin-mediated endocytosis contributes to toxicity in *Caenorhabditis elegans*, specifically by impairing its reproductive capacity. The involvement of the endocytic pathway in SiNP internalization was validated using inhibitors targeting different endocytosis types. The reproductive toxicity was attributed to clathrin-mediated endocytosis, as confirmed through studies involving endocytosis-defective mutants. Notably, the presence of SiNPs did not result in elevated ROS levels (Eom and Choi 2019).

Nanoparticle toxicity can also result from damage to the cell membrane, either through adsorption onto the cell surface or diffusion into the membrane. Recent studies have revealed that nanoparticle-induced toxicity may be linked to the unfolding of fibrinogen. The binding of nanoparticles to fibrinogen has been associated with various physiological and pathological effects, including macrophage uptake, blood coagulation, and protein aggregation. However, the underlying mechanisms leading to these abnormalities remain poorly understood (Wisniewska et al. 2023). Figure 5.4 provides a simplified overview of the most common mechanisms of nanoparticle toxicity. While the mechanisms discussed above are relatively

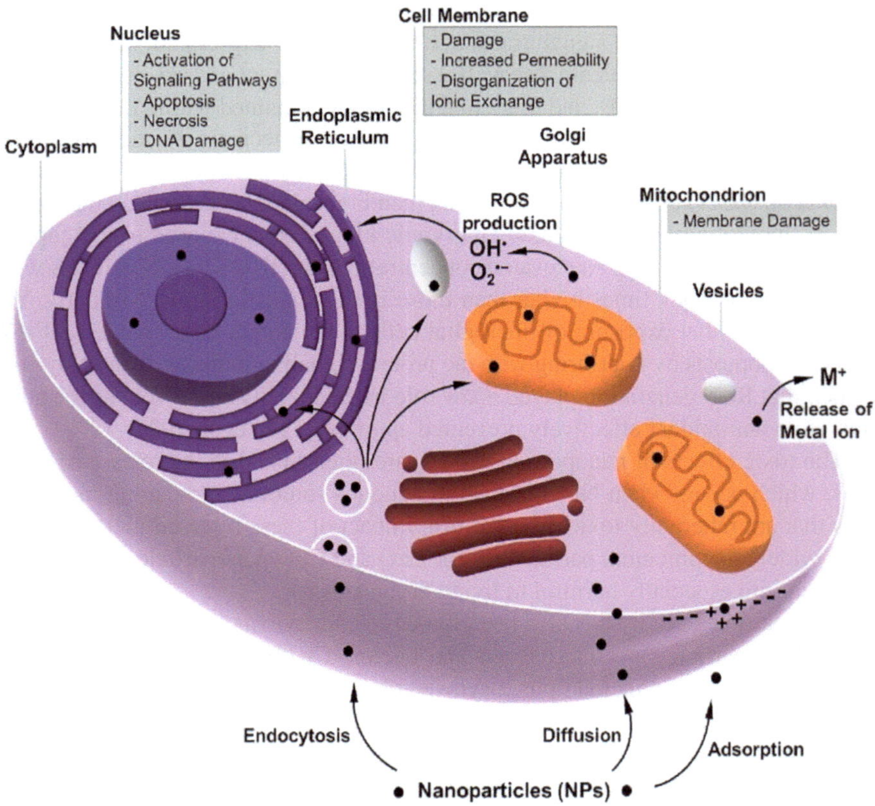

Fig. 5.4 Schematic illustration of the main mechanisms associated with MOF-induced toxicity. MOF-related toxicity can occur during the cellular internalization stage (endocytosis, diffusion, adsorption), resulting in membrane damage. Due to potential ROS production, internalized MOFs such as MOF NPs can also pose a risk to other cell structures (e.g., mitochondria, nucleus) (Wisniewska et al. 2023). Reprinted from Wisniewska et al. (2023), Copyright 2023, with permission from Elsevier

well-documented and, in some cases, clarify how toxicity occurs, numerous other potential mechanisms remain unexplored.

5.6.1 *Key Factors Affecting the Toxicity of MOF Nanoparticles*

Recent research has highlighted that the physicochemical properties of MOFs are key factors in determining their toxicity. Crucial parameters, including (1) the chemical composition of precursors, (2) particle size, (3) morphology, and (4) zeta potential and aggregation behavior, significantly impact the intrinsic toxicity of MOFs.

5.6.1.1 Chemistry of Precursors

The chemical composition of MOFs, determined by the metal ions and organic ligands used in their synthesis, directly influences their toxicity.

The toxicity of Co- and Zn-based MOFs was evaluated by Hao and Yan (2022), who specifically examined the hematotoxicity of ZIFs due to their potential contact with blood in biomedical applications. Their findings revealed that ZIF-67, which contains cobalt, caused significant hemolysis of red blood cells, whereas ZIF-8, composed of zinc, did not exhibit the same effect. The membrane rupture observed was attributed to the production of superoxide anions and hydroxyl radicals, which then led to hemoglobin binding. In contrast, Zhang et al. reported similar phytotoxicity for both ZIF-67 and ZIF-8. At concentrations of 0.1–1 mg/L, they observed inhibition of algal growth (by 1–32%), a reduction in chlorophyll content (by 7–30%), and damage to membranes and chloroplasts. Interestingly, the study also found that these effects were reversible within 72 h after the removal of the MOF, suggesting a promising outlook even for nanoparticles with moderate toxicity (Zhang et al. 2021c).

MIL-100 has been the most extensively studied MOF for its biocompatibility (Zhang et al. 2021b). Grall et al. assessed and compared the in vitro toxicity of three MIL-100 variants containing FeIII, AlIII, or CrIII. Their findings indicated that none of the MOFs caused toxicity in A549, Calu-3, or HepG2 cells, even at high concentrations (64 μg cm^{-2}). However, a mild toxicity (around 39% cell mortality) was observed in Hep3B cells exposed to 64 μg/cm^2 of MIL-100(Fe), which corresponded with a 1.8-fold increase in oxidative stress. The authors suggested that the ROS generation in the presence of MIL-100(Fe) might be linked to the redox properties of iron (as Fe^{3+} has a higher standard potential than Al^{3+} and Cr^{3+}) or its ability to produce hydroxyl radicals through the Fenton reaction (Eq. (5.2)). Furthermore, recognizing that oxidative stress can lead to potential genotoxicity, the study on Hep3B cells was expanded to include DNA damage assessment. This was done by counting the number of γ-H2Ax foci per nucleus. The results showed that only the Fe-based MOF, at a concentration of 64 μg/cm^2, caused an increase in the median foci from 3 to 5 per nucleus ($0.01 < p < 0.05$). While the observed DNA damage appeared relatively minimal, it is important to note that Hep3B cells lack p53 (unlike

HepG2 cells), and even a single DNA double-strand break can severely compromise cell integrity (Grall et al. 2015).

Other research groups have reported similar toxicity profiles for MIL-100 compounds. Hidalgo et al. found that neither MIL-100(Fe) nor MIL-100(Al) caused toxicity in J774.A1 cells, even at high concentrations (1200 μg/mL). However, both compounds led to an increased secretion of pro-inflammatory cytokines. The most significant changes were observed in the TNF-α levels, which showed a marked increase (4 ANOVA stars) in peripheral blood mononuclear cells exposed to 25 μg/mL MIL-100(Fe) or 250 μg/mL MIL-100(Al) (Hidalgo et al. 2022).

In conclusion, metal ions play a crucial role in determining the toxicity of MOFs, as demonstrated and confirmed by various studies. Consequently, when designing MOFs for biomedical applications, it is essential to carefully select the types of metals used.

Several studies have attempted to explore how the type of organic linker influences the toxicity of MOFs. However, understanding this relationship remains challenging, as the available literature presents diverse and conflicting results. Figure 5.5 offers a comprehensive toxicity assessment of various organic linkers and their corresponding Fe-based MOFs, based on reported IC_{50} values. Notably, it is evident that the cytotoxicity toward HeLa and J774 cells varied significantly, with some structures being compatible with HeLa cells but highly toxic to J774 cells. It is important to note that the toxicity of different MOFs does not always align with the toxicity of their organic components, even when the metal ion remains the same. In other words, the characterization of a MOF does not simply reflect the properties of its precursors. For example, even when MOFs are made from identical precursors, such as Fe ions and 2-amino-1,4-benzenedicarboxylic acid, their effects on J774 cell growth inhibition can vary. This serves as strong evidence that toxicity is a complex trait influenced by numerous factors that remain largely unexplored. We choose to focus on specific linkers, such as fumaric acid and certain terephthalic acid derivatives (BDC-NO_2, BDC-4CH_3, BDC-NH_2), which may contribute to the increased toxicity of MOFs. Figure 5.5 illustrates a general trend in toxicity from a chemical perspective. It is important to note that, although all the examined ligands are carboxylic acids, their biosafety profiles varied. Furthermore, incorporating different functional groups into 1,4-benzenedicarboxylic acid appears to offer little improvement in terms of the MOF's biosafety. Incorporating various functional groups into 1,4-benzenedicarboxylic acid appears to have little impact on improving the biosafety of MOFs. The authors propose that MOF toxicity may be influenced by the hydrophilic or hydrophobic nature of the organic linker. Specifically, the most toxic effects on J774 cells were observed with hydrophobic compounds, such as BDC-CH_3, BDC-NO_2, except for fumaric acid. In contrast, the hydrophilic 1,3,5-benzenetricarboxylic acid (BTC) caused the least harm to J774 cells. This hypothesis is supported by the fact that hydrophilic compounds are more easily excreted, reducing their potential negative effects on biological systems (Wisniewska et al. 2023).

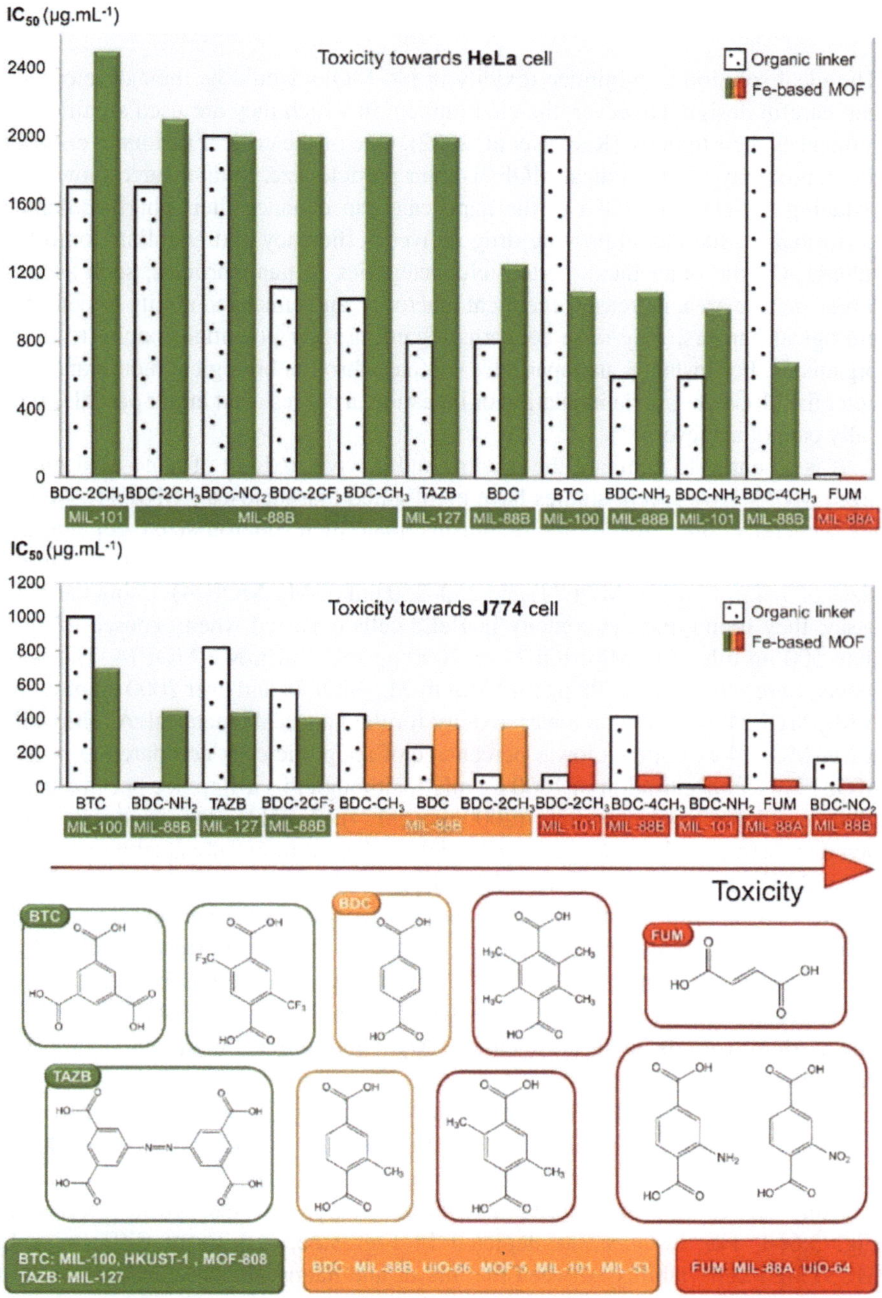

Fig. 5.5 Toxicity of organic ligands and corresponding Fe-contained MOFs toward HeLa and J774 cells based on reported IC_{50} data. Chemical names of the structures: *BTC* 1,3,5-benzenetricarboxylic acid, *TAZB* azobenzenetetracarboxylic acid, *BDC* 1,4-benzenedicarboxylic acid, *FUM* 2-butenedioic acid (Wisniewska et al. 2023). Reprinted from Wisniewska et al. (2023), Copyright 2023, with permission from Elsevier

5.6.1.2 Particle Size

The ideal solution to minimize toxicity in bio-MOFs would be their development and careful design. However, the environment in which they are used significantly influences their toxicity (Rabiee et al. 2022). One of the critical factors in ensuring the nontoxicity of biomedical MOFs is their particle size. Studies have shown that reducing the size of MOFs to the nanoscale can enhance their pharmacological performance, such as improving drug delivery efficiency and enabling controlled release. On the other hand, certain characteristics of nanoparticles, such as their larger surface area, increased chemical reactivity, and enhanced ability to penetrate biological barriers, may raise concerns regarding their potential toxicity to living organisms. For instance, nanoparticles can pass through biological membranes and enter the bloodstream via inhalation or ingestion, a process that larger particles typically cannot achieve.

It is essential to compare the in vitro and in vivo toxicity of nano- and micro-sized MOFs. Two recent studies have highlighted that nanoscale MOFs (n-MOFs) are generally safer for living organisms than their micron-sized counterparts (m-MOFs). For instance, Zhu et al. synthesized Mg-MOF74 with average particle sizes of 3–4 μm (m-Mg-MOF74) and 250–350 nm (n-Mg-MOF74). Using the MTT assay, they found that cytotoxicity in HeLa cells occurred when exposed to more than 500 μg mL of m-Mg-MOF74 or 1000 μg mL^{-1} of n-Mg-MOF74. The IC50 values were estimated at 798 μg mL^{-1} for m-Mg-MOF74 and over 2000 μg mL^{-1} for n-Mg-MOF74, indicating a lower toxicity for the nanoscale material. Additionally, n-Mg-MOF74 exhibited a lower percentage of apoptotic cells compared to m-Mg-MOF74 at a concentration of 1000 μg mL^{-1}, although this difference became negligible at higher concentrations (2000 μg mL^{-1}). Notably, the nanoscale form demonstrated superior osteogenic potential, attributed to its enhanced bioavailability of Mg^{2+}. For instance, n-Mg-MOF74 promoted greater collagen secretion in BMSC cells at equivalent concentrations. In vivo studies further revealed slightly reduced cardiotoxicity and less impact on the growth of rats treated with n-Mg-MOF74 compared to m-Mg-MOF74. However, no significant toxicity was observed for either form. The authors suggest that the nanoscale Mg-MOF74's favorable effects stem from its balanced release of Mg^{2+} both intracellularly and extracellularly (Rabiee et al. 2022).

Jiang et al. similarly observed reduced toxicity in n-MOF compared to m-MOF. They synthesized bio-MOF-1 with dimensions ranging from 400–600 nm for the nanoscale version and 50–80 μm for the microscale version (Fig. 5.6a–d). Key findings from the in vitro and in vivo toxicological analyses are summarized in Fig. 5.6e-l. For instance, as depicted in Fig. 5.6e, significant differences in MC3T3-E1 cell viability between the micro- and nanoscale structures emerged at concentrations exceeding 100 μg mL^{-1} (2–3 ANOVA stars). The IC50 value of n-bio-MOF-1, calculated at 599.3 μg mL^{-1}, was notably higher than the 248.3 μg mL^{-1} recorded for m-bio-MOF-1. These findings align with biochemical blood tests in rats, which suggested moderate biosafety at concentrations above 100 μg mL^{-1}. Furthermore, histological observations revealed that nanoscale MOF exhibited superior biocompatibility compared to its microscale counterpart (Jiang et al. 2022).

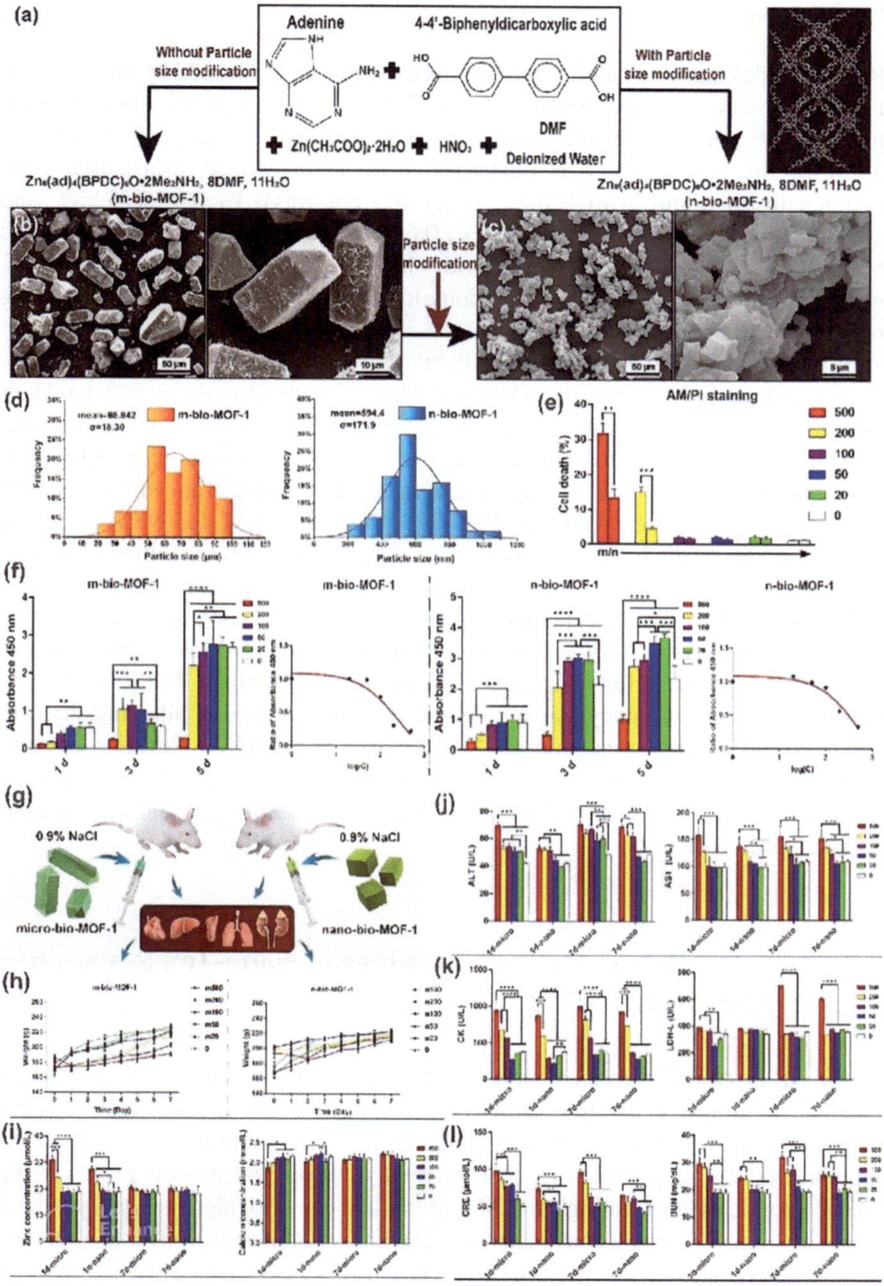

Fig. 5.6 Preparation and characterization of m/n-bio-MOF-1. (**a**) Synthesis. (**b, c**) SEM images of MOFs. (**d**) Particle size distribution. (**e**) Cytotoxicity assay. (**f**) CCK-8 results and corresponding curves of half inhibitory concentration. (**g**) Schematic depicting in vivo measurements. (**h**) Body weight. (**i**) Serum analysis of zinc and calcium. (**j**) Hepatotoxicity. (**k**) Cardiotoxicity. (**l**) Renal toxicity (Wisniewska et al. 2023). Reprinted from Wisniewska et al. (2023), Copyright 2023, with permission from Elsevier

5.6.1.3 Morphology

Recent studies have indicated that MOF toxicity may be influenced by morphology. Liu et al. observed that the viability of *E. coli* and *S. aureus* decreased under irradiation as MOF concentration increased. The greatest reduction was seen with rod-like TCPP-MOFs, followed by cubic, spherical, and spindle-shaped forms. After incubation with 100 μg·mL^{-1} MOF, the viability of *E. coli* was 0.3%, 13.7%, 32.3%, and 35.3%, while for *S. aureus*, it was 0.2%, 0.6%, 5.0%, and 9.3%, respectively (Liu et al. 2021). These findings aligned with singlet oxygen ($^1O_2$) production levels, with the authors attributing higher toxicity to larger pore sizes and greater spacing between TCPP activity sites.

In contrast, Zhou et al. identified a slightly different relationship between TCPP-MOF topology and cytotoxicity in HeLa and MCF-7 cells (Zhou et al. 2021b). Their findings revealed significantly lower cell viability after exposure to rod- or spindle-shaped MOFs compared to cubic or spherical structures under identical concentrations and irradiation conditions. The photodynamic IC50 values ranged 7.4–10.9 μM for PCN-222, 12.5–18.8 μM for PCN-223, 27.8–43.4 μM for MOF-525, and 40.4–52.8 μM for PCN-224, depending on the cell line. Flow cytometry analysis also indicated a higher proportion of necrotic and apoptotic cells induced by irregularly shaped TCPP-MOFs. The authors attribute this phenomenon to the selective endocytic pathways that MOFs can undergo based on their topology, which influences their internalization efficiency. Specifically, longitudinal TCPP-MOFs primarily utilize macropinocytosis as their main endocytic route, allowing them to penetrate HeLa cells more effectively than regular-shaped MOFs, thereby leading to higher cytotoxicity. Notably, no significant histopathological abnormalities or inflammatory responses were observed for any type of TCPP-MOF during examinations conducted on mice.

5.7 New Challenges for the Next Stage of Nano-Tox Research in Nanomedicine

Numerous firms and research institutes are developing a wide range of nano-enabled medical solutions, generating employment opportunities, raising money, filing investigational new drugs, and conducting clinical trials. From the viewpoint of the pipeline, there is ample evidence that the rapid development process results in the flexible application of biological NMs. Metal and metal oxides, polymers, silica-based materials, carbonaceous materials, 2D nanostructures, upconversion materials, metal-organic frameworks (MOFs), etc., are all covered by this. However, a higher level of uncertainty in biocompatibility and potential safety concerns presented new difficulties for them during their translation process. The degree of complexity requires a paradigm change from the traditional toxicology ("what happens" approach) to how mechanistic nano-/bio-interactions can help identify hazards and evaluate risks, ultimately resulting in nanomedicine that is safe by design (Fig. 5.7). To tackle safety concerns in clinical applications, it is essential to adopt existing experimental strategies. These include physicochemical characterization, high-throughput and high-content screening, nano-quantitative structure–activity relationship (nano-QSAR) analysis, and the study of nano-associated molecular patterns (NAMP) and adverse outcome pathways (AOP). These approaches are supported by nearly 15 years of progress in the fields of nanotoxicology and environmental health and safety (nano-EHS), as illustrated in Fig. 5.9. These elements form the foundation of "predictive toxicology," which was promoted as a paradigm change for evaluating the risks associated with nanomaterials. A good number of mechanistic endpoints that are commonly involved in the safety of NMs are covered by the assays that are currently in use, which were compiled in Fig. 5.7. These endpoints include oxidative stress, particle dissolution, surface membrane injury, lysosomes, inflammasomes, autophagy activation/blockade, mitochondrial damage, pro-inflammatory response, etc. The utilization of mechanistic cellular damage responses that are connected to disease pathogenesis is the most effective strategy that may plausibly handle the complexity of nanomaterials (Fig. 5.8). A quantitative

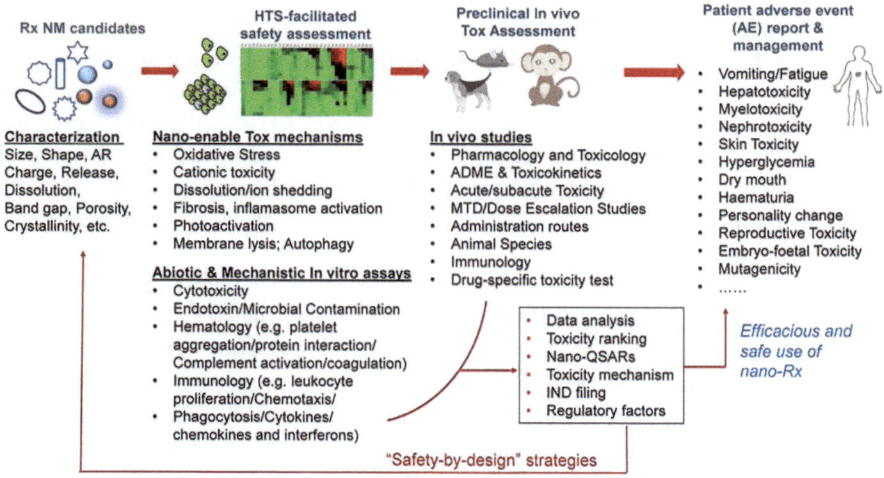

Fig. 5.7 Overall blueprint of safety assessment of nanomedicine. Therapeutic nanomaterials of various chemical compositions demand comprehensive characterization, such as size, morphology, charge, surface area, porosity, crystal structure, API concentration, release profiles, etc. It is necessary to consider HTS assays and response readouts in cells to obtain mechanistic toxicity insight with a view to plan and prioritize the animal assessments for the IND filing. The in vivo toxicity assessment should be designed and implemented with necessary consideration on ADME, toxicokinetics, administration routes, and acute versus chronic toxicity, which may not be fully covered in the in vitro assays. The output of these assays needs to be analyzed to facilitate the decision-making with respect to the technology translation. If necessary, a safe-by-design approach could be helpful to improve nanomedicine's safety features. The preclinical data can be used as a reference for the clinical safety assessment, i.e., patient adverse event report and management, for the efficacious and safe use of nanomedicine in patients (Reproduced from Liu et al. 2020). Reproduced from Liu et al. (2020) by permission of John Wiley & Sons Ltd

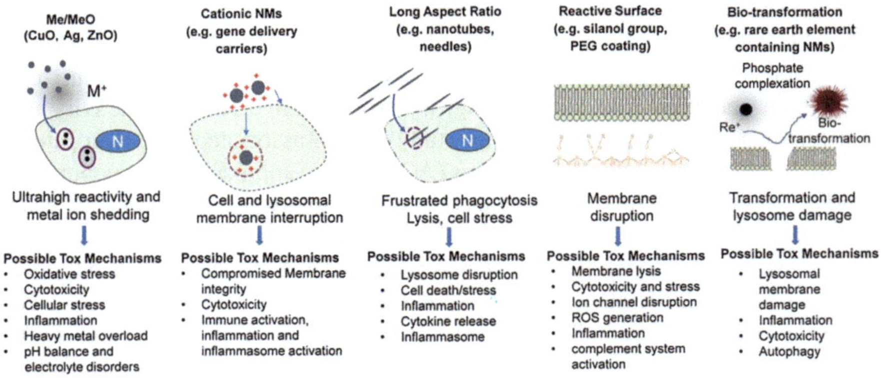

Fig. 5.8 Mechanistic injury responses that are frequently involved in nanomaterials. This includes dissolution and release of toxic metal ions from metal or metal oxide nanoparticles, cationic injury to surface membrane and organelles, frustrated phagocytosis and inflammasome activation by long aspect ratio nanomaterials, reactive surface that may lead to membrane disruption, and bio-transformation response in rare earth nanoparticles (Reproduced from Liu et al. 2020). Reproduced from Liu et al. (2020) by permission of John Wiley & Sons Ltd

analysis of these cellular toxicological endpoints would enable high-throughput screening (HTS) to be used for predicting in vivo safety, thus reducing the reliance on animal testing as the primary method for evaluation (Fig. 5.7). Examples of mechanistic injury at this stage include heavy metal ion shedding (in metal and metal oxide NMs), membrane damage (from cationic nanoparticles), lysosomal damage (due to long-aspect-ratio NMs), cell membrane disruption (from NMs with reactive surfaces), and biotransformation (in rare earth NMs), among others (Fig. 5.8) (Liu et al. 2020).

The key question is whether the current knowledge in nano EHS (environmental, health, and safety) and nanotoxicology is sufficiently advanced to ensure the safety of nanomedicine. Although valuable, the existing body of knowledge must be integrated into a more structured framework that supports further data generation and informs regulatory decision-making. This framework should rely on quantitative assessments of nanotoxicological mechanisms that are tied to both acute and chronic side effects, as well as adverse reactions in patients. Despite the potential benefits of a mechanistic approach for assessing nanomedicine safety, additional studies are needed to facilitate Investigational New Drug (IND) filings and comprehensive safety evaluations in patients. We believe that the next stage of research should focus on the following key areas (Fig. 5.9).

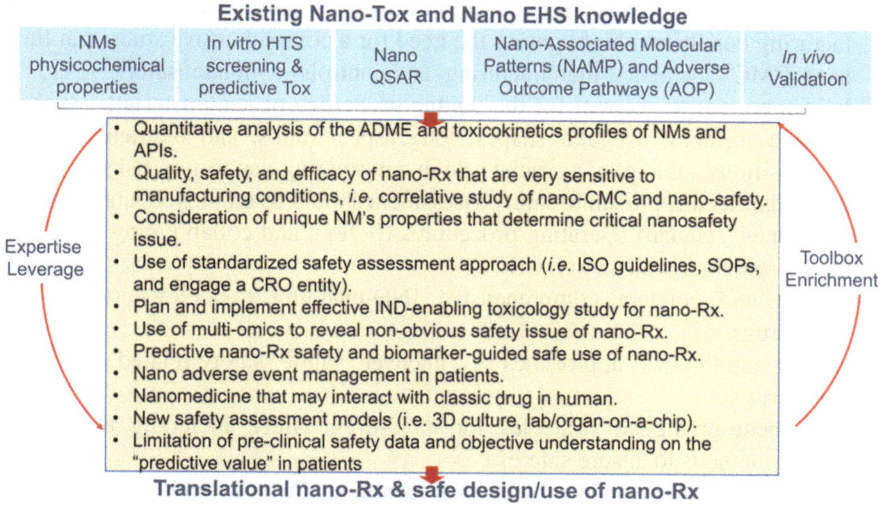

Fig. 5.9 Critical research aspects that are required to strengthen nanomedicine nanosafety investigation (Reproduced from Liu et al. 2020). Reproduced from Liu et al. (2020) by permission of John Wiley & Sons Ltd

5.8 Conclusions and Perspectives

Various cancer nano-biomaterials have the remarkable potential to revolutionize contemporary approaches to cancer prevention, imaging, early diagnosis, and treatment in a safe and effective manner. However, significant NP-associated toxicity, such as low biocompatibility and cytotoxicity, limits their utilization for cancer theranostics and can be hazardous to cancer patients up to life-threatening conditions. The adverse interactions between NPs and biological entities can lead to life-threatening toxicities. So, assessment of NP-associated toxicity is necessary. However, several challenges need to be addressed, such as developing standardized toxicity testing protocols and evaluating their long-term biocompatibility and stability in vivo. More efforts should be made to allow the generation of investigational new drug-enabling safety data that can be incorporated into a rational infrastructure for regulatory decision-making.

- It is essential to consider the absorption, distribution, metabolism, excretion (ADME), and toxicokinetic profiles of nanomaterials (NMs) and active pharmaceutical ingredients (APIs). New safety concerns may arise, as nano-drugs could improve the delivery of payload molecules to specific tissues where the free drug has limited access.

- The quality, safety, and effectiveness of nano-drugs are highly sensitive to manufacturing conditions, highlighting the need for a comprehensive study that links nano-CMC (chemistry, manufacturing, and controls) with nanosafety.
- It is important to account for the unique properties of nanomaterials—such as reactive surfaces, irregular shapes, long aspect ratios, and biotransformative compositions—that play a crucial role in determining nanosafety risks.
- A standardized approach to safety assessment should be employed, utilizing ISO guidelines, standard operating procedures (SOPs), and collaborating with contract research organizations (CROs).
- Design and conduct comprehensive IND-enabling toxicology studies for nano-drugs.
- Utilize multiomics approaches to uncover subtle safety-related issues in nano-drugs.
- Implement predictive strategies for nano-drug safety, including biomarker-guided methods to ensure safe use.
- Manage adverse events associated with nanomedicine in patients.
- Assess potential interactions between nanomedicine and traditional drugs in humans.
- Develop new experimental models with enhanced predictive capabilities to quickly assess the safety profile of nanomedicines, such as lab-on-a-chip or organ-on-a-chip technologies.
- Limitations of preclinical nanosafety assessments: While animal safety data is often used to predict human outcomes, there are notable differences between animal and human responses. These differences are influenced by various factors, such as differing levels of cytochrome P450 enzymes and other detoxification systems. When it comes to dosing, using body weight for linear dosimetry extrapolation often results in inaccurate conversion between animal and human doses. Allometric scaling based on body surface area is generally considered more accurate. Additionally, animal studies are unable to detect subjective effects like headaches, dizziness, nausea, or mental health issues, which often involve unique, individual reactions and varying responses within a diverse patient population.

References

Cao Y, Wu T, Dai W et al (2019) TiO_2 nanosheets with the au nanocrystal-decorated edge for mitochondria-targeting enhanced sonodynamic therapy. Chem Mater 31:9105–9114

Chen L, Li J, Chen Z et al (2020) Toxicological evaluation of graphene-family nanomaterials. J Nanosci Nanotechnol 20(4):1993–2006

Chen Y-C, Lin K-Y-A, Chen K-F et al (2021) In vitro renal toxicity evaluation of copper-based metal–organic framework HKUST-1 on human embryonic kidney cells. Environ Pollut 273:116528

Ding B et al (2020) Manganese oxide nanomaterials: synthesis, properties, and theranostic applications. Adv Mater 32(10):1905823

Eom H-J, Choi J (2019) Clathrin-mediated endocytosis is involved in uptake and toxicity of silica nanoparticles in Caenohabditis elegans. Chem Biol Interact 311:108774

Ettlinger R, Lächelt U, Gref R et al (2022) Toxicity of metal–organic framework nanoparticles: from essential analyses to potential applications. Chem Soc Rev 51(2):464–484

Fadeel B, Bussy C, Merino S et al (2018) Safety assessment of graphene-based materials: focus on human health and the environment. ACS Nano 12:10582–10620

Gazzi A, Fusco L, Khan A et al (2019) Photodynamic therapy based on graphene and MXene in cancer theranostics. Front Bioeng Biotech 7:295

Grall R, Hidalgo T, Delic J et al (2015) Horcajada in vitro biocompatibility of mesoporous metal (III; Fe, Al, Cr) trimesate MOF nanocarriers. J Mater Chem B 3(42):8279–8292

Greene A, Hashemi J, Kang Y (2020) Development of MnO_2 hollow nanoparticles for potential drug delivery applications. Nanotechnology 32(2):025713

Hao F, Yan X (2022) Nano-sized zeolite-like metal-organic frameworks induced hematological effects on red blood cell. J Hazard Mater 424:127353

Hidalgo T, Simón-Vázquez R, González-Fernández A et al (2022) Fingerprint of metal-organic frameworks. Chem Sci 13(4):934–944

Jiang S, Wang J, Zhu Z et al (2022) The synthesis of nano bio-MOF-1 with a systematic evaluation on the biosafety and biocompatibility. Microporous Mesoporous Mater 334:111773

Lan G, Ni K, Veroneau SS et al (2019) The versatile biomedical applications of bismuth-based nanoparticles and composites: therapeutic, diagnostic, biosensing, and regenerative properties. Am Chem Soc 14:4204–4208

Li J, Liu H, Ming J et al (2019) The biobehavior, biocompatibility and theranostic application of SPNS and Pd@au nanoplates in rats and rabbits. Chem Sci 10:1677–1686

Liu X et al (2020) Safety considerations of cancer nanomedicine—a key step toward translation. Small 16:36

Liu Y, Chen L, Zhao X et al (2021) Effect of topology on photodynamic sterilization of pophyrinic metal-organic frameworks. Chem Eur J 27(39):10151–10159

Loret T, de Luna LAV, Fordham A et al (2022) Innate but not adaptive immunity regulates lung recovery from chronic exposure to graphene oxide nanosheets. Adv Sci 9:e2104559

Martín C, Kostarelos K, Prato M et al (2019) Biocompatibility and biodegradability of 2D materials: graphene and beyond. Chem Commun 55:5540–5546

Mayer LD, Tardi P, Louie AC (2019) CPX-351: a nanoscale liposomal co-formulation of daunorubicin and cytarabine with unique biodistribution and tumor cell uptake properties. Int J Nanomedicine 14:3819

Mei L, Zhang X, Yin W et al (2019) Translocation, biotransformation-related degradation, and toxicity assessment of polyvinylpyrrolidone-modified 2H-phase nano-MoS2. Nanoscale 11:4767–4780

Miao Z, Chen S, Xu C et al (2019) PEGylated rhenium nanoclusters: a degradable metal photothermal nanoagent for cancer therapy. Chem Sci 10:5435–5443

Peng F, Setyawati MI, Tee JK et al (2019) Nanoparticles promote in vivo breast cancer cell intravasation and extravasation by inducing endothelial leakiness. Nat Nanotechnol 14:279–286

Rabiee N, Atarod M, Tavakolizadeh M et al (2022) Mashhadzadeh green metal-organic frameworks (MOFs) for biomedical applications. Microporous Mesoporous Mater 335:111670

Raja IS et al (2020) Two-dimensional theranostic nanomaterials in cancer treatment: state of the art and perspectives. Cancer 12(6):1657

Rodrigues AF, Newman L, Jasim D et al (2020) Size-dependent pulmonary impact of thin graphene oxide sheets in mice: toward safe-by-design. Adv Sci 7:1903200

Shahbazi M-A, Faghfouri L, Ferreira MP (2020) The versatile biomedical applications of bismuth-based nanoparticles and composites: therapeutic, diagnostic, biosensing, and regenerative properties. Chem Soc Rev 49:1253–1321

Taheriazam A, Abad GGY, Hajimazdarany S et al (2023) Graphene oxide nanoarchitectures in cancer biology: nano-modulators of autophagy and apoptosis. J Control Release Off J Control Release Soc 354:503–522

Wang Z, Liu Z, Su C et al (2019) Biodegradable black phosphorus-based nanomaterials in biomedicine: theranostic applications. Curr Med Chem 26:1788–1805
Wang H, Bremner D, Wu K et al (2020a) Platelet membrane biomimetic bufalin-loaded hollow MnO_2 nanoparticles for MRI-guided chemo-chemodynamic combined therapy of cancer. Chem Eng J 382:122848
Wang X, Fan L, Cheng L et al (2020b) Biodegradable nickel disulfide nanozymes with GSH-depleting function for high-efficiency photothermal-catalytic antibacterial therapy. iScience 23:101281
Wang X, Zhong X, Cheng L (2021) Titanium-based nanomaterials for cancer theranostics. Coord Chem Rev 2021:430
Wei R, Gong X, Lin H et al (2019) Versatile octapod-shaped hollow porous manganese(II) oxide nanoplatform for real-time visualization of cargo delivery. Nano Lett 19:5394
Wisniewska P et al (2023) Mitigating metal-organic framework (MOF) toxicity for biomedical applications. Chem Eng J 471:144400
Xu W et al (2022) Hollow mesoporous manganese oxides: application in cancer diagnosis and therapy. Small 18(15):2106511
Yan L, Zhao F, Wang J et al (2019) A safe-by-design strategy towards safer nanomaterials in nanomedicines. Adv Mater 31:1805391
Yan X, Yang Q, Fang X et al (2021) Co (ii)-based metal–organic framework induces apoptosis through activating the HIF-1α/BNIP3 signaling pathway in microglial cells. Environ Sci Nano 8(10):2866–2882
Zanoni TB et al (2015) The oxidation of p-phenylenediamine, an ingredient used for permanent hair dyeing purposes, leads to the formation of hydroxyl radicals: oxidative stress and DNA damage in human immortalized keratinocytes. Toxicol Lett 239(3):194–204
Zeng W, Zhang H, Deng Y et al (2020) Dual-response oxygen-generating MnO_2 nanoparticles with polydopamine modification for combined photothermal-photodynamic therapy. Chem Eng J 389:124494
Zhang X, Hu X, Wu H et al (2021a) Persistence and recovery of ZIF-8 and ZIF-67 phytotoxicity. Environ Sci Technol 55(22):15301–15312
Zhang H, Li M, Zhu X et al (2021b) Artemisinin co-delivery system based on manganese oxide for precise diagnosis and treatment of breast cancer. Nanotechnology 32:32. https://doi.org/10.1088/1361-6528/abfc6f
Zhang L, Ouyang S, Zhang H et al (2021c) Graphene oxide induces dose-dependent lung injury in rats by regulating autophagy. Exp Ther Med 21:462
Zhou H, Ge J, Miao Q et al (2020) Biodegradable inorganic nanoparticles for cancer theranostics: insights into the degradation behavior. Bioconjug Chem 31:315–331
Zhou J, Li Y, Wang L et al (2021a) Structural diversity of nanoscale zirconium porphyrin MOFs and their photoactivities and biological performances. J Mater Chem B 9(37):7760–7770
Zhou Z, Liang S, Zhao T et al (2021b) Overcoming chemotherapy resistance using pH-sensitive hollow MnO_2 nanoshells that target the hypoxic tumor microenvironment of metastasized oral squamous cell carcinoma. NanoBiotechnology 19:157
Zhu P, Chen Y, Shi J (2018) Nanoenzyme-augmented cancer sonodynamic therapy by catalytic tumor oxygenation. ACS Nano 12(4):3780–3795
Zhuang H, Zhao M, Ding S et al (2020) (2020) multifunctional smart yolk-shell nanostructure with mesoporous MnO_2 shell for enhanced cancer therapy. ACS Appl Mater Interfaces 12:38906

MIX
Papier aus verantwortungsvollen Quellen
Paper from responsible sources
FSC® C105338

If you have any concerns about our products,
you can contact us on
ProductSafety@springernature.com

In case Publisher is established outside the EU,
the EU authorized representative is:
Springer Nature Customer Service Center GmbH
Europaplatz 3, 69115 Heidelberg, Germany

Printed by Libri Plureos GmbH
in Hamburg, Germany